MAINSTREAMING OUTSIDERS:
The Production of Black Professionals

MAINSTREAMING OUTSIDERS:
The Production of Black Professionals

James E. Blackwell
University of Massachusetts/Boston

GENERAL HALL, Inc.
Publishers
23–45 Corporal Kennedy Street
Bayside, New York 11360

MAINSTREAMING OUTSIDERS
The Production of Black Professionals

GENERAL HALL, INC.
23–45 Corporal Kennedy Street
Bayside, New York 11360

Publisher: Ravi Mehra
Editor: Philip Eisen
Associate Editor: Marie I. Corsalini
Composition: *Graphics Division,* General Hall, Inc.

LIBRARY OF CONGRESS CATALOG CARD NUMBER: 81-82121
ISBN: 0-930390-39-3

Manufactured in the United States of America

For my parents, Edward and Celia, whose love, motivation, strengths, and dreams were the sources of an enduring inspiration...

CONTENTS

LIST OF TABLES

ix

List of Figures

Acknowledgments of Data Sources

Although their names are not cited below, the author is deeply indebted to the administrators and institutional research staff of participating graduate and professional schools for their responses to the survey inventory which were a rich source of immensely important data. A special note of gratitude is expressed to those organizations and individuals listed below because of their generous assistance in providing so much of the data upon which this study is based:

Association of Colleges of Pharmacy – Dr. Lars Solander
American Association of Dental Schools – Dr. James K. Walker
American Bar Association – Dean James White
American Dental Association – Dr. James W. Graham
Association of American Law Schools – Dr. Millard Rudd and Prof. Derek Bell
Association of American Medical Colleges – Dr. Davis G. Johnson, Mrs. Jewel Hodge-Jones and Mrs. Lillye Mae Johnson
Association of American Veterinary Medical Colleges – Dr. W. M. Decker
Association of Schools and Colleges of Optometry – Mr. Lee Smith
Committee on Minorities in Engineering – Mr. Levoy Spooner
Council on Legal Education Opportunity – Mr. Wade Henderson, Esq.
Council on Social Work Education – Mrs. Ruth Posner
Engineering Manpower Commission – Mr. Patrick Sheridan and Ms. Barbara Ann Supple
National Architectural Accrediting Board – Dr. Hugo Blasdel
National Fund for Minority Engineering Students
National Medical Fellowships – Ms. Kate Harris
National Optometric Association – Dr. Edwin Marshall and Dr. Charles Comer
National Research Council – Ms. Dorothy Gilford and Mr.

1

Peter D. Syverson
The National Center for Educational Statistics
The U.S. Office of Civil Rights
Tuskegee Institute – President Luther Foster and Dr. Ellis Hall
Xavier University of Lousiana – President Norman Francis and
Dean Warren Mckenna

Preface

Mainstreaming Outsiders is a product of my long-standing professional interest in the dynamics and complexities of inter-group relations, and of the conditions in the social structure which perpetuate patterns of group dominance and group subordination. One of the most effective strategies employed by dominant groups to protect their control over the highly prized values and rewards within a society is to manipulate and limit access to these values and rewards. High educational attainment is one example of a highly prized value in a modern, industrialized, and technological society since the society places a great premium on expert knowledge, competence, training and skills. Inequities in access to education generate other forms of structural inequality. As a result, it becomes even more difficult for subordinate groups, who are characteristically outside that reward structure, to alter their status within the social system.

This book is an examination of the efforts to fulfill what appeared to have been a national commitment during the civil rights era to eliminate structural inequalities and to equalize educational opportunities between all segments of the American population. Although the primary focus here is on increasing access of black Americans to graduate and professional schools and their graduation from them, much of what is said is generally applicable to the experiences of other minority groups such as Puerto Ricans, Mexican Americans and Native Americans. Indeed, just as black-white comparisons are drawn in the analysis of data, so are occasional distinctions made between other minority groups and the white population regarding access and production.

I am indebted to several persons who assisted me in the various stages of this project. I express my profound gratitude to: (1) my colleagues, Siamak Movahedi, Russell Schutt and Philip Hart for their advice on methodological issues raised during the research, and T. Scott Miyakawa for his invaluable and constructive criticisms made after thoroughly reading the manuscript in its early stages, (2) Kathleen Leon, Charlotte Craig and Jeanne O'Neill who served as research assistants, (3)John Murphy, Dianne

Horabik, and Tom Marx of the Computer Center of the University of Massachusetts/Boston for special assistance in data processing, (4) Patricia Garrity for typing assistance, and to Barbara Saulnier and Anne Knight who were so diligent in typing the entire manuscript and its revisions, (5) Mrs. Lynne E. Brown, of the Southern Education Foundation, for her very careful reading of the manuscript and thoughtful suggestions, and (6) I particularly appreciate the assistance provided by the Southern Education Foundation for its sponsorship of the study. Special appreciation is extended to those scholars whose meticulous reading of the manuscript and insightful analysis of pertinent issues helped to strengthen the final product. Included among these scholars are Marie Haug of Case Western Reserve University and Edgar Epps of the University of Chicago.

Above all, I am especially grateful to Myrt, my wife, for the myriad roles she played throughout this project. Her contributions as a research assistant, reader, and critic, and her abiding faith in the importance of this undertaking, as well as her humaneness and understanding, were of inestimable importance to me.

James E. Blackwell

Foreward

The book, *Mainstreaming Outsiders: The Production of Black Professionals,* grew out of the mutual and special interest of the author and the Southern Education Foundation in the problem of the underrepresentation of black Americans in graduate and professional education. As a result of that sustained interest, and consistent with the research goals of the Southern Education Foundation, the Foundation agreed to sponsor the survey research upon which this book is based. This study is one of nine research projects sponsored by the Southern Education Foundation within the last two years under its Higher Education Program. A major function of this program is to monitor desegregation efforts in states which historically have operated dual systems of public higher education, and ultimately to inform the public of the issues that affect equal opportunities for minorities in higher education.

SEF, since its incorporation in 1937 and the establishment of its initial fund by George Peabody in 1867, has been concerned with issues related to race and education. Specifically, the issues which surfaced in the *Adams V. Richardson* case, filed in 1970 charging that the U.S. Department of Health, Education, and Welfare had defaulted on its obligation to enforce Title VI of the Civil Rights Act of 1964, began to shape the Foundation's program in higher education. The Foundation recognized that the issues obviously affected black access to and graduation from post-secondary institutions in the South. Consequently, funds were provided to support black coalitions, region-wide conferences, the commissioning of studies and publishing of other materials. As stated by a former SEF Board member, the *Adams* case "will affect hundreds of institutions, tens of thousands of employees, faculty administrators, hospital workers, workers in experiment stations, cafeteria employees, etc., and millions of students. . ." It was within this context, that the Foundation made a decision to allocate resources to explore further the status of blacks in public higher education institutions. This research effort was informed in great part by this earlier decision. MAINSTREAMING OUTSIDERS offers data in one document for the purpose of analysis. It provides a graphic picture of the state-of-the-arts in the various professional fields and can be a reference book for students in their exploration of career and job opportunities. It refutes in many cases the notion that blacks are the recipients of preferential treatment in being admitted and retained in special admissions programs in existence at several professional schools. It brings into focus the importance of employing and retaining black faculty if successful role models are to be set for black students, and it highlights effective programs at black institutions that reflect the concept of expanded missions for the traditionally black colleges on a statewide level. Despite statements from various quarters of the education community that significant progress has been made for blacks to gain access into higher education institutions of their choice, the author continues to flag in MAINSTREAMING OUTSIDERS troubling areas. It is our hope that readers will look closely at the data and use the information as benchmarks while attempting to resolve the issue of educational equity for minorities. Morever, MAINSTREAMING OUTSIDERS serves as a challenge to those in policymaking positions to effect change where needed in the present decade. Dr. James E. Blackwell is not new to the Southern Education Foundation. We have enjoyed a close working relationship over the last seven years. Dr. Blackwell has assisted the Foundation with several of its conferences addressing desegregation of public higher education systems and has prepared several publications for the Foundation. He has concentrated in recent years on analyzing data on the participation of blacks in graduate and professional schools, and this most recent publication is a more comprehensive assessment of specific fields.

The Board and staff of the Southern Education Foundation applaud Dr. Blackwell's outstanding efforts that *Mainstreaming Outsiders* represents. We are confident that this timely and significant document will be useful to those who continue to be involved with the critical issues addressed in this book.

Eldridge W. McMillan
Executive Director
Southern Education Foundation

CHAPTER 1

Historical Context of Mainstreaming

The mainstreaming of outsiders is deeply rooted in American history. The association of the concept with non-European racial and ethnic groups is, however, a relatively recent phenomenon. If mainstreaming is conceptualized as one of the many dimensions of assimilation into American life, the differences in the applicability of the idea to the actual experiences of European white ethnic groups and as compared with those of third world minority groups are apparent. Almost from the inception of the first wave of European immigrants during the colonial period, it was assumed that these groups, as well as those Europeans who followed them would be assimilated into the mainstream of American society after an unspecified period of time. With each successive group, it may be argued, the time frame for its becoming mainstreamed was longer but the inevitability of its attainment of equal access to the opportunity structure of American life was never in doubt.

This was not the case for third world groups. Most assuredly, during the antebellum days of the American Deep South, the prevailing sentiment, due primarily to the human characteristics attributed to black persons, was that they could not and should not be brought into the mainstream of American life. After slavery, in all parts of the nation, all sectors of the society heatedly debated whether or not blacks could be brought into the mainstream and if they, indeed, had a right to those privileges reserved for the dominant group. Today, racist individuals and groups and ultra-right wingers in American society deny such rights; others demand them, while others raise the issue of mainstreaming at whose expense.

What do we mean by mainstreaming? How is the process applied to the experiences of black Americans in graduate and professional education? And, how is it related to the concept of blacks as "outsiders" in American society? What social, political, and economic conditions or factors have facilitated or restricted the mainstreaming of outsiders in recent years? They are some of the salient questions addressed in this chapter.

7

A Definition Of Mainstreaming

The term *mainstreaming* refers to a process by which an unfavored racial or ethnic group is provided opportunities to fulfill occupational roles in the same manner as members of the more favored or dominant group in a desegregating American society. The unfavored group is generally regarded as outsiders or peripheral to the American society because of previous conditions of servitude, or ascribed lower status based on racial identification, or because of their disfavored position in the overall social structure which has been relatively frozen by law and social custom. This group is subjected to categorical discrimination and exclusionary policies.

The primary mechanism for mainstreaming is an expansion of the educational opportunity structure. However, broad educational opportunities as a social process are contingent upon the degree to which such persons either have suffcent economic means to take advantage of new educational opportunities or the degree to which financial resources are made available to them from external sources. Access to the widening educational opportunity structure also depends upon the internal and external political climate of an institution, the dynamics of institutional behavior, the quality of social and political pressure brought to bear on a state, on the nation as a whole, particularly in terms of responses to pressures for enduring social change which elevate the overall status of the unfavored group.

It should be noted here that it is not always a compelling necessity for the expansion of educational opportunities to take the form of new positions or seats for unfavored members of the society. This fact applies to institutions previously closed to them and to those institutions that offered only token representation to members of outside groups. It may mean locating special programs or strengthening those already present in institutions established primarily for members of the outsider group because of de jure or de facto segregation and customary policies of discrimination and exclusion.

Hence, to mainstream black Americans is to remove restrictions and barriers constructed against their entry, individually or collectively, into the central fabric or into educational, economic and political life of a desegregating American society.

As a process, mainstreaming requires all of the characteristics that social scientists have come to label cultural assimilation.[1] At the outset, implicit in mainstreaming is the notion that individuals partaking of the process can and do speak the language of the dominant group. They have incorporated much of the dominant group's cultural norms, values, and social expectations as their own. Or, at least, these persons submit to both expediency and reality since behaving, as dominant group mandates require, is the most efficacious method of being mainstreamed.

However, persons who seek to enter the mainstream of American society confront, at minimum, two options. In the first instance, they can attempt to behave precisely as the mainstream of the dominant group behaves; that is, they must use the dominant group life patterns as a model or as normative guidelines and, as anglo-conformity philosophy demands, in the process, divest themselves of their original culture. In the second instance, such persons may compartmentalize their actions as do individuals and groups who practice cultural pluralism. In this way, they follow dominant group expectations in language, values, social expectations, and customs regarding the world of work and in the educational process but adhere to the cultural mandates of their ancestral culture in other aspects of daily life. In either case, the quid pro quo involved in mainstreaming exacts a price that those who seek to be mainstreamed must be willing to pay. Depending upon the degree of compatibility between the culture and the values of the outsiders and of the dominant group, a certain amount of cultural loss is involved for the outsiders. However, it is assumed that the cultural, economic, social and political gains achieved by being mainstreamed more than adequately compensate for cultural loss. In fact, being mainstreamed is a sine qua non for upward mobility, social acceptance by the dominant group and members of the influence establishment of the outsider group. It is the key to access, privilege, and highly prized rewards of the total society.

In the case of black Americans as outsiders, the process of mainstreaming is continuous and has a long but episodic history. This situation reflects the fact that black Americans have remained a minority group longer than any other group in the United States except the indigenous, Native American population. A brief examination of that historical experience is instructive with respect to current efforts to incorporate blacks into the American mainstream.

Mainstreaming During The Reconstruction Period

The importance of higher education as a mechanism for mainstreaming blacks was recognized several years before the Civil War began. Law excluded blacks who sought a college or university education in the southern portions of the United States and they were restricted by social custom in all other sections of the country. Occasionally, a few institutions such as Berea

College and Oberlin College in Ohio, admitted blacks on a selective basis. Before the Civil War began and ended, two black colleges were established to Christianize blacks and to provide them with an education in the liberal arts. These were Wilberforce University in Ohio and Lincoln University in Pennsylvania.[2] Between 1854 and 1952, some 123 colleges were established to serve black students who did not have access to "white institutions" because of de jure and de facto discrimination.

As the Civil War was drawing to a close, and as Reconstruction began, attention was given to the responsibility of the federal government to provide educational programs for the newly emancipated blacks. That responsibility was implicit in President Lincoln's Emancipation Proclamation of September 22, 1862, the passage of the Fourteenth Amendment which in 1868 assured "due process," and the Fifteenth Amendment which in 1870 extended the franchise to black males. The federal government's role in accelerating access to education for blacks was made even more explicit in 1865 with the establishment of the Freedman's Bureau. This agency used its authority to found Howard University in 1867 as the only federally sponsored college or university for black Americans in the United States.[3]

Federal intervention in efforts to insure the expansion of educational, economic, social, and political opportunitites for the newly emancipated black Americans was enhanced by the enactment of Civil Rights Bills in 1866 (passed by Congress over President Andrew Johnson's veto), the Civil Rights Bill of 1875 and by the radical reconstruction program, in general, that Thaddeus Stevens constructed. As a result, blacks entered grade school and secondary education programs at a comparatively rapid pace while Howard University became one of an increasingly large number of colleges established for black students in a continually segregated society.

Almost all the black colleges founded during the Reconstruction Period (1865-1877) were established by or with the assistance of Northern white philanthropists who committed themselves to the educational advancement of 4 million newly freed slaves and about one-half million free blacks classified as "free men of color" prior to the Civil War. Many of these philanthropists were affiliated with religious bodies that were also instrumental in the founding of black colleges. Among such religious organizations and missionary groups were the American Missionary Association and the Congregational Church, the Baptists, Methodists, and Presbyterians. Hence, the majority of the earliest black colleges were private institutions, founded either by blacks or by and with the assistance of white citizens.

Among the earlier private black colleges were Fisk University (Tennesee, 1866); Morehouse College (Georgia 1867); Talladega College

(Alabama, 1867); Atlanta University (Georgia, 1867); and Hampton In- stitute (Virginia, 1864).[4] According to Frank Bowles and Frank A. DeCosta, sixteen of the thirty-four historically black public colleges now in existence were established in the period between 1866 and 1890. Only Cheyney State College (1837) in Pennsylvania was founded prior to the Civil War.[5] However, all but two of these institutions were listed as normal or industrial schools and did not confer bachelor degrees. Four of them were affiliated with a religious body and came under public control after 1890. The two public institutions permitted to offer a college degree did not confer their first degrees until 1884 and 1895.[6]

Seventeen public black colleges, now in existence, were established under the so-called second Morrill Act of August 30, 1890. This act paved the way for the development of legally separated black and white land-grant public institutions in various states. As a result, within a nine year period between 1890 and 1899, one land-grant college for black students was either established or planned in each of the seventeen Southern and border states.[7] They were separate, unequal and, for the most part, could not award baccalaureate degrees at this time. The legacy of industrial, mechan- cial and agricultural education, often associated with publicly supported historically black colleges, can be traced to this period. Liberal arts educa- tion was presumed by many to be the domain of the private institutions for black college students. Importantly, despite the predominance of normal schools, elementary and secondary programs at institutions called colleges, between 1865 and 1895, more than 1,100 black Americans received col- lege degrees.[8] Hence, black institutions of higher education were in the forefront of efforts to bring black Americans into the mainstream of American life. Their first graduates, like many of those who followed, were the vanguard of leadership in the segregated black community at the turn of the twentieth century.

Dual systems of education for black and white Americans were stimulated by the 1896 U.S. Supreme Court decision in the case of *Plessy v. Ferguson*. This case clearly established the principle of separate-but-equal regarding all aspects of American life. Under this decision, states were not compelled to construct even the rudimentary components of an integrated and desegregated society. In fact, they were permitted to rigidify segregated structures since they had only to satisfy the mandate of establishing separate facilities that could be labeled equal. Suffice it to say, such facilities were in- deed separate but never equal. White privilege prevailed in every aspect of American life and notably in higher education. The proliferation of both private and public colleges for black students was a reaction to the stark reality that excluded indigenous black Americans from the rights enjoyed by all other citizens of the United States and by European foreigners who desired to attend American colleges and universities.

The advent of *Plessy v. Ferguson* was certainly signaled by the Compromise of 1877 followed by the withdrawal of federal troops from the South and the reestablishment of "home rule," and state actions to circumvent the Civil Rights Bill of 1875. Nevertheless, the successes of the black colleges during the Reconstruction period were the first systematic attempts to assure the mainstreaming of a very selected group of black Americans.[9]

The Dubois – Washington Controversy

Ideological and philosphical differences among black Americans over what were assumed to be appropriate strategies for mainstreaming surfaced in the DuBois – Washington controversy. Booker T. Washington, educated at Hampton Institute and eminent President of Tuskegee Institute in Alabama, had become the most widely known and influential black American in the United States by 1895. Thus, the entire nation listened with special attention when in 1895 at the Atlanta Exposition he gave his famous "Let down your buckets where you are" address. In this speech, Washington pointedly advocated racial separation, the cultivation of friendship with white Southerners, and industrial, agricultural, and practical education. As John Hope Franklin points out in *From Slavery to Freedom,* this speech pleased especially Southern whites who were suspicious and distressed over the classical education advocated for blacks by various Northern white persons and groups.[10] It also pleased accommodationists among the Northern white leadership undisturbed by new social boundaries between the races. Clearly, Booker T. Washington saw industrial education not as a means of propelling blacks into the mainstream of the total American society but as a mechanism through which some could be of special service to the growing black communities around the nation. Washington then, became an apostle of accommodation and separation. W.E.B. DuBois, on the other hand, was cut from an entirely different intellecutal frame. Born in Massachusetts, educated at Fisk, Harvard and Berlin, he envisioned a qualitatively different future for American blacks. In 1903, he published one of his several important books, *The Souls of Black Folk.* One of the essays in this book, entitled, "On Mr. Booker T. Washington and Others", DuBois leveled severe criticisms of Washington's "Tuskegee machine," Washington's role in what he viewed as an apostle for injustice to black people, his disregard of the debilitating consequences of the caste structure in American society, and of his relegation of higher training for the brighter minds among black youth to a level of impracticality, and of secondary importance. DuBois advocated training what he called a "talented tenth" of black intellectuals who could not only assume positions of academic and professional responsibility but who could provide sufficient leadership for rescuing blacks from the pitfalls of racial segregation and discrimination. This "talented tenth" could organize blacks to attack racial injustice as well as the devastating impediments that the caste system imposed on any effort to mainstream black Americans.[11]

These were two diverse approaches to mainstreaming; one centripetal, the other centrifugal. One advocated turning inward toward the black population itself while accommodating the harsh realities of an openly segregated and overtly discriminating society. The other admonished blacks to recognize their twoness of being a black person and an American and to prepare to claim those rights and privileges assured and safeguarded under the Constitution of the United States. It is in DuBois that the idea of preparing black Americans for occupational roles and leadership positions in the American society as a whole takes form. Blacks nurtured and developed this idea in parallel institutions of a segregated society even as laws, social customs, and categorical discrimination prevented them from open access to the institutional fabric of American society.

Black Professional Schools and Mainstreaming Roles

In the period between 1896 and 1954, black professional schools assumed the responsibility of providing professional education for black Americans. Once again, their role in this endeavor was imperative since blacks could not receive professional training at most historically white institutions in the country. Obviously, there were major exceptions, exclusively in the North, since some black students received graduate and professional training from institutions such as Harvard University, the University of Wisconsin, Yale University, and others during that period.

In the early part of this period, especially at the turn of the century, a half dozen medical schools were established at historically black institutions. While they existed, they trained almost all black physicians, dentists, nurses, pharmacists and other health care professionals. The implementation of Plessy V. Ferguson's mandate for separate-but-equal assured re-segregation, pernicious discrimination, and racial isolation in higher education. Philanthropic contributions declined during the first two decades of the twentieth century. Consequently, such instititions as Shaw University in North Carolina and Virginia Union University, both Church-dominated, were compelled to discontinue their professional schools in medicine and law.[12] Negative recommendations from the Flexner commission resulted in the demise of others. By 1915, Meharry Medical College (Tennessee) and Howard University in the District of Columbia remained as the two historically black professional schools which offered medical and dental degrees. These institutions provided excellent training in the health professions and scientific fields for black students. Attestation to this fact is evident in the observation that for a better part of this century, a distinct majority of blacks in positions of significance in medical and dental fields, either as practitioners in all-black or in predominantly white settings, in all parts of the nation, received their medical and dental school training at one of these institutions. More than 80 percent of all black physicians and dentists up until 1968 were graduates of these two institutions.

Inasmuch as no historically black college offered a doctoral degree or very much of a graduate education program, in general, prior to 1954, blacks who lived in the South were forced to travel North or West for this type of training. Similarly, Black residents of Northern States confronted quota systems that seriously restricted the number of blacks who could be admitted at any given time. This limitaiton helps to explain the paucity of blacks who received doctoral degrees, as well as other professional degrees, prior to 1970. For example, only 57 doctorates were conferred on black Americans between 1930 and 1939.[13]

Legal Attacks Against "Separate-But-Equal" in Professional Education

The underrepresentation of blacks in the professions not only meant the inability of blacks to take advantage of whatever opportunities for professional roles that opened to them in unsegregated situations, it also meant that they could not possibly meet the needs of a segregated black community.

However, as more and more black lawyers were trained, sufficient legal resources were mounted for attacks against de jure segregation in higher education. The legal route was arduous, time-consuming, and laden with psychological stress for plaintiffs involved in each suit filed before local, district, appeals and Supreme Court. In the forefront of this attack against racial injustice and in an effort to attain equality of opportunity for black Americans were the NAACP and the NAACP Legal Defense and Educational Fund (hereafter referred to as the LDF). The NAACP had already attacked discrimination in housing and in the political arena with notable success, prior to the intiation of its work during the mid-1930s on graduate and professional education. Individuals had also sued for access to professional education as early as 1933 when Thomas Hocutt unsuccessfully sued officials of the University of North Carolina for entrance into that institution's School of Pharmacy. The technicality which led to the dismissal of his case against the University was his inability to establish "eligibility."[14]

Relevant Supreme Court Cases

Given the de jure restrictions against admission of black students to graduate and professional schools in the seventeen Southern and Border States, as well as only token inclusion in Northern institutions, it is not without significance that the NAACP and the LDF directed primary and initial attention to discrimination in public institutions. To combat the assault against "separate-and-unequal," it should be noted, several of the seventeen states responded first by constructing " separate-but-equal" professional schools for black students, and, second, by establishing out-of-state tuition grants for black students. Both of these measures were calculated to circumvent efforts by blacks to gain admission to historically white public institutions. By 1933, black college graduates increasingly demanded access to professional school education. During that same year, some 97 per cent of the approximately 38,000 black students enrolled in colleges were studying at historically black institutions.[15]

Between 1935 and 1954, five Supreme court cases, which addressed the problem of equality of educational opportunity for black Americans, received national attention and illustrate the issues raised by the LDF. These five cases were: (1) *University of Maryland v. Murray,*165 Md. 478 (1935), (2) *Missouri ex rel Gaines v. Canada* (1938),(3) *Sipuel V. Board of Regents of the University of Oklahoma* (1948), (4) *Sweatt V. Painter* (Texas, 1950), and (5) *McLauren V. Oklahoma Regents* (1950). All these cases were a prelude to *Brown v. Board of Education of Topeka, Kansas* (1954). Yet, each is significant for special reasons and all are important for their collective attack against categorical discrimination in graduate and professional schools.

The Murray case in Maryland attacked both the principle of "separate-but-equal" and the policy of out-of-state tuition grants to black students as a means of excluding them from the publicly-supported professional schools within the state of their residence purely because of their race. Murray had applied to the school of Law at the University of Maryland but was offered out-of-state tuition to a law school of his choice in another state. He refused; ultimately his case reached the Supreme Court where the Court ruled in his behalf. The U.S. Supreme Court stated that, since no separate School of Law was established for blacks in Maryland and, because attending a law school other than in Murray's home State of Maryland would result in undue hardship on him, the University of Maryland was compelled to admit him to its School of Law.

Similarly, the University of Oklahoma was ordered to enroll Ada Lois Sipuel in its law School. The State also was told to desist in compelling its black students, who desired graduate and professional education, to accept out-of-state tuition grants in lieu of admission to a State-supported graduate or professional school.

Perhaps, the most significant aspect of the Gaines case in Missouri was the success realized by the LDF in establishing the doctrine that separate-but-equal could not fully satisfy the demands of dual protection; that only actual *identity* of facilities and shared use of the *same* public institutions, would suffice.[16] A hurriedly constructed Law School at all-black Lincoln University was not equal. As a result of this case, the U.S. Supreme court once again voided out-of-state tuition grants as a mechanism for excluding blacks from all-white law schools. Although the State was compelled to admit Lloyd Gaines to the University of Missouri School of Law, he disappeared before he was formally admitted and was never heard from thereafter.[17]

The attack on separate facilities achieved a new significance and added momentum in the Sweatt case in Texas. Heman Sweatt applied for admission to the all-white University of Texas school of Law and was denied. Instead, he was expected to attend the make-shift law school that the State of Texas had established for black students under the auspices of all-black Texas Southern University. He sued and when his case reached the United States Supreme Court, the Court agreed that these make-shift facilities were indeed unequal and absolutely unacceptable. Therefore, the University of Texas was ordered to admit Mr. Sweatt to its law school. However, the fundamental question underlying the *Plessy v. Ferguson* decision was not addressed: Are state mandated, racially separate educational facilities and programs *inherently* unequal?

That question was not resolved to any degree of satisfaction even when the University of Oklahoma admitted George McLauren to its doctoral program in Education. The University had subjected him to demeaning segregated facilities *within* the institution; however, the U.S, Supreme Court ordered the University to desist in these practices. In so doing, the court articulated the principle that when black students are admitted to a traditionally all-white institution, the right to equal and fair treatment within these institutions must not be violated. Blatant re-segregation of students *within* desegregated institutions was intolerable.

These five cases brought national attention to state and institutional behavior which collectively impeded efforts by black students to enter the mainstream of educational opportunity as well as to the legal successes of the LDF in attacking institutional discrimination. However, forcing the LDF to file state-by-state suits time and time again seemed to have been the strategy of resistance to desegregation and for the preservation of the policy of "whites only" at the major public graduate and professional schools. As a result, the practice of out-of-state tuition grants continued in several states not previously subject to litigation.

In some states, trepidations arose over the litigation successes of the LDF—a concern heightened by every new evidence that the Supreme Court was becoming impatient with each state's calculated resistance to equal educational opportunities for black residents. Nine years before the famous Brown decision, the conference of Deans of Southern Graduate Schools concluded in one of its studies that the black Americans' demand for equal educational opportunities within their own home state was exceptionally strong and that positive response was imperative. The conference suggested the construction of "regional" graduate and professional schools for black students and that each state would contribute its share to cover operational, management, and educational costs of these institutions. Again, this was a strategy to prevent blacks from entering the all-white state-supported institutions within the seventeen Southern and border states. The Conference report was issued in 1945 and by 1947 this idea was widely accepted by the white educational leadership but had garnered little support from black Americans.[18] Inasmuch as the the Supreme Court decisions in the Murray, Sipuel, Sweatt, Gaines and McLauren cases during this period were favorable to blacks, there seemed to be no need to think seriously about regional schools for blacks. They did not need to embrace a diversion from the ultimate goal of a fully integrated society.

On May 17, 1954, the U.S. Supreme Court ruled in the case of *Brown v. Board of Education of Topeka, Kansas* that the principle of "separate-but-equal" was unconstitutional. The Court pointed to injustices inherent in racial separation in public institutions and in 1955 ordered the various jurisdictions to embark on desegregation in education "with all deliberate speed." The relentless pressure by the NAACP and the LDF had finally paid off since that greatly needed sweeping decision against the "separate-but-equal" principle had been won. Now, the problem was implementation. What was meant by "all deliberate speed"? By its reluctance to provide more specific implementation guidelines for its unanimous decision in the

Brown case, The Supreme Court left the doors of resistance open. Philosophies of "Nullification" and "Gradualism," among others, followed. Nevertheless, a few blacks enrolled in the graduate schools of the seventeen Southern and border States immediately affected by the decision. However, their representation in the all-white professional schools of these States was miniscule and tokenism in its worst form.

Although no State-controlled medical schools were established by the Southern states for black students prior to this decision, this was not the case regarding law schools. Faculties or Schools of Law had been established at all-black Texas Southern University, Southern University (Louisiana); North Carolina Central University, Florida A & M College, and at South Carolina State Agricultural and Mechanical College.[19] Black students continued to be excluded from the all-white pharmacy, optometry, veterinary medicine, social work, architecture, and other professional schools. These exclusionary policies were the major factor in the underrepresentation of blacks in the high status professions.

In order fully to comprehend the mounting demand for inclusion and mainstreaming of black Americans in the late 1960s and 1970s, the issue of "underrepresentation" must be addressed. The basis for determining both representativeness and underrepresentation in enrollment is the proportion of blacks in a state's population. Therefore, in this view, blacks are not adequately represented in a graduate or professional school until their enrollment approximates their proportion in the state's population. There is, of course, a controversy over parameters of "underrepresentation" and "representation ." In fact, other measures may be utilized.

Representation is inextricably related to all of the processes that influence access as well as to efforts to meet a compelling social need. The latter is tied to proportionality. Hence, the issue is not simply a question of the available pool of blacks which usually implies those who have completed a baccalaureate degree with certain majors. Strict reliance on this factor is frequently a rationalization for inaction. It is a justification for doing so little to recruit more black students at college and pre-college levels. It is a position that underscores a compelling urgency to train a sufficient number of black students at every level of education and to eliminate all social, economic, political, and institutional barriers to access so that black students can be admitted with the same regularity and expectations of accomplishments as other students who are admitted to graduate and professional schools. Hence, underrepresentation is a reflection of horrendous political, economic, academic, and institutional behavioral barriers which continually exclude black students from equal access to graduate and professional schools.

Although some of these barriers will be discussed in greater detail in a following section, the point should be reiterated here that the barriers of economic deprivation, racism, and particularly racist admissions policies helped to foment the accelerating demand for social change witnessed during the sixties and seventies. Unquestionably, the collective force of such barriers explain the underrepresentation of blacks in the major professions.

Previous studies on this subject are instructive at this point. For instance, Dietrich C. Reitses reported that the proportion of blacks in medical schools remained at about 2.5 per cent of total enrollment for the decade, 1947-1956. In 1947-48, the 588 black medical school students in U.S. medical schools comprised only 2.59 per cent of total enrollment in all U.S. medical colleges. Of that number, all but 93 were enrolled at either Meharry or Howard University Medical colleges.[20] At the same time, A. Sorenson pointed out that at least twenty-six medical schools officially excluded blacks from admission.[21]

There was a twenty-four per cent decline in the proportion of black students enrolled at predominatly white medical schools between 1955 and 1962. Up to 1969, black student enrollment in U.S. medical schools never exceeded 2.8 per cent of total enrollment in all U.S. medical schools and the overwhelming majority of those students continued to be enrolled at the two historically black medical colleges. However, by 1969, the proportion of blacks in first year medical school classes had reached 4.2 per cent in that category.[22]

In the field of law, black students comprised a mere two per cent of the 65,000 law students enrolled in ABA approved schools of law in 1964—a decade after the historic Brown decision. Black student enrollment in graduate schools and in such professional degree programs as Architecture, Engineering, Pharmacy, and Optometry was one per cent or less in 1964. Graduate and professional school educational opportunities for black students were literally a sham. Admissions policies and practices were antithetical to the spirit of the Brown decision.

Impact Of The Civil Rights Movement On Access

The Civil Rights Movement of the 1960s was undoubtedly the most significant social protest movement for racial justice; for social, political, and economic equality, and for change in the pattern of race relations that this country has ever known. Consequently, special attention is given to the role of students in stimulating social and educational change, the impact of Federal government policies and Civil Rights legislation, and external factors that influenced the major educational institutions to respond in positive ways to the pressures and demands for change.

Prior to the first publicized sit-ins staged by college students at a luncheon counter in Greensboro, North Carolina, the wall of racial segregation in higher education had already been cracked by court ordered admission of a few black students to professional schools in a handful of Southern states. Similarly, a few, all-white universities had permitted black undergraduates to enroll. For instance, the University of North Carolina admitted a black student for the first time in its 162 year history in 1951. The University of Tennessee followed a year later with its first black student. Under court pressure, the University of Alabama admitted its first black student, Autherine Lucy, in 1956 but suspended her four days later following a campus riot and finally expelled her on February 29, 1956.[23] These were responses to external pressure and political forces rather than an expression of positive institutional behavior.

The first sit-ins were a demand for social rights and privilege such as eating without discrimination at lunch counters or swimming in the same pools as those used by the white population. The Student Nonviolent Coordinating Committee, formed at Shaw University as an inter-racial body in 1960, took the leadership in organizing this aspect of the civil rights movement. However, the objectives of the SNCC-sponsored sit-ins changed over the course of the decade. So did the fundamental aspirations of black students in general change whether located on the campus of the historically black institutions or the historically white colleges.

For example, in 1962, demonstrators at the University of Chicago charged that university with operating 100 segregated apartment houses. During that same year, the Federal government had to deploy federal troops to the University of Mississippi to safeguard James Meridith who had enrolled in the School of Law under a court order. The Governor of Mississippi, Ross R. Barnett, in defying the federal government, stated, "There is no case in history where the Caucasian race has survived social integration.[24] This fear of social integration connoted the racism involved in resistance to equality of access to higher education. However, black students showed as much concern about overall quality of life on black college campuses as they did about gaining access to the historically all-white professional schools.

The shift toward other concerns than opposition to segregated housing and the ability to drink at the same water fountains or eat at the same lunch counters as whites was first observed on the campuses of black colleges and universities in 1968. During that year, black students either demonstrated or seized buildings and hostages at such predominantly black institutions as

Alcorn A & M College in Mississippi, Cheyney State College in Pennsylvania, Tuskegee Institute in Alabama and Bowie State College in Maryland. Their demands included campus reforms and a black-oriented curriculum.[25]

These demonstrations, a political and social statement as they were, provided models for similar activities at such predominantly white institutions as San Francisco State College, Boston University, Northwestern University, Trinity College, Columbia University, Brandeis University, University of California/Berkeley, Cornell University and Queens College — all of which were struck by major student revolts during 1968 and 1969. By this time, the impetus for student sit-ins had assumed a character altogether different from their thrust in 1960. The students of 1968 and 1969 demanded courses in Afro-American Studies, the establishment of Black Student Unions, changes in admissions policies concerning the use of standardized tests, the recruitment of more black students, and the hiring of more black faculty on the predominantly white college campuses. In effect, not only did these students demand curriculum reforms but significantly greater access to diverse programs of study than those in which many black students traditionally enrolled. As Harry Edwards points out, these students sought "freedom. . . educational opportunity," and it did not matter whether it was gained through integration or through separation within the institutions.[26]

However, they seemed ambivalent about "mainstreaming." A part of them cried out for the educational skills needed to be competitive in the larger society and a part of them insisted upon a curriculum that would help them be more effective in serving the black community. And some of them observed that these same kinds and quality of training that would make them more effective in the black community were precisely the same things required for wider participation in the mainstream of American society. Albeit, student pressures from the inside of universities and colleges were also instrumental in extracting positive responses from institutions to increase the presence and participation of black students in graduate and professional schools. Their movement helped to transform the racial composition of colleges and universities throughout the nation.

The Civil Rights Act of 1964, which many claim the U.S. Congress enacted as a final tribute to the late President John F. Kennedy, contained the legal basis for "affirmative action" in higher education. Other legal justification were based in Executive Orders issued by President Lyndon Johnson and requirements for federal contracts mandated by the U.S. Department of Labor and Equal Employment Opportunity Commission (EEOC). Its Title VII challenged the constitutionality of many requirements for access to jobs and admissions policies then in effect in various institutional structures. Title VII provided the statutory prohibition against discrimination in admissions on the basis of race, religion, or creed. The power of the federal government to cut off funds to educational systems not in compliance with federal policy regarding desegregation was provided by the Secondary Education Act of 1965. The Equal Employment Opportunity Commission began to issue various guidelines for confor- mance to federal statutes and brought pressure to bear on educational institutions to expand educational opportunity to blacks and other dis- advantaged minorities. Clearly, some faculty members and administrators demonstrated a commitment to the democratization of graduate and pro- fessional education. Consequently, they helped to reformulate recruitment and admissions policies that would lead to changes in the racial composi- tion of many institutions in all parts of the nation.

As will be amplified in succeeding chapters, the private sector was par- ticularly influential in efforts to increase the presence of black students in graduate and professional schools during the sixties and into the 1970s. Major corporations, responding to pressures from the federal government to hire more blacks, realized that the paucity of blacks in such fields as engineering and acrhitecture could only be relieved by dramatic changes in the rate of participation of black students in academic programs which prepared them for occupational roles in the corporate structure. Groups like the Association of American Medical Colleges, the Association of American Law Schools, and the Council on Social Work Education, among others, enunciated policies that demonstrated a firm commitment to *in- clude* blacks rather than drive them away from these professions.

Career Choices of Black Students

The success of these efforts was and continues to be determined in part by the shifting career choices of black students as a greater array of options become available to them. The black population, regardless of social class, has always stressed education as the primary avenue to upward mobility. The black middle-class, particularly, has consistently given high priority to the acquisition of post-secondary education not only for its economic benefits but for its social rewards in terms of status, prestige, and presumed cultural attainment. As long as de jure segregation was supported and as long as de facto segregation and systematized patterns of discrimination persisted, the skills obtained through post-secondary and professional education largely served a black population in a segregated racial community. The primary career open to black college graduates was teaching. A selected few could enter medicine and dentistry; fewer entered law because the opportunities for practicing that profession were seriously wanting. Their entry into such professional fields as optometry, pharmacy, and veterinary medicine is a relatively recent phenomenon compared to their entry into the teaching profession. But, orientation to social welfare services is also of long standing.

Career choices made by black Americans, therefore, reflect their historical experiences with racial isolation and oppression, exclusionary policies of higher education, economic deprivation, and discouragement from considering the possibilities of other options. Given these conditions, the black community was limited in the variety of professional role models that it could provide to black youth. However, because of the proportionately higher number of educated blacks who lived in this closed community, it is important to observe that the segregated southern black communities had a far greater number of highly visible role models than did black communities outside the South. Nevertheless, even before 1954, when desegregation was constitutionally sanctioned, innumerable poor, working-class black fathers and mothers impressed upon their children the importance of "being better educated than they were," of "making something out of yourselves," in order to obtain higher status jobs, a significantly improved economic condition and more effective means of coping with racial injustice and discrimination.

Shifts in career choices of black students began in the late 1950s and have continued into the present time. As previously indicated, prior to the fifties, black college students focused almost exclusively on teaching and social service as career choices. With the implementation of desegregation, black students were exposed to considerably more information about career possibilities. They encountered counselors who had the information to impart and who could advise them on how to obtain financial resources when needed. They were in contact with academic advisers who could suggest courses appropriate for their new career objectives, They also benefitted from social encounters with peers interested in a full range of career possibilities that black students dared not dream about prior to desegregation. Their level of aspiration rose appreciably in this context. Consequently, distribution patterns in career choices began to change.

Studies by such researchers as Robert L. Crain and Patricia Gurin and Edgar Epps substantiate the thesis of shifts in career patterns consequent to desegregation. Crain observed that black students in a desegregated school setting increased their level of aspiration for a variety of careers far beyond the level of aspiration for careers demonstrated by black students in predominantly black institutions.[27] In a survey of 535 black college seniors from six institutions in 1964 and 536 black college seniors in 1970, Gurin and Epps found a continuing commitment to the professions but notable shifts in career aspirations. For instance, the proportion of black males and females who in 1964 aspired to be physicians declined by 50 per cent in 1970 (i.e., from 8% to 4% and from 2% to 1%, respectively). The proportion of men who aspired to be public school teachers dropped from 22% of the total in 1964 to 13% in 1970. On the other hand, the proportion of black college seniors who aspired to be accountants doubled among men and tripled among women.[28] Similarly, increasing numbers of black college seniors aspired to be lawyers, engineers, architects, and public advisors, and to acquire other types of professional careers.

In the late 1960s and into the 1970s. the career choices of black college students, as well as black high school students, were undoubtedly influenced by environmental cues over and beyond those found in the immediate family or school situation. Information obtained from the printed and electronic media provided secondary role models and subtle cues, if not subliminal suggestions about career possibilities. The United Negro College Fund's advertising campaign showed black students in a variety of settings; commercials from oil companies, engineering and computer firms, and business and industry in general began to depict blacks in diverse occupational roles. Some commercial television programs utilized black actors in what was for the industry "nontraditional" occupational characterizations. Hence, some blacks were physicians dentists, nurses, lawyers, and atheletes. All of these factors contributed heavily to the changing career choices of black students. While they may help to explain the entry of a greater number of blacks into diverse professional programs, they do not explain the continuing dominance of education as a major among black recipients of the doctoral degree. That choice is apparently explained by other factors. Notwithstanding, the evidence shows that black students entered into more diverse professional fields in the late sixties and seventies than at any time in their history. They did so despite their continuing condition of economic deprivation, discrimination, and suspicious institutional behavior.

Economic Barriers

The five economic barriers to higher education are: (1) high rates of unemployment, (2) inadequate income to support students in higher education programs, (3) low occupational status, (4) limited access to scholarships and fellowships and (5) fear of excessive indebtedness resulting from loan programs. Specific attention is given in this section to the first three barriers while the remaining two, which are more specifically student-related, will be discussed in subsequent chapters.

For approximately thirty years, the unemployment rate among black Americans has been consistently about twice that of the white population. As indicated in Table 1, the unemployment rate among white Americans was 4.9 per cent in 1950 but 9.0 per cent for blacks and other races. In 1954 white unemployment rose to 5.0 per cent while the rate for blacks went to 9.9 per cent or a black to white ratio of 2:1. That ratio remained at 2:1 or higher for every year thereafter except for a few years in the 1970s when it dipped to 1.8:1 among known job-seekers. However, when the number of discouraged workers is added to this figure, the ratio climbs far beyond 2:1. In addition, black teenagers are three times more likely to be unemployed than white teenagers; consequently, they cannot save as much to support a college education.

Unemployment inevitably means low income and this situation indicates that impoverished people who barely have sufficient income to meet daily basic and survival needs will have little or no funds to support a college or professional education. As shown in Table 3, black families are paid less than sixty per cent of the income paid to white families in the United States. Only in 1969 did the median family income of black families exceed more than 60 per cent. However, black families are receiving proportionately less income today than they did a decade ago. Even though the dollar income of blacks and whites is increasing, the gap between the median income of whites and blacks is also widening at an alarming rate. (Table 3)

Low income and unequal income reflect low occupational positions held by black Americans compared to positions held by whites as well as the persistence of institutional racism in hiring practices. Although substantial improvements have taken place in the occupational position of black Americans between 1950 and 1980, blacks continue to be found in the lower status positions compared to white Americans. According to the latest available data from the U.S. Department of Labor, white males, for instance, outnumber black males in white collar jobs by an almost two-to-one margin (42 per cent and 23 per cent, respectively). White males outdistance black males in professional and technical positions by more than two-to-one (i.e., 15 per cent white compared to 7 per cent black). The white male percentage in managerial and administrative positions is three times that of black males in such positions. It is only in blue collar and service jobs that black males outnumber white males by substantial margins.

Table 1. Unemployment Rates for Black and White Americans: Aged 16 and over, 1950 - 1980* (Annual Averages)

Year	Unemployment Rate		Ratio: Black and Other Races to White
	Black and Other Races	White	
1950	9.0	4.9	1.8
1951	5.3	3.1	1.7
1952	5.4	2.8	1.9
1953	4.5	2.7	1.7
1954	9.9	5.0	2.0
1955	8.7	3.9	2.2
1956	8.3	3.6	2.3
1957	7.9	3.8	2.1
1958	12.6	6.1	2.1
1959	10.7	4.8	2.2
1960	10.2	4.9	2.1
1961	12.4	6.0	2.1
1962	10.9	4.9	2.2
1963	10.8	5.0	2.2
1964	9.6	4.6	2.1
1965	8.1	4.1	2.0
1966	7.3	3.3	2.2
1967	7.4	3.4	2.2
1968	6.7	3.2	2.1
1969	6.4	3.1	2.1
1970	8.2	4.5	1.8
1971	9.9	5.4	1.8
1972	10.0	5.0	2.0
1973	8.9	4.3	2.1
1974	9.9	5.0	2.0
1975	13.9	7.8	1.8
1976	13.8	7.0	1.9
1977	13.1	6.8	1.9
1978	12.4	5.5	2.3
1979	11.5	5.6	2.1
1980	14.2	7.8	1.8

*Adapted from Table 47 in *The Social and Economic Status of the Black Population in the United States: An Historical View, 1790 - 1978* (Current Population Reports, Special Studies Series P-23. No. 80). Washington, D.C.: U.S. Department of Commerce, 1979 *and* from Robert Hill, *The Illusion of Black Progress.* Washington, D.C.: The National Urban League. 1979, p. 33.

Table 2. Distribution of Families by Race and Income for Selected Years: 1953 to 1974*

Percentage of Families in Each Income Bracket by Year and Race

Income $$$	1953 Black	1953 White	1959 Black	1959 White	1964 Black	1964 White	1969 Black	1969 White	1974 Black	1974 White
Under $1,500	16	7	14	4	8	3	5	2	4	2
$1,500- 2,999	16	7	19	6	13	5	10	3	10	3
$3,000- 4,999	23	11	19	10	20	9	15	7	17	7
$5,000- 6,999	18	16	16	12	18	10	15	8	14	8
$7,000- 9,999	16	26	17	23	18	18	19	14	16	14
10,000-11,999	6	13	7	16	8	13	9	11	8	11
12,000-14,999	9	3	5	12	7	15	12	16	11	15
15,000 & over	2	11	4	19	8	28	17	40	19	42

*Adapted from Table 15 in *The Social and Economic Status of the Black Population in the United States: An Historical View, 1970-1978*, (1979), p. 32.

Table 3. Ratio of Black Median Family Income for Selected Years: 1953-1979

Year	Income Black (per cent of White Income in parenthesis)	White
1953	$ 4,547 (56%)	$ 8,110
1959	$ 4,931 (52%)	$ 9,547
1964	$ 5,921 (54%)	$10,903
1969	$ 8,074 (61%)	$13,175
1974	$ 7,808 (58%)	$13,356
1976	$9,242 (59%)	$15,537
1979	$11,650 (57%)	$20,520

Adapted from Table 19 in *The Social and Economic Status of the Black Population in the United States: An Historical View, 1790 - 1978*, p. 36, and data supplied by the Boston office of the U.S. Department of Commerce.

Fifty-eight percent of black males are in blue collar jobs, compared to 45 per cent of white males in this category. The percentage of black males in service positions is more than double that of white males in these positions: 17 per cent of black males, compared to 8 per cent of white males.[29]

Black males earn only two-thirds of the income received by white males when all occupations are combined. According to Robert Hill's analysis, black male professionals earn about 73 per cent of the income received by white professionals while black managers are paid 77 per cent of the amount of income received by white managers. On the other hand, black laborers are paid approximately ten per cent more than white laborers.[30] The fact that black females are currently achieving greater parity with white female workers in the same occupational areas does not obfuscate the reality that black workers, in general, are in lower status occupations with lower incomes. This situation results in untoward economic disabilities which dramatize the need for greater financial assistance to black students who wish to enter college and graduate and professional schools. It is estimated that over three-quarters of all black students come from families whose median family income is less than $12,000 at a time when the cost of graduate and professional education continues to climb out of the reach of lower income families.

The mainstreaming of outsiders is not solely dependent upon the removal of economic barriers; it is also conditioned by the systematic removal of serious eductional barriers to college, and graduate and professional school enrollment.

For instance, increasing the available pool necessitates reduction in the high school drop-out rate and increasing the number of both college-going and college-graduating black students. The median number of years of schooling for blacks (between the ages of 25 to 34) is 12.5 years compared to 12.8 for the total population in the same cohort. Although the proportion of black eighteen year olds who are high-school drop-outs continues to decline, the rate still hovers about eighteen percent compared to approximately ten per cent for whites. However, between 1950 and 1975, the proportion of black males who had completed four years of high school increased by 390 per cent and the proportion of black women high school graduates increased by 416 per cent. It is estimated that approximately 72 per cent of all black Americans (compared to 82 per cent of white Americans), in the 20-24 year age bracket have completed high school. Among the college trained, between 1950 and 1975, the proportion of black males who had completed one to three years of college rose by 388 per cent while the proportion of black males who had completed four or more years of college increased by 380 per cent. The increases for black females in these two categories were 333 per cent and 315 per cent, respectively.[31] One must also keep in mind that the rates for the white population increased substantially.

These figures point to dramatic changes in the available pool of black students for admission into graduate and professional schools, even if we assume that many will not meet academic requirements for acceptance in these programs. Further, the reality that more and more black youths are college bound is indicated by some 426,000 more blacks enrolled in college in 1975 than there were in 1970. By 1979-80 the number of blacks enrolled in college exceeded 1,200,000. However, the enrollment profile is complicated by the relatively low persistence rate for black college students and by the 50 per cent in Community Colleges—in technical programs. This low persistence and retention rate explains why blacks comprise only 8 or 9 per cent of all college seniors. Nevertheless, the percentage of blacks who have graduated from college increased from four per cent of the total black population in 1960 to six per cent in 1970 and then to eleven per cent of all blacks in 1975. (The figures for white Americans were 12, 17 and 22 per cent, respectively.) As a result, the pool from which candidates for admission to graduate and professional degree programs has also expanded significantly.

Characteristics of The 1970s

Many of the conditions previously described continued to affect the overall participation of black students in graduate and professional schools in the 1970s. In addition, other social, political, and economic conditions have exacerbated efforts to stimulate increased enrollment and graduation rates among black students at these levels.

Three sets of litigation during the 1970s are important for clarifying the access issue but for entirely different reasons. These cases are the *Adams v. Richardson* case, the *DeFunis v. Odegarrd* case, and *The Regents of the University of California v. Allan Bakke* case.

The *Adams v. Richardson* case, also known as *Adams v. Califano,** dates back to 1969 when a suit was filed with the assistance of the LDF against the Office of Civil Rights and the Department of Health, Education and Welfare for not enforcing those provisions of Title VI of the Civil Rights Act of 1964 applicable to dual systems of higher education. (The suit takes its name in part from the fact that Elliot Richardson was HEW Secretary when the suit was initiated and Joseph Califano was HEW Secretary throughout most of the important decisions regarding resolution

---

*In 1981, this suit was re-named to become Adams v. Bell; thus, it recognizes a new Secretary of Education.

of the conflict between plaintiffs, the Federal Government, and the State systems, educational policies of which were challenged). Nine Southern states and the State of Pennsylvania were charged with continually operating dual systems of higher education that effectively channeled black college students into predominantly or historically black colleges and white students into the historically white institutions.

On June 12, 1973, the Department of Health, Education and Welfare was ordered by Judge John Pratt of the U.S. District Court of the District of Columbia to terminate federal funds to those states that continued to violate the Civil Rights Act of 1964. The ten States involved were Arkansas, Florida, Georgia, Louisiana, Maryland, Mississippi, North Carolina, Oklahoma, Pennsylvania, and Virginia. Collectively, these States became know as the "Adams States." (Tangentially, the ruling affected all seventeen Southern and border States in which an historically black college is located.) These States were required to submit plans for dismantling their dual systems, and for demonstrating substantial movement toward assuring equality of opportunity in access, distribution and retention of students, faculty and administrators in state systems. Since 1973, because of various legal challenges, focus of attention has been on seven States; Arkansas, Florida, Georgia, Maryland, North Carolina, Oklahoma, and Virginia. The unresolved question is the degree to which these and other affected States have complied with specific aspects of the court order pertaining to the enrollment of black students in their graduate and professional schools. That determination is one of the objectives of this study.

Both the DeFunis and Bakke cases challenged the ligitimacy of special programs designed to bring relief to black students through rectification of the consequences of past patterns of discrimination. Both cases arose in non-Southern states and both raised questions about the utilization of subjective criteria, such as racial and ethnic designation, as opposed to the so-called neutral or objective criteria in admission decisions. Both argued that the utilization of subjective criteria in ways that granted special favor to black applicants for admission to the professional school of law and medicine constituted what has come to be characterized as "reverse discrimination." The plaintiffs further challenged admission quotas for minority students in general and "set-aside" programs that guaranteed a specific number of seats to minority students. There did not appear to be any concern in these cases for rectification of past policies of categorical discrimination which excluded black students from graduate and professional school programs.

These two suits, DeFunis in Washington and Bakke in California, focused national attention on the myriad problems inherent in the assumption that objective or neutral criteria for admission, such as MCAT, LSAT and GRE test scores, grade point averages, and rank in class, are free from bias and select only the most meritorious from among any group of students. The arguments heard in these cases, especially the Bakke case, illuminate the many instances of exceptionality regarding merit when these standards are applied to white students as well as their abuses when applied to black and other minority students. This debate provided disturbing evidence that many persons had now taken the position that "blacks had had their chance," that "they should not get ahead at the expense of whites," and, that whatever commitment that any number of individuals, organizations, and to some degree the federal government, had had in the past to aggressive equal educational opportunity programs had all but dissipated. While the Bakke case was in litigation, between 1974 and 1978, institutions seemed to have been in a holding pattern. The question here is to what degree is this period characterized by changes in recruitment and enrollment patterns of black students in graduate and professional schools?

The decision reached by the supreme court in the DeFunis case did not resolve the issue since by the time the case reached this Court, it was moot: DeFunis was already enrolled in law school. On the other hand, the decision reached in the Bakke case was significantly more far-reaching in its implications and it will be discussed in the Chapter on the Medical Education of Black Students.

In addition to allegations of reverse discrimination and challenges to special admissions and recruitment programs, the 1970s were characterized by fluctuations in financial aid and scholarship programs, the development of retention programs to influence educational outcomes, and escalating attack on standardized tests that culminated in the enactment of a "truth in testing law" in New York State. The debate on the efficaciousness of coaching to raise test scores reached the U.S. Congress. Admonitions that retrogression was occurring in efforts to safeguard what limited access to graduate and professional schools that blacks had experienced were sounded in all parts of the nation.

Clearly, the political climate has changed from the upheavals of the 1960s and the progressive orientation toward the resolution of social, economic, and political injustices that these upheavals spawned. A new conservatism has indeed appeared and misconceptions about access of black students to the detriment of white applicants are encouraged by both the printed and electronic media. Our evidence in succeeding chapters will demonstrate that just as enrollment among black students increased during the 1970s, overall enrollment rose in significant leaps for American students as a whole. Succeeding chapters will address these issues with greater depth and specificity.

In addressing the problems elaborated in this book, it must be stressed that all black and all minority students do not fall into the category of "underprepared, under-financed and problem-inflicted" students. A significant proportion of these students are more than adequately prepared, often exceptionally well-prepared for graduate and professional school programs regardless of the field of study. It is a tragic mistake to assume that blacks and other minority students represent a homogeneous type and it is decidedly worse to treat them as if all required special, and all too frequently, demeaning paternalistic attention. These students simply want equality of opportunity, a fair and equal chance, respect for themselves for the individuals they are, the same quality of training offered to others and unbiased evaluations of their performance.

Footnotes

1. Milton Gordon, *Assimilation in American Life.* New York: Oxford University Press, 1964.
2. James E. Blackwell, *The Black Community: Diversity and Unity. New York: Harper & Row, 1975.*
3. Lerone Benett, *Before the Mayflower: A History of Black America.* Chicago: Johnson Publishing Company, 1969, pp. 400-403.
4. *Ibid.*
5. Frank Bowles and Frank A. DeCosta, *Between Two Worlds: A Profile of Negro Higher Education,* New York: McGraw—Hill Co., 1971, p. 32 and Appendix B-34.
6. *Ibid*
7. *Ibid*
8. Blackwell, *Op. Cit.,* pp. 5-10.
9. John Hope Franklin, *From Slavery to Freedom.* New York: Alfred Knopf Co., 1952, pp. 385-387.
10. W.E.B.DuBois, *The Souls of Black Folk.* Chicago: 1903.
11. *Op. Cit.,* p. 538
12. Commission on Human Resources, *Minority Groups Among United States Doctorate Level Scientists, Engineers and Scholars,* 1973. Washington, D.C.: National Academy of Sciences, 1974.
13. John Hope Franklin, *Op. Cit.,* p. 541.
14. *Ibid.,* p. 538.
15. Richard Bardolph, *The Civil Rights Record.* New York: Thomas Y. Crowell, 1970, p. 271.
16. *Ibid.*
17. *Op. Cit.,* p. 542.
18. Cf. Lerone Bennett, *Op. Cit.,* p. 204. It is somewhat ironic that South Carolina had moved so deeply into rigid segregation since its major all-white university had been

desegregated during the Reconstruction period. In 1873, Richard T. Greener, the first black graduate of Harvard University, was named professor of metaphysics at the University of South Carolina—a university which then had a racially integrated student body and Board of Trustees. By the turn of the century, with the re-establishment of home rule and the advent of Plessy v. Ferguson, the nation had once again experienced a triumph of white supremacy in all aspects of day-to-day living. The actions of South Carolina, therefore, in establishing a make-shift law faculty for black students was characteristic of the time.

19. Dietrich C. Reitzes, *Negroes and Medicine*. Cambridge, Mass.: Harvard University Press, 1958.

20. A.A. Sorenson, "Black Americans and the Medical Profession, 1930-1970", *Journal of Negro Education.* 41 (Fall, 1972), 337-342.

21. Cf. James E. Blackwell, *Access of Black Students to Graduate and Professional Schools.* Atlanta: The Southern Education Foundation, 1975 and James E. Blackwell, *The Participation of Blacks in Graduate and Professional Schools: An Assessment.* Atlanta: The Southern Education Foundation, 1977.

22. Lerone Bennett, *Op. Cit.,* pp. 420-423.

23. *Ibid.*

24. Harry Edwards, *Black Students.* New York: The Free Press, 1970.

25. *Ibid.*

26. Robert L. Crain, "School Integration and Occupational Achievement Among Negroes." *American Journal of Sociology.* 75 (1970), 593-606.

27. Patricia Guren and Edgar Epps. *Black Consciousness. Identify and Achievement.* New York: John Wiley & Sons, 1975. pp. 39-43.

28. *The Social and Economic Status of the Black Population in the United States: An Historical View,* 1790-1978. Washington, D.C.: The U.S. Department of Commerce Bureau of the Census, Special Studies Series p. 23, No. 80, 1979, p. 218.

29. Robert Hill, *The Illusion of Black Progress.* Washington, D.C.: The National Urban League, 1978, p. 36.

30. James E. Blackwell, Social Factors Affecting Educational Opportunity for Minority Group Students," Chapter 1 in The College Board, *Beyond Desegregation: Urgent Issues in the Education of Minorities.* New York: The College Entrance Examination Board, 1978, pp. 1-12.

31. *Ibid.*

CHAPTER 2

Methodological Issues and Reference Points

This study has two basic components. This first component is based upon an analysis of exploratory survey data on 743 cases (i.e., professional schools) covering eight professional fields of study offered at one or more of 427 institutions in the United States. The eight professional fields are (1) Dentistry, (2) Engineering, (3) Law, (4) Medicine, (5) Optometry, (6) Pharmacy, (7) Social Work and (8) Veterinary Medicine. The disparity between cases and numbers of institutions is attributed to the fact that some institutions offered multiple opportunities for professional education. The second component of the study is a trend analysis of first year enrollment, total enrollment and graduation rates of black students from professional schools during the 1970's. In fact, the time period covered by both components of the study is 1970 through 1979. In addition to the eight professional fields included in the first component, the trend analysis also includes selected data on architectural education and the production of doctoral degrees by race. In both instances, comparative data, especially between Blacks and Caucasians, are presented.

Objectives of The Study

The principle objective of this study is to ascertain answers to the following questions:

1. What changes occurred between 1970 and 1979 in the access of black students to graduate and professional schools? Has access improved, remained static, or retrogressed?

2. To what extent were the observed changes influenced by such factors as: special recruitment and special admissions programs, quality of financial aid and scholarships offered to black students, the number of black faculty in the professional school, the presence of a minority affairs officer, the presence of a retention program, the number of professional schools within

34

the State, the proportion of blacks in the State population, the presence or absence of a black college in the State, whether or not the State is under litigation to dismantle a dual system of higher education (i.e., the "Adams States"), the location of the institution, and the type of institution (e.g., private, public or State-assisted).

3. What happened in the "Adams States" with regard to increasing access of black students to graduate and professional schools? Were the observed changes in access and production in those States different from those patterns noted in other parts of the nation? If so, what factors appear to account for those differences?

4. To what degree can role model theory be tested through an empirical analysis of recruitment, admissions, enrollment and graduation rates at selected institutions?

5. What appears to be the major impediments to increasing equality of educational opportunity in graduate and professional schools? How can these problems be alleviated?

6. To what degree did the U.S. Supreme Court decisions in the DeFunis and Bakke cases have a "chilling effect" on the recruitment and enrollment of black students in graduate and professional schools?

7. How salient are such factors as institutional commitment and behavior, assistance from the private and corporate structures, and governmental intervention for facilitating matriculation of black students as well as their graduation from graduate and professional schools?

8. How can the responses to the above sets of questions instruct us as to what factors appear to be the most salient predictors of enrollment and graduation and what policies appear to be successful in fostering minority student enrollment and graduation from graduate and professional schools? And, finally, what do responses to such questions as those raised above suggest as future courses of action in the 1980's and beyond for achieving the goal of equality of educational opportunity for all Americans?

A Theoretical Viewpoint

This research was not undertaken to garner support for a specific theoretical perspective. Nor was it intended to justify the overarching utility of a single factor theoretical approach as opposed to employing several marco-theories to explain social phenomena. The relative strengths of both approaches are demonstrated in sociological literature.

A case can be made for approaching the experiences of black students in attempting to gain access to graduate and professional schools from either a functionalist or from a Marxist/conflict theoretical perspective. Functionalist proponents would likely argue that changes in access of minority students to professional education occurred through a series of necessary evolutionary stages beginning with a denial of opportunity to these students, toward a more advanced state of limited and imperfect access and finally, if the system required it, the attainment of a mature stage of complete equality for black Americans. Since functionalists would probably perceive some form of discrimination as a functional imperative in order to make the social system work, it is doubtful if blacks or any other racial minority would ever reach the final or mature stage of complete equality. In my view, this approach is not only static and supportive of the status quo, as so many of its critics have maintained, it underestimates the potential influence of disruption in the social system generated by both internal and external forces in creating major changes in the educational subsystem.

Neo-Marxists and class conflict theorists would undoubtedly conceptualize the problem strictly in terms of a struggle between classes. In this view, inequality of access may be perceived as evidence of the purposeful utilization of education as a means to solidify and rigidify class inequality. In this view, since black Americans are overrepresented in the lower income classes and since upward mobility for them proceeds at an intolerably slow pace in a racist society, the continued denial of equality of access to the educationl opportunity structure assures a prolonged state of occupational servility and overall subordination to the dominant group. Although both models, functionalism and conflict theory, contain attractive elements, neither is totally adequate nor sufficiently encompassing of the critical factors which explain inequality of access as does power theory![1]

Consequently, for purposes of this study, certain sociological concepts have a higher degree of salience in facilitating understanding of access as a process (e.g., *equalizing* educational opportunities between minorities and majorities). Of special importance are such concepts as structural inequality, norms, authority, power and social status. As a personal choice greater attention is given to the theoretical notions of power to explain the process of access and outcomes (e.g., enrollment through graduation from graduate and professional schools) than to other perspectives.

Chapter 1 cast efforts to propel black Americans into the mainstream of higher education in terms of legalistic attacks against discrimination that culminated in the Civil Rights Movement of the late 1950s and of the 1960s. A central goal of that movement, which the student movement subsequently re-articulated, was to eliminate structural inequality and to change institutional norms. Pressures for these changes in higher education arose from evidence that prevalent institutional policies resulted in discrimination against blacks and other minorities and did not promote democratic partcipation in decision-making processes. Hence, it was imperative for colleges, universities, and their graduate and professional schools to overhaul or transform substantially their racial composition and to revolutionize those methods of making decisions which impacted on the lives of matriculated students and on the very nature of the labor force itself. In this context, concepts such as "shared authority" and "participatory democracy" achieved widespread popularity.

However, these movements also sought to break through the demeaning stereotypes about black Americans and other minorities which were sufficiently prevalent in graduate and professional schools as to influence decisions regarding the admissibility of these students. For example, one may conclude that some selection or admissions officers held stereotypes concerning the "educability"of black students or about persumed intellectual differences between blacks and whites. If those stereotypes were strongly believed, then decisions on admissions would reflect that belief system. Those officials were likely to consider only those blacks who were clearly superior to all candidates for that was the only way that they could be regarded as "equal" to white candidates whose objective scores were at the lower end of admissions cutoff points. Assuming a certain ubiquity of these types of stereotypes as normative in educationl institutions, we need to focus also on institutional behavior regarding the access process.

Institutional behavior is an intervening variable of special significance for any theory advanced to explain the process involved in creating greater equality of educational opportunity. It is a multi-dimensional variable which embraces such components as commitment, authority, power, sanctions, and a concern for status and image within the network of institutions of higher education. Strong and positive institutional behavior regarding access of black students is indicated by forceful leadership, the allocation of resources to achieve articulated goals, and modifications in policies proven to be impediments to equality of opportunity.

The authority validating institutional behavior may be either centralized or decentralized. Centralized authority is likely to be more effective in inducing sustained change under certain circumstances. For example, it is a function of the degree of commitment of the decision-makers in the hierarchy of authority to the primary goal and the willingness of that authority to invoke institutional sanctions against those units within the university which do not comply with mandates to recruit, admit, and enroll more black students.

It is also highly probable that certain decentralized units within the institutional community (e.g., a school, college, or a department) may be considerably more inclined to take steps in the direction of educational change than are the leaders at the top of the authority structure. In such instances, it is necessary for them to have a least a tacit agreement that the top echelons of the leadership structure will not undermine or subvert their efforts to achieve this goal. Support or lack of support for this endeavor may be a function of the leaders' perceptions as to whether or not a significant change in the racial composition of the department, college, school, or university will alter their status or ranking within the network of educational institutions or in the eye of the public.

Similarly, formal organizations and associations, such as accrediting bodies and coordinating functionaries, have a great deal of both manifest authority and power as well as implied authority and latent power. Transformations in institutional behavior may be related to the type of pressure brought to bear upon institutions by related associations as well as by an even more external structure—the federal government. The former may use their influence, persuasive powers, and overall organizational support to encourage changes in institutional norms. Some are in a position to invoke sanctions to assure compliance, such as loss of accreditation or some forms of financial support which may be obtainable only though associational approbation.

In the case of the federal government, informal strategies of persuasion may be employed as a means of fostering greater equality of educational opportunity. However, the federal government alone has at its disposal the ultimate weapon of power—cut-off all financial support from the government—against institutions whose behavior violates established governmental norms and policies. That action is, perhaps, its most obvious manifestation of power. However, power may take many forms and routes to social change. Just as the government has maximum power in certain critical areas, the power and authority of institutions and of their units may also be quite broad regarding the processes of access and graduation from graduate and professional schools. Institutions may use their discretionary authority in the area of admissions as manifestations of their power to either *exclude* of to *include* certain groups of students.

Power may be conceptualized as the ability to control decision-making processes regarding access to and distribution of social, educational, economic, and political rewards offered within a social system. In societies characterized by dominant and minority groups, that ability to control is vested primarily in the hands of the dominant group. Members of the dominant group, irrespective of social class and position in the social structure, regard themselves as more worthy and more entitled by virtue of their membership in the dominant group to a greater share of this control than are members of subordinate or outsider groups. Dominant group members come to believe in their entitlements and actualize their claims over them. Consequently, they are not prepared to share their access to social rewards, economic gains or opportunities in the educational and political sub-systems without pressure.

Dominant group elites, leaders and decision-makers are in a unique position since they have the power to deny outsiders sufficient access to scarce values which foster social, political, and economic equality, as well as educational parity with dominant group members. They not only institutionalize their own standards of conduct as normative but they use their influence and authority to regulate normative behavior for members of the social system as a whole. However, a subordinate group is not entirely powerless since it may possess one or more of those three sources of power described by Robert Bierstedt. These sources are (1) the size of the group, (2) the degree of social organization within a specific group, and (3) the group's resources.[2]

The size of a group is not a necessary precondition for gaining access to scarce resources or highly prized values. The quality of a group's social organization and its ability to capitalize on its own sense of social cohesion may be of far greater salience for changing power relations. Possessing resources alone is not adequate; it is the ability to mobilize successfully those resources—ultimately a manifestation of sophisticated social organization—that becomes paramount. As Hubert M. Blalock asserts, the resources are not only expressed in terms of money, status, and property but they may be of one of two kinds—pressure or competitive resources. The pressure resources possessed by a group refer to its ability to use coercive techniques to exact rewards from another (i.e., a boycott in order to obtain jobs) whereas competitive resources are exemplified in expertise that is needed by the dominant group.[3]

Just as dominant groups have control over opportunities to improve their member's conditions in the economic structure through provisions of better jobs, equal wages and a reduction in unemployment, they also have control over decisions on admissions to graduate and professional schools. During the 1960's, blacks employed both pressure and competitive resources in order to transform the traditional admissions practices and to alter institutional norms. Even though the ultimate responsibility for admissions decisions in graduate and professional schools remained in the hands of the dominant group, the pressure of blacks on admissions committees did result in positive changes. These outsiders discovered first-hand evidence of the ability of dominant group decision-makers to modify traditional admissions criteria and to make "adjustments in objective assessments" for both purposes of selectivity and preferential treatment for members of the *dominant* group.

Whenever dominant group members feel threatened by the encroachments of subordinate group members on their perceived entitlements, they increasingly resist the organized efforts undertaken by subordinate group members to achieve greater gains. That resistance takes any number of different forms, such as new legislation that regulates "normative behavior" as defined by the dominant group, systematized attacks on subordinate group member in terms of attributed negative traits and on their presumed inadequacies, denial of privilege or alterations in requirements for admission to the mainstream itself, and allegations of "reverse discrimination." Resistance, as evidenced perhaps in legal assaults, may result in retrogression or retardation of subordinate group advances. Consequently, the struggle for equal opportunity through changes in power relationships seems continuous for outsiders.

The Universe of the Study

An initial goal for the first component of this study was to include all institutions in the United States that offer a professional degree in the fields selected for analysis. This objective would have meant including all 102 schools of architecture; 124 schools of medicine; 168 American Bar Association approved schools of law; 59 schools of dentistry; 13 colleges of optometry; 80 colleges of pharmacy; 287 schools of engineering; 22 colleges of veterinary medicine; 88 schools of social work and some 220 institution listed by the American Council on Education as Ph.D. granting institutions. This would have yielded a statistical universe of 1,162 professional and graduate degree programs or cases.

The constraints of manageability, feasibility and the likelihood of institutional responses ultimately required modifications in the original goal for an inclusive universe of the study. Preliminary testing indicted major difficulties in obtaining data directly from institutions on doctorate students by field primarily because of the tremendous amount of labor required by each institution. That idea was abandoned and selected institutional data were obtained from other sources. Interviews with key individuals knowledgeable about architectural education suggested extreme difficulty in obtaining data of sufficient quality from schools or colleges of architecture for the time frame covered by the study. Hence, the attention given to architectural education was reduced. Consequently, the major fields of study covered by the survey research portion of this analysis included a potential total of 840 cases. Selected institutional data, coupled with information obtained through personal interviews, and data from governmental sources comprise the limited data sets on architectural education.

Data Collection Procedures

Description Of The Survey Inventory:

Data for this study were collected in two overlapping phases. Phase I of the study consisted of data collection through a mailed survey. A survey inventory was constructed for distribution to the deans of the colleges or schools of specific professional programs (e.g., the Dean of the School of Law, the Dean of the School of Medicine, the Dean of the School of Dentistry, and so on, were each forwarded a copy of the instrument even though all may have been physically located at the same university). Each dean was requested to complete only those sections of the twenty-two items in the inventory which were pertinent to his/her specific professional school.

The survey instrument asked for: (1) Institutional identification data, including the name location, and type of institution; (2) Numerical data by race on applications, acceptances and enrollment for the study period; (3) Place of undergraduate degrees of full-time black graduate or professional degree students (i.e., whether received from an historically black college or from a predominantly white college, aggregated by year); (4) Number of students graduated by year, profession, and race for each year during the study period; (5) Number of students graduated by year, profession, race, and sex for each year of the study period; (6) Number of doctoral degrees conferred by year, field, and race for each year during the study period; (7) Number of Masters degrees conferred by year, field, and race for each year during the study period; (8) Number of full-time faculty by year, degrees, and race; (9) Number of part-time faculty by year, degree, and race; (10) Tenured faculty by year, degree, and race; (11) Distribution of black faculty by teaching fields or discipline; (12)Description of minority recruitment program; (13) Description of special admissions program; (14) Description of program admission requirements; (15) Costs of graduate or professional education by year; (16) Description of academic support services; (17) Description of non-academic support services; (18) Explanation of goals for enrollment of black students; (19) Discussion of the impact of the Bakke decision on admissions policies, financial aid programs and retention; (20) Financial support programs by year, type and racial distribution; (21)

Views of major problems in the recruitment and enrollment of black graduate and professional school students; and (22) General comments about the project and its goal.

Two mailings of this inventory were sent to the deans of each program. These two mailings yielded responses from 121 programs and/or institutions which indicated inability to participate in the study and from 131 active participants in the study. The pattern of participation and non-participation by field is described in Table 4.

Table 4

Participation Profile by Field in Response to Survey Inventory

Program/Field	Number of Participating Schools	Number of non-Participating Schools
Dentistry	13	3
Engineering	20	14
Law	15	16
Medicine	30	21
Optometry	6	1
Pharmacy	16	2
Social Work	8	9
Veterinary Medicine	11	2
Architecture	2	6
Ph.D Granting Institutions: Arts & Sciences	20	43
Totals	131	121

Even though returns from the mailed survey were considerably less than anticipated, usable data were obtained from both the correspondence indicating inability or unwillingness to participate and from the completed inventories. Much of the correspondence was supportive, thoughtful, and explicit in explanations for non-participation. The responses to the completed surveys tended to be careful, analytical, and of inestimable value in understanding problems of access.

Non-participants offered a variety of explanations. In general, reasons for refusing to participate can be categorized, in descending frequency: (1) information requested is not available for the time period needed; (2) to complete the inventory would be too time-consuming; (3) available personnel is too limited to complete the inventory acceptably; (4) most of the information has already been reported to professional associations and can be obtained directly from them; (5) budgetary resources are too limited to do the computer search and analysis required by the inventory; (6) State approval is required to release some of the data requested; (7) it is against institutional policies to participate in studies of this type, and (8) "other" reasons.

Selected comments from the refusals' correspondence illustrate this range of responses and reveal some of the problems in collecting this type of institutional data:

One writer stated:	"After examining the specific information being sought, I am afraid we are not able to be of much help. We have only very recently been recording many of the items about which you seek information at all or in a form which would allow reasonable retrieval."
Another wrote:	"The ADA has been collecting identical data to that which you request for many years. It has been tabulated and widely published. If after contacting the ADA you still find that you need additional information from . . . , I will be more than happy to see if I can gather it."
Another wrote:	"We are making headway to attract more black undergraduate students for our various engineering programs. Hopefully, in years to come, some of the better graduates will decide to continue their education at the graduate level. The prospects are dim, since graduate stipends of some $460.00 per month can not at all compete with lucrative salaries of $1,500—$1,800 offered by industry."
Still another:	"This questionnaire is not relevant to . . . College since we do not have a separate graduate school with that name generally used."
Another wrote:	"This college does not maintain records on minorities, the university does but not by major department. We do not actively recruit any students. All students meeting minimum admission standards are admitted . . . The questionnaire is not applicable to this college."
Another wrote:	"Our records are inadequate."

Still another: "The data you request has not and is not maintained. I
                      am sorry we cannot assist you with your research in
                      this matter."

And, another: "I regret to inform you that we will not be able to col-
                      lect the data you need since it would involve a major
                      amount of time and effort on the part of this office. I
                      simply do not have the resources available."

And, finally: "If, however, in your project funding you have the
                      resources which could be used to provide us with a
                      graduate assistant to help us with the project on our
                      campus we would be delighted to coordinate this ac-
                      tivity for you."

This range of response also illuminates some of the limitations of the
survey data which will be discussed in greater detail under limitations of the
study. They also underscore the necessity of continuing to obtain data from
other sources, especially in projects of this type.

Other Sources of Data

Due to the limited number of institutional responses to the survey inventory, it was necessary to obtain data from other sources. Hence, the second phase of the data collection process consisted of interviews with administrators, data analysts, and other key informants of the professional
associations with whom the various institutions and their programs had
some type of formal affiliation. Through these key informants, the major
portion of the data utilized in the statistical analysis of this study were collected. These formal organizations collect data on an annual basis from all
institutions which have specialized programs in one or more of the professional fields selected for study here. The institutional data reported to these
professional associations form the primary data source for this study.

These sources are as follows:

1. *The Council on Social Work Education,* located in New York, collects
    and publishes statistical data on social work education on an annual
    basis. These data are included in a volume, *Statistics on Social Work
    Education.* All schools of social work are included in the statistical
    analysis. The volume includes descriptive statistics on the following
    areas: student enrollment; institutional characteristics with special at-

tention to student enrollment; ethnic characteristics of students and faculty; admissions data; tuition charges; degrees awarded; applications for admissions to first year Master's degrees programs; number of full-time students receiving financial grants in graduate school by institution; and characteristics of Doctorate degree students.

The Council also publishes additional documents which provide data on special programs and services offered by schools of social work.

2. *American Dental Association (Chicago) and the American Association of Dental Schools* (Washington, D.C.). The American Dental Association publishes an annual volume called *Dental School Trend Analysis; Minority Report: Supplement to the Annual Report Dental Education,* and *Annual Report: Dental Education.*

These volumes provide comprehensive statistical data, both historical and current, on dental education, advanced specialty education, levels of pre-dental education, and dental licensure results. They also contain enrollment data for each institution by sex, minority status, graduation data by sex and minority status, tuition data, academic programs and admissions policies, faculty by institution (but not disaggregated by race), minority recruitment, financial assistance, special programs, if any, and such admissions requirement data as Dental Aptitude Test, grade point average, residency as well as general information about the institution's curriculum. General characteristics of entering class by race and sex for institutions are also given. Selection factors are discussed in one or more of these volumes.

3. *Association of American Medical Colleges* (Washington, D.C.). This association publishes annually *Minority Student Opportunities in United States Medical Schools* which contains statistical data on minority applications, acceptances, enrollees, and faculty by institution. It has a thorough description of medical school programs for disadvantaged students in the areas of recruitment, admission policies and procedures, and financial assistance.

Another publication is *Minority Student Information'on Individual Medical Schools.* This volume identifies the contact person for minority students at each medical college and depicts first year and total enrollment data on specific minority group students by institution.

The Participation of Women and Minorities in United States Medical School Faculties contains aggregate data on minority faculty at U.S. medical schools. In December of each year, the *Journal of Medical Education* publishes "Medical Education in the United States" that provides graduation statistics which can be disaggregatd for minority group descriptions.

The Divisions of Students Services of the Association of American Medical Colleges, under the leadership of Dr. Davis G. Johnson, publishes annually *U.S. Medical School Enrollments by Race and Ethnicity.*

The National Medical Fellowships provided financial data on scholarship awards by institution and year.

4. *The Engineering Manpower Commission* (New York City). This association[5] has two major annual publications: *Engineering and Technology Enrollments* and *Engineering and Technology Degrees.* A major function of the Commission is to collect, analyze, and publish significant data on engineering manpower. It collects data on institutions accredited by the EPCD and these data are displayed by school, curriculum, enrollment, degrees, and race. The enrollment documents provide enrollment by year, program, and race for each institution so that an examination of these data will enable one to construct a profile of how many blacks, for instance, are enrolled in the first, second, third, or fourth year at a given institution. The annual publication of *Engineering and Technology Degrees* cites the number of degrees conferred, first professional, M.S., and Ph.Ds., by institution, program, and race.

    The National Research Council publishes information on the activities of the Committee on Minorities in Engineering. Among these activities are the collection and dissemination of information on special recruitment and admissions programs for minority students. The Conference Board's publications also provide qualitative and quantitative data on engineering programs and on the work of the Committee on Minorities in Engineering. The College Board has sponsored a Minority Engineering Scholarship Program and an Upper Division Scholarship Program designed to increase minority student access to colleges of engineering. Major sources for financial data were publications and other materials provided by the National Fund for Minority Engineering Students.

5. *A Review of Legal Education* is published annually by the Section on Legal Education of the American Bar Association. This document contains:

    (a) Law School admissions requirements;

    (b) A list of ABA—approved law schools; and

    (c) For each approved law shcool, there are enrollment statistics, a survey of minority group students enrolled in J.D. programs, admissions requirements, number of degrees conferred, law school enrollments since 1950, and tuition and fees and related data. Its annually up-dated report on minority student enrollment by year and class permits the reader to develop an immediate perspective on enrollment trends for each minority group.

In addition to data obtained from interviews at the national headquarters of the Association of America Law Schools (Washington, D.C.), considerable data were obtained from an examination of the *Proceeding: Report of Committees and Projects* of its annual meetings. Important committees for purposes of this examination were the AALS Special Committee on Effective Use of Legal Education, Committee on Admissions Standards, Council on Legal Education Opportunity (CLEO), and the Committee on Minority Groups. These reports contain data on recruitment, admission, financial support, enrollment and on minority faculty in law schools. Persons associated with those committees were also interviewed.

Other information on the distribution of black and other minority faculty in law schools was obtained from Professor Derek Bell of the Harvard University School of Law.* Data on CLEO were obtained from Atty. Wade Henderson, Director of CLEO in Washington, D.C. Educational Testing Service Publications were invaluable for their analysis of institutional data, performance on the Law School Admissions Test by race, and in the presentation of impediments to the growth and retention of minority students in Schools of Law.

6. *The Association of American Veterinary Medical Colleges* (Washington, D.C.) This Association collects enrollment data from all colleges of veterinary medicine in the nation. Prior to 1974, these data were categorized by race and disaggregated by minority group. Beginning in 1975, an agreement was made that these data would no longer be published in a way that identifies specific institutions. Instead, aggregate data on minority groups, foreign students, and on Caucasion students would be reported annually. Hence, the data from the AAVMC are useful for developing a trend of total Caucasion, minority and comibined enrollment for the period since 1973. These had to be supplemented by data from other sources, such as the HEGIS Reports, which provide racial composition by school and personal interviews. Data for the profile on Tuskegee Institute's School of Veterinary Medicine were provided by Dr. Ellis Hall who also made available a number of conference Reports which contained institutional data.

Interviews were conducted by phone with the Deans of fourColleges of

---

*Professor Bell has since become Dean of the Law School at the University of Oregon.

Veterinary Medicine who also chair important committees of the Association that have an impact on the enrollment and graduation of black students.

7. *The Association of Collegiate Schools of Architecture* (Washington, D.C.). This Association publishes *Architectural Schools of North America* which includes information on its 102 schools. For example, it provides descriptions of departments and colleges of architecture, type of program offered, admissions requirements, enrollment data by race, degrees conferred, ethnicity and sex of students, and financial data. However, these data were supplemented by interviews with key informants in the Washington office and by data from the National Center on Educational Statistics and Office of Civil Rights Publications.

8. *The American Association of Colleges of Pharmacy (AACP)* (Bethesda, Md.). This Association has a number of publications used in this research.

   The more relevent ones were the annual *Reports on Graduate Enrollment Data and Graduate Study in Member Colleges.* These documents contain yearly enrollment and graduation data by race and by first professional and graduate degree programs. They provide information on teaching assistantships, non-service stipends, general financial support for Pharmacy students, and admission requirements. Other reports provide faculty distribution data and information on recruitment and special programs.

   The material used for the development of the Profile on the Xavier University School of Pharmacy was provided by Warren McKenna, Dean of the School of Pharmacy and by Dr. Norman Francis, President of Xavier University of New Orleans.

9. *The Association of Schools and Colleges of Optometry,* (Washington, D.C.). This Association annually publishes *A Survey of Optometric Educational Institutions.* This survey includes institutional data on student characteristics, enrollment by year, retention and withdrawal data, minority group student enrollment characteristics, faculty characteristics, annual student expenditures, financial aid data, and admissions/selection factors.

   The Association also publishes *Information for Applicants to Schools and Colleges of Optometry* each year. This document includes a general survey of the academic requirements for admission to each of the thirteen schools and colleges of optometry in the United States, and financial aid data.

   As in all of the above professional degree programs, these data were enriched by personal interviews with key informants of the Association, additional brochures and publications (e.g., journal articles), and by data obtained from the National Center for Educational Statistics, the Higher Education General Information Survey (HEGIS), and publications of the U.S. Office of Civil Rights.

10. *Graduate Education in the Arts and Sciences.* Enrollment and degree data were obtained from the National Center for Educational Statistics; publications from the Department of Health, Education and Welfare, especially its *Racial and Ethnic Enrollment Data From Institutions of Higher Education* (published biennially), which now provides institutional data by ethnicity, (HEGIS Reports), from the National Academcy of Science and the National Research Council.

11. The National Academy of Sciences publishes an *Annual Summary Report: Doctorate Recipients from United States Universities.* This report presents tables which depict ethnic characteristics of doctoral degree recipients by field of study. It also contains historical data as well as data on methods of financial support and plans for employment after the completion of the doctoral degree, and time lapse between the undergraduate degree and the completion of the doctorate.

12. General:
    In addition to these basic data sources, a major search of the literature on graduate and professional education was undertaken. This revealed support for the specific observation made through the analysis of primary and secondary data. Other agencies and bureaus in Washington, such as the Department of Health, Education and Welfare, the U.S. Office of Education, the Health Manpower and Resources Agency, and others provided significant data on financial aid and scholarship programs.

    Data from all of these sources were matched and discrepancies in statistical presentations were reconciled, re-organized for presentation with the survey inventory described in phase one, and prepared for computer processing. As a result, data were obtained on 427 institutions and 743 professional school programs (cases) to be used in analysis. The following distribution of these cases by field shows the actual number of schools in that field plus the percentage that number is of the total N of 743 computerized cases:
    (1) Dentistry = 59 cases and 8%; (2) Engineering = 235 cases and 32%; (3) Law = 161 cases or 22%; (4) Medicine = 108 cases and 15%; (5) Optometry = 13 cases or 2%; (6) Pharmacy = 74 cases or 10%; (7) Social Work = 84 cases or 11%; and Veterinary Medicine = 9 cases or 1%.*

---

*per cents exceed 100 due to rounding off.

Limitations Of Data Sources

Just as no statistical technique for data analysis is infallible, likewise, no data collection strategy is without problems. This study is based upon data collected directly from institutions through a survey inventory and, principally, institutional and State data collected from a variety of primary and secondary sources. The major problem presented by this approach lies in the tremendous effort required to reconcile differences and to match data reported from diverse sources. This difficulty involves some institutions whose records are sketchy and whose data have never or only recently been computerized, and for which reconstructuring time sequences of data becomes problematic. This problem is overcome in those cases in which reporting of institutional data to their accrediting agency or to the Office of Civil Rights in the Department of Health, Education and Welfare has become regularized for a substantial period of time.

A problem could arise from the tendency of most institutions not to collect data by race prior to 1972. Despite legal constraints to the contrary, some institutions, especially for their professional programs, collected such data and reported them with regularity to their respective national association as early as 1964. Matching data from a variety of sources reconciles whatever degrees of inconsistency that may exist in quantitative data and provides greater numerical precision.

However, even with the most careful reconstructions, the data are uneven for some fields for certain years. The reason is that some institutions either did not report data for those specific years or they omitted selected items for that particular year. As a result, there are missing data for some years which reduce the number of cases below the total N of 743 for certain specific years and items. However, this is no different from the occurrence of usable and non-usable responses to forced—choice and open—ended questions in survey research. As long as a sufficient number of cases are present in the sample to render statistical analysis appropriate, these omissions do not present unremediable problems.

The availability of data from the professional associations and from both the National Center for Educational Statistics as well as the Office of Civil Rights assures more effective reconciliation and checks on the accuracy of institutionally reconstructed data reported in a survey inventory. These sources enabled the collection of usable data on 743 programs and most professional schools. In all, more than 98 per cent of the institutions are presentated in the discussion of specific professional fields.

Operational Definitions

1. The term *access* refers to the enrollment of black students in graduate or professional schools. However, access may be conceived as existing in one of three possible levels. Level one is called "weak or limited access", in which a very small number of black students are enrolled in relation to their proportion in the state or national population. The second level of access is labled "approximate access" in which the enrollment is in excess of one-half the proportions. The third and most desirable level or form of access is referred to as "equal access". In this final stage, which happens to be the goal of equality of opportunity proponents, the proportion of black students enrolled is the same or slightly higher than the proportion of black persons in the state or national population.

2. *Distribution* refers to actual dispersion or concentration of black students in various graduate or professional fields. This term encompasses a quantitative spread or concentration relative to enrollment and graduation rates.

3. *Outcomes* is a term that applies to the number or proportion of black students who actually receive degrees in a given field. It is expressed quantitatively as the actual number of black graduates and/or the proportion of black students graduating with graduate or professional degrees in a field.

4. *Retention program* means any organized or formally constituted activity, the specific purpose of which is to retain students until they successfully complete degree requirements (e.g., tutorials, counselling and guidance, academic advising, financial aid programs, and so forth), as well as informal associations (e.g., Black Student Association) and black faculty who may provide psychosocial support for black students during various stages of the educational process.

5. *Quality of financial aid* and *quality of scholarship aid* refer to the dollar amounts awarded to students from all sources and to the proportion of black students receiving financial assistance. On the basis of total information available, (e.g., dollars awarded, financial resources, etc.), these variables were trichotomized as follows: good, adequate, or little or no assistance provided.

6. *Presence or absence of special admissions programs* is self-explanatory. However, it means that a determination was made as to whether or not an institution had ever used any form of special admissions program for minority students during the 1970s. Similarly, a determination was made as to whether or not an institution had established a *special minority recruitment program* and a *minority affairs office.*

7. *The presence or absence of a Black College in the State* simply refers

to the State locations of the 107 historically black colleges and whether or not a specific professional or graduate school is located in one of the States in which these institutions are located.

8. Use was made of the latest Bureau of the Census data to make a determination of the *proportion of black persons in the total population* of the State. The precise percentage of black persons was entered into the computations.

9. The actual *number of black faculty* employed full-time in a given institution was also entered into the data base for analysis. Hence, this definition does not refer solely to whether or not black faculty are present but to the size of the black faculty cohort as well.

10. *Litigation status* refers to whether or not the institution or professional school is located in a State that is either currently covered by litigation in the *Adams vs. Califano* case, *or* may be covered by this litigation because of the presence of an historically black college within the State, *or* is not likely to be under litigation due to the presumption of innocence regarding the perpetuation of a dual system of higher education. Although ten States are as of 1980 involved in the *Adams* case, at varying stages of litigation, it is possible that at least eight additional States could be sued by the Justice Department along the same lines as the claims specified in the *Adams* case.[a]

11. *Location of the institution* is a trichotomized variable. For purposes of this study, institutions are located in one of the three following locales: (1) rural area or small town, (2) city, and (3) large metropolitan city.

12. *First year Black student enrollment** refers to the number of black students who entered the first year classes of a graduate degree or professional school class in each specific year of the decade.

13. *Total Black student enrollment** means the combined enrollment of all black students in a given professional degree program for each year.

14. *Total black students graduated* refers to the actual number of black students who were reported as degree recipients for each year of the study.

---

*Note: enrollment data only in schools of law were not always disaggregated by race. It is estimated that black students comprise about 90 per cent of minority enrollments for this field reported in survey data.

[a] By January 1981, Alabama, Texas, West Virginia, Delaware, Ohio, Kentucky, South Carolina, and Missouri were added to this group by the U.S. Department of Justice.

Data Analysis

The first stage of data analysis concerns the exploratory survey research which characterized phase I of the study. It involves regression analysis between one or more of three dependent variables (i.e., first year enrollment of black students, total enrollment of black students, and graduation rates of black students) and the independent or predictor variables (i.e., quality of financial aid, quality of scholarships, presence or absence of a special recruitment program, presence or absence of a special admissions program, presence or absence of a minority affairs office (officer), number of black faculty, presence or absence of a retention program, type of institution (e.g., public, private or State-assisted), number of professional schools within the State, the proportion of blacks in the State population, presence or absence of a black college in the State, whether or not the State is currently or may be under litigation in the future to dismantle a dual system of higher education, and the location of the institution. The Statistical Package for the Social Sciences (SPSS) was used for this analysis.

The second stage in the data analysis involves the development of trends in enrollment and graduation by race. These trends are presented in six separate ways: (1) total professional school enrollment or graduation by field and race; (2) enrollment of first year black professional school students by field and by year for the study period; (3) enrollment and graduation data by sex on a selective basis; (4) the participation of black students by field and State (in order to plot changes and to make comparisons between States in the access of black students to professional schools); (5) enrollment and graduation activity in the "Adams States" by selected fields of study, and (6) trends in the production of black students with doctorate degrees in selected fields.

In succeeding chapters, the findings will be presented first in terms of the 743 cases; that is as a summary of findings for all cases and for all fields combined, and for all measures. This analysis will be followed by specific chapters for the major professional fields and doctorate level education.

Footnotes

1. For further discussion of one or more of these positions, cf., Randall Collins, "Some Comparative Principles of Educational Stratification", *Harvard Educational Review* 47:1 (February 1977), l-29, and footnotes 2, 3, and 4.
2. James E. Blackwell, "The Power Basis of Ethnic Conflict in American Society", in Lewis A. Coser and Otto N. Larsen (eds.), *The Uses of Controversy in Sociology*. New York: The Free Press, 1976, Chapter 10.
3. Robert Bierstedt, "An Analysis of Social Power", *American Sociological Review*. 15 (December 1950), 730-738.
4. Cf., Hubert M. Blalock, "A Power Analysis of Racial Discrimination", *Social Forces* 39 (1960) 53-69 and _____, *Toward A General Theory of Intergroup Relations*. New York: John Willey & Sons, 1967.
5. The engineering Manpower Commission works cooperatively with a number of agencies, societies and associations, located in the same building, who have collective interest in facilitating broader participation of minority groups in the Engineering profession. Among these are the Engineering Joint Council and the American Society of Engineers.

CHAPTER 3

Data Analysis Of Survey Findings

The mainstreaming process discussed in this book is based upon an analysis of two sets of data. The first type of data refers to information obtained using the institutional profiles constructed both from survey reports from the participating institutions and supplemental data on those institutions gained from other sources. This survey and supplemental process yielded 743 professional school programs or cases which could be subjected to statistical analysis utilizing the criterion and predictor variables.

The second type of data is constructed for a trend analysis of the process of mainstreaming black professional school students and for ascertaining the kinds of factors that appear to be associated with these trends. Hence, this second level of analysis builds upon the first presentation. The trend analysis is integrated into the discussions within subsequent chapters covering the full range of processes involved in mainstreaming black students, from access to graduate and professional schools through the actual number who are graduated, and it will be addressed in Chapters IV through XII.

Our primary concern here is an analysis of the findings of data obtained on the 743 cases resulting from the institutional surveys. In order to address the issues posed at the beginning of Chapter II, three dependent variables and as many as thirteen predictor variables were identified.

The three dependent variables employed in this exploratory study are (1) the first year enrollment of black students, (2) the total enrollment of black students, and (3) the total number of black students graduated. These three variables were subjected to statistical analysis for every year of the 1970s. the variables were selected on the basis of the current assumption that they are the primary manifestation of access and of outcomes (graduation rates).

The following variables were selected as possible predictors of one or more of the three dependent variables:
 (1) Quality of financial aid was selected because the literature is

55

replete with *assertions* that black students require substantial amounts of financial assistance due to the impoverished economic backgrounds of the majority of black students (Assigned score: 2 = good; 1 = adequate, 0 = poor or none).

2) For basically the same reason, quality of scholarship aid was selected as a possible predictor variable for determining access and graduation (Assigned score: 2 = good; 1 = adequate, 0 = poor or none). As with the quality of financial aid, these scores were arbitrarily assigned based upon available information from all sources. This information consisted of institutional survey and catalog data, foundation reports, governmental data, and the like.

3) *Special Admissions Program* was selected because of the suggestions made that these programs have enabled a significant number of black students to matriculate in graduate or professional schools. It has also been suggested that it would not have been possible for many of them to enroll had it not been for the establishment of special admissions programs. (Assigned score: 1 = yes; 2 = no).

4) The same can be said of *special recruitment programs*. The essential question here is to what extent has the presence or absence of special minority recruitment programs either facilitated or retarded the enrollment of black students in graduate and professional school? (Assigned score: 1 = present; 2 = absent).

5) Conventional wisdom has similarly permitted many to assume that the access of black students is a function of the presence or absence of either a minority affairs officer or a minority affairs office which presumably is the conscience of the institution regarding the implementation of equality of opportunity programs in higher education through a broad range of affirmative action mandates. (Assigned score: 1 = present; 2 = absent).

6) The number of black faculty in the graduate or professional school was employed to test the various assumptions of role model theory. One of the central assumptions, explicit in the socialization literature, is that role models provide a standard of possible achievement of conduct and serve as sources of inspiration regarding alternative career choices or avenues for upward mobility. An important question emerges: Does the presence or absence of black faculty serve either as a facilitator of black student matriculation or as a barrier to access? (The assigned score

is the actual number of black faculty reported).

7) Presence or absence of Retention program is presumed to be a significant correlate of outcomes (graduation). It is also assumed to be a factor upon which some students may rely in making a decision as to whether or not to enroll in graduate or professional degree programs. (Score: 1 = present; 2 = absent).

8) The number of professional schools in the State may be a determinant of the range of opportunities for gaining access available in a given state. It can be assumed that the greater the number of professional schools within a State the greater will be the enrollment of black students in that type of professional degree program within the same State. On the other hand, a small number of professional schools may be restrictive and, therefore, serve as an obstacle to access. (Score: 1 = 0 – 3; 2 = 4, 5; and 3 = 6 + professional schools).

9) Another assumption that may be made on the basis of historic patterns of college enrollment is that the presence or absence of a black college in a State may be associated with success in increasing black students' access to post-college programs. Underlying this assumption is the notion that these institutions may serve as quasi-feeder institutions to the graduate and professional schools within the same State and may be a primary site for recruitment. (Score: 1 = present; 2 = absent).

10) Inasmuch as enrollment may be a function of the students in the available pool, an important question that arises is the degree to which the number of black students enrolled in these programs is a function of the per cent of black persons in the State's population. Is it appropriate to assume that the larger the black population in the State the greater will be the per cent of black students matriculated in the graduate and professional schools of that State? (The assigned score here was the actual percent of blacks in the population).

11) The most fundamental rationale for including "litigation status" as a predictor variable was to attempt a determination of whether or not being under a court order to desegregate higher education actually makes a difference in the type of access that black students experience. (Score: 1 = yes; 2 = maybe; 3 = no or not likely).

12) The black population in the United States is predominantly an urban population. Hence, an interesting question arises as to whether or not the physical location of an institution makes a

difference in its success in recruiting and enrolling black graduate or professional degree students. Are black students more prone to select institutions located in large, metropolitan communities similar to their homes or are they attracted in this type of study to institutions located in smaller cities or communities? (Based upon U.S. Census definitions, this variable was scored: 1 = up to 50,000; 2 = 51,000 to 999,999; 3 = 1 million and above).

13) Finally, there has been some suggestion that private institutions, especially in the early part of the seventies, were more aggressive in the recruitment of black and other minority students. There is also some evidence that private institutions draw the lion's share of financial resources and can award more financial aid and scholarship assistance to deserving black students. Therefore, it has been assumed that private institutions are more successful in both enrolling and in graduating black students. Hence, this variable was selected as a predictor in order to ascertain possible associations between institutional types and the criterion variables. (Score: 1 = private; 2 = public; 3 = state-assisted).

Correlates of Total Black Students Enrollment

In order to ascertain possible relationships between the criterion and the predictor variables, and in the interest of parsimony, a decision was made to test these relationships by examining "total black students enrollment" and the predictor variables in two ways. First, Pearsonian correlations were calculated for three different points in the decade: year number one, year number 5 or the middle part of the decade, and year number 9, the last year on which the most complete data are available. The second step was an examination of the Pearsonian correlation matrix for a single time frame: year number nine was selected for this purpose.

According to Table 5, in year number I, 1970-71, the most powerful predictor of total black student enrollment in graduate and professional schools is the presence of black faculty; (r = .82; p <.001). Explaining 65 per cent of the variance in enrollment, the magnitude of this correlation lends support to the theoretical premise that the presence of role models serves to attract black students to institutions and that presence may raise the level of aspiration among this group of students for a professional career.

Other important correlations observed for the first year were between

Table 5. Pearsonian Correlations of Total Black Student Enrollment With All Predictors For Selected Years: 1971, 1975 and 1979.

Year						Variables							
Criterion	Fin/ Aid	Sch/ Aid	Sp/ Admiss.	Sp/ Recruit.	Min/ Office	Blk Fac.	N/Prof Sch	% Blk Pop	Blk Col/State	Litigation Status	Loc/ Inst.	Inst. Type	Reten-tion
Total Enrollment 1971	r=.21 **	r=.21 **	r=-.25 **	r=-.18 **	r=-.15 **	r=-.82 **	r=.08	r=.32 **	r=-.10 *	r=-.00	r=.15 **	r=-.01	r=-.25 **
Total Enrollment 1975	r=.33 **	r=.31 **	r=-.26 **	r=-.19 **	r=-.13 *	r=.84 **	r=.11 **	r=.32 **	r=-.08 *	r=.01	r=.22 **	r=-.00	r=-.19 **
Total Enrollment 1979	r=.37 **	r=.35 **	r=-.14 **	r=-.15 **	r=-.08 *	r=.83 **	r=.13 **	r=.29 **	r=-.10 *	r=-.01	r=.16 **	r=-.04	r=-.17 **

* p < .05
** p < .01
Not significant when no star is given.

total black student enrollment and quality of financial aid, (r = .21, p < .001); quality of scholarship assistance, (r = .2l, p < .001); special admissions programs*, (r = −25, p < .001); *special recruitment programs, (r = −.18, p. < .001); per cent of black persons in the State's population, (r = .32, p. < .001); location of the institution (r = .15, p < .001), and *retention programs, (r = −.25, p < .001). Taken singly, only the percent of blacks in the State population explains more than ten per cent of the variance. (See Table 5).

These correlations lead to the conclusion that the greater the financial and scholarship assistance provided to black students the larger will be the number of black students enrolled in graduate and professional schools. The *absence* of a special admissions and recruitment program is a deterrent to enrollment. Similarly, the *absence* of retention programs does not persuade black students to enroll in graduate or professional schools. On the other hand, the larger the number of black persons in the State population the larger is the total enrollment. This suggests that the most successful institutions in attracting black students are those institutions located in States having a significant black population. The correlation between location of the institution and total enrollment suggests that black students are more frequently attracted to institutions located in large metropolitan areas presumably similar to those of the place of their home residence. The absence of a black college in a State seems to have made a difference in 1970-71. This may be explained by the tendency of institutions to draw upon black colleges as the primary source of their recruitment for graduate and professional education. However, these findings are cautioned by the generally weak associations observed.

According to these data, being under litigation to desegregate higher education did not make an apparent difference in the total number of black students enrolled. In other words, those institutions that were to be subjected to this litigation were not any more successful than those institutions that were not under litigation in enrolling black students. Neither did the type of of institution make an apparent difference. Private institutions did not appear to be any more successful in attracting black students than either public or state-assisted institutions in 1970-71. Private institutions may have been more successful in this endeavor in the late 1960's but this did not appear to be that case by 1970. It is possible that in a more rigorous

*The negative correlations refer to "absence" of these special services.

assessment of possible relationships between these two variables, type of institution might emerge as a significant variable. That is not the situation at this point.

In the middle of the 1970s and in the last year of the analysis, the same forms of correlation emerged. Of primary importance here, however, is the power of the black faculty variable as a predictor of total enrollment. In every year, the correlation between these two variables is substantially stronger than is the correlation between total enrollment and any of the remaining predictor variables.

The Pearsonian correlation matrix tables present intercorrelations between the predictor variables. In some instances, these intercorrelations are both strong and significant. Although this particular group of variables taken collectively, appears to have a high degree of salience in influencing access of black students to professional schools, these findings are tentative and must await more refined analysis.

The high correlation between some of the variables suggest that they may, in fact, be measuring the same phenomenon. That is to say, for instance, the correlation between quality of financial aid and quality of scholarships awarded to black students may reflect the strength of institutional commitment to provide a sufficient amount of financial assistance that enables black students to pursue a professional degree. Similarly, the correlation between "being under litigation to dismantle dual systems of higher education" with the "presence of a black college within the State" is not unexpected. It is precisely in States under litigation that all of the historically black colleges are located. (Table 6).

Equally important to observe is the predictive power of the "presence and size of the black faculty" in relation to this criterion variable. While high correlations between this predictor and this criterion variable may, indeed, be particularly supportive of role model theory, the probability that something else is at work in this relationship is certainly worth noting. The presence of a substantial number of black faculty and a fairly large black student enrollment in the same professional school may also be strong indicators of an institutional commitment and a racial climate within the institution sufficiently positive as to facilitate both the hiring of black faculty as well as to promote the matriculation of black students beyond token levels.

Conversely, the absence of black faculty and black students or their token approximations in professional schools may suggest lack of institutional commitment to equality of employment and educational opportunity as well as a negative institutional climate.

Given these considerations, it can be argued that the factors treated as

Table 6. Pearsonian Correlation Matrix: Year 9
Total Black Student Enrollment With All Variables

Variables	Total Blk Enroll.	Fin/Aid	Sch/Aid	Sp/Admiss.	Sp/Recruit.	Min/Office	Blk Fac	N/Prof Sch	% Blk Pop	Blk Col/State	Litiga. Status	Loc/Inst.	Inst. Type	Retention
Total Blk Enroll.	1.000													
Fin/Aid	r=.37**	1.000												
Sch/Aid	r=.35**	r=.59**	1.000											
Sp/Admiss.	r=-.14**	r=-.12**	r=-.21**	1.000										
Sp/Recruit.	r=-.15**	r=-.23**	r=-.19**	r=.25**	1.000									
Min/Office	r=.08	r=-.18**	r=-.16**	r=-.06	r=.18**	1.000								
Blk Fac	r=.83**	r=.21**	r=.30**	r=-.27**	r=-.11**	r=-.09	1.000							
N/Prof Sch	r=.13**	r=.08	r=.07	r=.12**	r=-.08	r=-.04	r=-.10*	1.000						
% Blk Pop	r=.29**	r=.11*	r=.11**	r=-.03	r=-.11**	r=.05	r=.34**	r=.00	1.000					
Blk Col/State	r=-.10*	r=-.02	r=.00	r=-.01	r=.10*	r=-.05	r=-.19**	r=.06	r=.48**	1.000				
Litigation Status	r=-.01	r=.01	r=.03	r=-.12**	r=.08	r=-.08	r=-.00	r=.05	r=-.27**	r=.86**	1.000			
Loc/Inst.	r=.16**	r=.16**	r=.18**	r=-.13**	r=-.04	r=.04	r=.08	r=.19**	r=-.14**	r=.03	r=.10*	1.000		
Inst. Type	r=-.04	r=.00	r=.04	r=-.02	r=-.08	r=.04	r=.10	r=.19**	r=.03	r=.01	r=-.00	r=.25**	1.000	
Retention	r=-.17**	r=.02	r=-.18**	r=.21**	r=.42**	r=.11*	r=-.13**	r=-.08	r=-.12*	r=.05	r=.03	r=-.12**	r=-.02	1.000

*p < .05
**p < .01
Not significant when no star is given

predictor variables in this research could be conceived as dependent variables if "institutional commitment" and "institutional climate" are regarded as dependent variables. Indeed, in subsequent chapters, such inferences do occur from time to time but without regard to a direct statistical analysis. In effect, enrollment and graduation of black students from graduate and professional schools do attest to degrees of institutional behavior and the willingness of black faculty and students to engage in the learning and teaching process of a given institution because the overall institutional milieu is not viewed as an impediment of significant consequence to them. Notwithstanding, for purposes of this analysis, first year black student enrollment, total black student enrollment and the number of black students graduated are treated as dependent variables. The probability that they, as a group of variables, might reflect special actions taken or not taken by an institution to promote equality of opportunity is amplified in subsequent chapters. It should also be reiterated here that all of the variables treated in this analysis as predictors may be regarded as explicit manifestations of institutional policy and behavior.

Again, it is important to note that these interactions occur within a dynamic system characterized by constant feedback between variables over time; consequently, the technical distinction between independent and dependent variables may, in the long run, be meaningless.

Retention programs appear to be most successful when they include a good quality of scholarship monies, are associated with organized special recruitment and special admission programs, and are located in institutions with minority affairs officers or offices. They are also more likely to be present in institutions located in large metropolitan communities and in States with a substantial proportion of black persons in the total population. Irrespective of location, the very existence of retention programs is a response to a clearly defined need. This suggests that many schools have not done a particularly good job in providing basic educational skills to their pupils and/or that the quality of the learning environment creates a need for various kinds of support services.

Identification of Most Powerful Predictors

The next step in this analysis was an attempt to isolate the most powerful set of predictors of first year enrollment, total enrollment, and of total black students graduated. To accomplish this task, a stepwise regression method was employed.[1] Given the magnitude of the data collected over nine or ten years time, the principle of parsimony was once again applied.

Hence, a decision was made to illustrate the outcomes of the stepwise regression by reporting on last step findings for years one, five and nine. In this manner, patterns or trends and inconsistencies in the predictive power of certain variables could be specified.

It should be stressed that this study is not confirmatory; rather it is exploratory. Therefore, the observations made regarding the power of predictor variables should be regarded as primarily heuristic in that the statistical analysis may shed light on relationships between variables which could impact on the professional education of black students. Although a stepwise regression technique was employed in the analysis of data, for purposes of this study, a decision was made to focus exclusively on the last step in the equation and to report Betas, R^2 for the equation as a whole, F scores, and level of significance. The general rule of thumb for including variables in the tables was that they appeared significant at least at the .05 level. Exceptions to this general criterion were made when the correlation between an independent variable and one or more of the criterion variables was significant at an earlier step at least at the .05 level and was only slightly different at the last step of the equation. Consequently, a table may sometimes include a variable with a relationship that is significant at the .07 level.

First Year Black Student Enrollment in Professional Schools

During the first year covered by the study, this analysis shows that the most important predictor of first year enrollment of black students in professional schools was "the number of black faculty" in the professional school. (Table 7). The salience of the black faculty variable for predicting first year black student enrollment was observed once again in the fifth year as well as in the ninth year. However, during the fifth year, "the proportion of blacks in the State's population" was also significantly correlated with first year black student enrollment. In both years, the correlations between these two independent variables and the dependent variable, first year black student enrollment, met the criterion for determining the most important or powerful predictors. This situation was observed when the Betas, R^2s, Fs and significance levels were examined. (Table 7).

In year nine, four variables met the criterion for selection as the most powerful predictors. These were: (1) number of black faculty in the professional school, (2) quality of financial aid, (3) number of professional schools within the State, and (4) litigation status. In addition, two variables were included in the table because they had appeared as significant at least at one earlier step in the stepwise regression analysis even though their level of significance at the last step varied slightly from the .05 level of confidence. These variables were the type of institution and the presence or absence of a black college in the state.

According to this analysis, the number of black faculty in the professional schools appears to be consistently correlated with first year enrollment of black students in professional schools. The persistence of this variable as the most powerful predictor suggests that those institutions seriously committed to improving the level of enrollment of black students in professional schools might be more successful in achieving that objective by hiring more black faculty. One can also speculate that a more detailed analysis which is directed to other minority groups might show a similar relationship between the presence of faculty members of those groups and first year professional school enrollment of minority students from the same groups. This is, of course, speculation; however, it is informed by impressionistic evidence observed in certain professions (e.g. law and social work).

The correlation between a good quality of financial aid and significant first year enrollment is not unexpected. As previously, discussed, a disproportionate number of black students come from a household with a median family income that is only about fifty-seven percent of that of the white population. We have also pointed out that a large proportion of these families subsist at or near the poverty level. Since the costs of professional school education often exceed the median family income of black Americans, and, given the extreme difficulties that blacks encounter in finding other sources of financial support, some form of financial aid is required to assure their enrollment and retention. This analysis suggests that those institutions with a better quality of financial aid programs do a better job of enrolling black students at the first year level of professional school training. Although this variable appears to be more powerful in the ninth year of this analysis, it did appear as an important variable at earlier steps in previous years.

Similarly, one should expect that the greater the number of professional schools within a State, the more likely are black students to be enrolled in professional schools. Having more professional schools means, obviously, an enlarged amount of space. In turn, these States provide more opportunities, at least in theory, for students of all races to be included. This expanded opportunity seems to be the case since it appears that those States with several professional schools are more likely to enroll a larger number of black students in their first year professional school classes.

The anticipated impact of the *Adams* vs. *Califano* case has not been achieved so far. According to these findings, the most effective States,

Table 7. Regression of First Year Black Student Enrollment In Professional Schools With Most Significant Variables: All Cases for Selected Years, 1, 5 and 9

Year	Variable	Beta	F	Significance
1	*Number of Black Faculty in the Professional School	.50	5.61	.03

R² for the equation as a whole = .32
F for the equation as a whole = 13.53, Significance = .001

| 5 | *Number of Black Faculty in the Professional School | .86 | 92.20 | .00 |
| | *Proportion of Blacks in the State's Population | .30 | 7.00 | .00 |

R² for the equation as a whole = .49
F for the equation as a whole = 73.13, Significance = .003

9	*Number of Black Faculty in the professional school	1.67	105.63	.00
	*Quality of Financial Aid	7.60	5.41	.02
	*Number of professional schools in the State	4.98	4.65	.03
	*Litigation Status: Re: "Adams States"	−13.12	4.57	.03
	*Type of Institution	−7.66	3.19	.07
	*Presence or absence of Black College in State	19.93	3.35	.07

R² for the equation as a whole = .62
F for the equation as a whole = 23.75
Significance = .000

*Most significant predictor variables in the equation

regarding the enrollment of black students in professional schools, are those States outside the seventeen southern and border states which are not likely to be included under the "Adams" litigation. It must be stated, however, that in some professional schools, significant increases in first year professional school enrollment have occurred in some States that are presently under litigation or in States likely to be included under such litigation. These increases may be a direct response to "Adams" litigation. This finding may also suggest that effective enforcement of the legal mandate to desegregate higher education is required to accelerate the process of enrolling larger numbers of black students in all types of professional schools in States under "Adams" litigation.

During the sixties, it was generally assumed that private institutions were more effective in enrolling black students in professional schools. While that conclusion may have been valid for the sixties, the situation during the seventies appears to be episodic. The findings in this investigation suggest that publicly-supported and state-assisted institutions have begun to enroll more black students than private institutions. But, in the last year of the decade private institutions appeared to have been more successful. One reasonable explanation for this change lies in the exorbitant costs of private school education. When that fact is coupled with the uncertainties of financial assistance programs, speculations about the correlations or relationships between the type of institution and black student enrollment achieve a new reality. However, the reported confidence level is weaker than the level reported for correlations between other variables in the equation.

Similarly, the confidence level of the correlation between the presence of a black college in the state and first year black student enrollment in professional schools is weaker than that for other correlations reported. Nevertheless, some support can be made for the position that, at least with regard to the 743 cases in this study, those States in which no black college is located were more effective in the seventies in enrolling black first year professional school students than those States in which black colleges were located. That is precisely why the "Adams" suit was necessary — to move those institutions beyond the levels of limited tokenism in the desegregation of higher education. This conclusion may be borne out by the trend analysis presented in subsequent chapters when institutional and State-by State data are discussed.

Total Black Student Enrollment in Professional Schools

To discover the most important predictors of total black student enrollment in professional schools, the same procedure was followed as that presented in the discussion of first year black student enrollment. The regressions for years one, five and nine are presented in Table 8. According to this analysis, the number of black faculty in the professional school was consistently the most powerful predictor of total black student enrollment in each of the years selected for analysis. In addition, three variables which were also correlated with first year enrollment were correlated with total enrollment. These were: (1) quality of financial aid (fifth and ninth year), (2) number of professional schools in the State (fifth year), and the proportion of blacks in the State population (fifth year). Still another predictor appeared (in both the first and fifth years) to be correlated with total enrollment. This variable was the presence or absence of special admissions programs. Only in the fifth year, however, did that correlation meet our criterion for unqualified inclusion as a powerful predictor. It appears that institutions with special admissions programs are more successful in enrolling black students than are those institutions without them.* The comparatively stronger level of confidence shown in the fifth year in relationship to the relatively weaker confidence level displayed in the first year may be important. For instance, it might suggest that in 1970 a larger number of institutions utilized special admissions programs to attract black students. However, by the fifth year, when organized attacks against special admissions programs proliferated and achieved nationwide publicity, professional schools were less likely to admit the utilization of these programs (Table 8).

Total Black Students Graduated from Professional Schools

When similar procedures for data analysis were applied to correlations between total numbers of black students graduated and the predictor variables, the number of black faculty once again appeared as the most significant predictor in every year selected for analysis (Table 9). It was the only variable which met the criterion for acceptability during the first year. In fact, only two predictor variables attain unqualified acceptance in subsequent years selected for analysis, namely; (1) presence or absence of a special admissions program and (2) the proportion of blacks in the State

---

*The minus signs for the Beta scores are simply a function of the way in which the variable is coded.

Table 8. Regression of Total Black Student Enrollment In Professional Schools With Most Significant Variables: All Cases For Selected Years, 1, 5 and 9

Year	Variable	Beta	F	Significance
1	*Number of Black Faculty in the professional school	2.51	52.93	.00
	*Presence of Special Admissions Programs	21.89	3.70	.07

R² for the equation as a whole
=   .74
F for the equation as a whole
= 18.30
Significance =   .000

Year	Variable	Beta	F	Significance
5	*Number of Black Faculty in professional school	1.33	22.83	.00
	*Proportion of Blacks in State population	.87	5.72	.01
	*Presence of Special Admissions Program	−17.79	4.31	.03
	*Number of professional schools in State	6.81	3.78	.05
	*Quality of Financial Aid	9.38	3.31	.07

R² for the equation as a whole
=   .35
F for the equation as a whole
= 15.70
Significance =   .000

Year	Variable	Beta	F	Significance
9	*Number of Black Faculty in professional school	1.95	16.15	.00
	*Quality of Financial Aid	20.30	4.10	.04

R² for equation as a whole =
.32
F for the equation as a whole
= 21.32
Significance =   .000

*most significant predictor variables in the equation

population. The presence of a retention program achieved a level of confidence only slightly varied from the previously stated criterion for acceptance. It should be pointed out that rigid adherence to the .05 level of confidence is not intended to communicate something sacred about that level. It is only a criterion established for the sake of statistical analysis. There are instances, of course, when conventional wisdom might be highly salient for understanding the relative importance of the effect of an independent variable on a dependent variable even though the correlation does not fall within narrowly defined parameters of acceptance. Such is the case with the retention variable and its correlation with black student graduation in the fifth year. Although most black students do not require "special retention" programs (e.g., tutorials, traditional academic support services), all benefit, either directly or indirectly from the presence of other components of a broadly defined retention program. That program includes the presence of black faculty, a critical mass of black students, formal or informal psychological support structures, and a positive learning environment. These features may not be reflected in simplistic correlational analysis.

In sum, the analysis of the data on the 743 cases suggests that the most powerful predictor of first year black student enrollment, total black student enrollment, and of the total number of black students graduated from professional schools is first and foremost, the presence of black faculty members in professional schools in the state. Others are the presence or absence of special admissions programs, proportion of blacks in the State population, presence or absence of a retention program, type of institution, and litigation status with regard to the *Adams* vs. *Califano* case. Once again, it is possible that all these predictors may be a manifestation of the same thing — institutional commitment. What is, therefore, needed is another construction that permits a more detailed investigation which specifies institutional commitment as the dependent variable.

The question may be raised as to whether or not the findings are repeated if each dependent variable is tested against all of the predictor variables employed in this study within the context of specific professional fields as opposed to the aggregate of all cases. An attempt at that analysis failed because of the paucity of cases per year in relation to the large number of predictors attempting to enter the regression equation. In other words, the number of cases, when disaggregated by professional field for every year of the study, was not adequate for the type of analysis desired. Hence, the findings reported in this exploratory study apply to the aggregate of cases. Without a more complete data set, it may be premature to make unqualified or grandiose generalizations about a particular field.

Albeit, some of the data utilized in the trend analysis are, in fact, more

Table 9. Regression of Total Black Students Graduated From Professional Schools with Most Significant Predictor Variables: All Cases for Selected Years, 1, 5, and 9

Year	Variable	Beta	F	Significance
1	*Number of Black Faculty in professional school	.15	7.8l	.01
	R^2 for the equation as a whole = .4l F for the equation as a whole = 19.48 Significance: = .000			
5	*Number of Black Faculty in professional school	.37	54.43	.00
	*Presence of a Retention Program	−1.8l	3.38	.06
	*Proportion of Blacks in State Population	.41	5.84	.01
	*Presence of Special Admissions Program	−3.38	4.16	.04
	R^2 for equation as a whole = .43 F for equation as a whole = 27.82 Significance: = .000			
9	*Number of Black Faculty in professional school	1.41	34.59	.00
	*Presence of Special Admissions Program	−5.79	6.18	.0l
	R^2 for equation as a whole = .4l F for equation as a whole = 31.80 Significance: = .000			

*most significant predictor variables in the equation

complete and do permit more definitive statements about the relationships between one or more of the dependent variables and such predictors as the number of black faculty and quality of financial aid. Clearly, there is a need for even more complete data on all of the variables included in this analysis and, perhaps, a sharper statistical model to move this analysis beyond its heuristic level.

Footnotes

1. The strengths and weaknesses of the stepwise regression technique is discussed in Maurice G. Kendall and A. Stuart, *The Advanced Theory of Statistics, Vol. II.* London: Charles Griffin, 1961.

CHAPTER 4

The Medical Education of Black Americans

Few issues in graduate and professional education created as much controversy, political enmity, personal dilemmas, and social strains among the nation's citizenry as did that of the admission of black and other minority students to medical colleges during the 1970s. Most of the litigation in the past decade regarding access to professional schools involved either medicine or law but the most celebrated cases were in medical education.

The dilemmas unearthed during the controversy over the medical education of black Americans in the 1970s revealed problems considerably more fundamental in nature and scope than suggested by specific observations. For instance, whether the nation listened or not, Americans were informed of the critical nation-wide shortage of black and other minority group physicians in general and, most especially, in the underserved areas. Paucity of physicians coincides with other health related problems. This shortage of physicians and other health care personnel dramatizes the gross inadequacies in the delivery of health care services to those groups in the population whose needs are most pronounced.

It is clearly not a problem of the quality of training but most fundamentally a manifestation of the limited number of individuals and appropriate facilities available to provide the services required. As a result, disproportionate numbers of black Americans needlessly die each year of hypertension, cardiovascular disorders, malignant neoplasms, early childhood diseases, lead poisoning, influenza, and cirrhosis of the liver. Inordinate numbers of black Americans continually suffer from serious and incapacitating mental health problems because of inabilities to develop the kinds and quality of coping strategies obtained through proper psychiatric care. This, too, is a manifestation of both shortage and maldistribution of physicians and other health care professionals.

This chapter focuses on those efforts mounted during the past decade to alleviate the problem of underrepresentation as well as on programs

73

designed to utilize medical education as one strategy for mainstreaming black Americans in the professions. It devotes attention to some of the major dilemmas and social conflicts of the seventies which affected the implementation of those programs and the ultimate realization of their goal —the production of larger numbers and proportions of black physicians.

Howard University and Meharry Medical College in Mainstreaming

In terms of history, no two institutions have played such a pronounced and vitally important role in the training of black Americans in the health care professions as have Howard University in Washington, D.C. and Meharry Medical College in Nashville, Tennessee. During Reconstruction and Post-Reconstruction periods, eight medical schools opened for the training of black physicians, dentists, and other health specialists. Of those groups, only the College of Medicine at Howard, established in 1867 and Meharry Medical college, founded in 1876, remain.[1]

These two institutions, Howard, entirely federally funded, and Meharry, heavily subsidized by the federal government although privately established, assumed the responsibility for medical education among black Americans at a time when as policy blacks were excluded from all other medical colleges. Their responsibility emcompassed far more than formal course work in medical schools. It was necessary for these institutions to provide clinical training, internships, and residencies for blacks. That type of training was unavailable to blacks throughout southern "whites only" facilities and only minimally and selectively open to them in other parts of the country.

Because of the deepening patterns of segregation, discrimination and rampant racism which followed the legalization of the "separate-but-equal" philosophy and its dispersion among all forms of institutional life, opportunities for medical training for black Americans were universally restricted. Clinical training was confined to historically black hospitals because of the refusal of white hospitals to accept even those black students who were graduated from historically white medical schools. Meharry and Howard graduates shared the same experience.[2] Racism, segregation, discrimination, and absolute necessity account for the establishment of the teaching hospitals associated with the two black medical colleges. Freedmen's Hospital was founded at Howard University and Hubbard Memorial Hospital was established at Meharry. They provided a large share of the spaces required for black interns and residents. Other historically black hospitals that played an early role in this endeavor included Homer G. Phillips (St. Louis), Provident (Chicago), Flint Goodrich (New Orleans), and others in Detroit, Philadelphia, Greensboro, and elsewhere.

Similarly, blacks were not permitted to hold faculty positions in the historically white medical colleges because white students objected. They were barred from administrative positions in medical schools because of institutionalized patterns of racism and racial discrimination. Few practicing physicians were permitted to join local, state, or the national medical associations and societies. Consequently, they were prohibited from influencing health care policy that affected the lives of American citizens. In essence, black Americans were relegated to what Professor Montague Cobb calls "The Negro medical ghetto".[3]

Howard University and Meharry Medical College assumed the responsibility of meeting the needs of black Americans not only in a medical ghetto but in preparing their graduates to fulfill a leadership role in the medical community of the larger American society. The magnitude of their success can be gleaned from the fact that as late as 1967, approximately 83 per cent of the 6,000 black physicians then in practice received their training at either Meharry or Howard.[4] Despite the increases in the absolute numbers of black students enrolled in historically white medical schools and the growing numbers of black faculty in these institutions, Howard and Meharry still account for 23 per cent of black first-year enrollees in U.S. medical schools; about 22 per cent of total black students enrollment and approximately 26 per cent of all blacks holding faculty positions either in the basic sciences or as clinical appointees.[5] And it is their graduates who hold leadership roles today as in the past in the National Medical Association* as well as in policy making roles in the broader medical community. Hence, Meharry and Howard continue to play a major role and to perform an important service in the American society as a whole.

The Role of Professional Associations in Mainstreaming Blacks in Medicine

The centerpiece of the Civil Rights Movement of the sixties was "access" — opening up opportunities to better jobs, public accommodations, political rights, and equality of educational opportunity. Since it encompassed all aspects of social life in the American society in which blacks were

*The National Medical Association is the black counterpart to the largely white American Medical Association (AMA). It was founded in 1895.

categorized as "outsiders," medical education could not escape its thrust. As the student sit-in movement shifted from public accommodations toward gaining other dimensions of structural equality, a primary target was defined as access to graduate and professional schools, especially in medicine and law. These were high status professions presumed to produce lucrative financial rewards and to be the source of tremendous power, influence, and prestige in the community.

Any assessment of the proportions of blacks or of their absolute numbers in the medical profession leads to the inevitable conclusion that black Americans not only are under represented but have been short-changed. Prior to the escalation of the Civil Rights Movement, there was no substantial evidence of serious commitment by the white medical establishment to the medical education of black and other minority students. The over-all shortage of black physicians is suggested by the fact that as the decade of the seventies began black Americans represented approximately 11.1 per cent of the total U.S. population but only 2.1 per cent of all practicing physicians in the nation.[6] The paucity of black applicants to medical schools further exacerbated the situation. This was indeed, an intolerable disparity — a situation that begged rectification. This dearth of black physicians, the recognition of the inadequacies in the delivery of health care services to underserved areas, pressures from the Civil Rights Movement, and consciousness-raising among some medical school faculties and administrative leadership coalesced to produce major policy decisions by the power structure of the medical profession.

In response to this situation, a number of efforts, aimed at rectification of past and current injustices and inequities in medical education, were initiated either jointly or independently by a variety of professional organizations in the health care field. Among these were the Association of American Medical Colleges (AAMC), the National Medical Association (NMA), the American Medical Association (AMA), the American Hospital Association (AHA), and others. Policy statements of major Foundations also signaled increasing commitment to accelerate the mainstreaming process of black students in the field of medicine. This list included Ford Foundation, Carnegie Foundation, the Alfred P. Sloan Foundation, the Josiah T. Macy Foundation, the Rockefeller Foundation, the Robert Wood Johnson Foundation and many others. As will be amplified, the federal government became more aggressive in its efforts to stimulate greater equality of access through a variety of enabling scholarship programs and direct financial assistance to medical and dental schools. The combined activities and programmatic developments fostered by these groups were directed at changing the racial and class composition of medical schools and toward reducing their long-standing characteristic as the bastions of white male privilege and opportunity.

One of the prime movers in these endeavors was the AAMC whose leaders, with the encouragement and support of its membership institutions and through policies enunciated at annual meetings, embarked upon a program of significant change in 1968. In that year, the AAMC publicly expressed its endorsement of a position previously taken by some of its member institutions to move toward increasing diversification of students of medical colleges. This move meant effectuating those changes that would lead to more aggressive recruitment, selection, admission, and enrollment of larger numbers of students from varied ethnic, racial, and economic backgrounds and from different geographic areas of the country.[7]

In 1969, in cooperation with the National Medical Association, the AMA demonstrated its changing commitment through the endorsement of resolutions which stressed wider participation of all outsider groups in the health care fields. They supported greater use of equivalency and proficiency tests in order to provide a more systematic method of evaluating previous education and experience. This policy assured recognition of the value of on-the-job experience for academic credit should such persons decide to seek professional training in medical colleges.[8]

The AAMC, in the same year, obtained its first major grant (approximately $1.5 million) from the U.S. Office of Economic Opportunity (OEO) for the purpose of facilitating equal access to educational opportunities for minority students in health care fields. As a result of this grant and others that followed, the AAMC established national projects that have proven indispensable for increasing access to medical schools. Among these projects are the AAMC's Office of Minority Affairs, the Medical Minority Applicant Registry (also known as Med-Mar), and the publication, *Minority Student Opportunities in United States Medical Schools*. The Office of Minority Affairs, located at the AAMC national headquarters, is an envaluable link in the recruitment process. The grants received from OEO were also a prime source of funds utilized for the development of about fifty minority programs at colleges and professional schools.[9] These concerted actions represented early manifestations of a strong commitment to attack head-on the problem of under-representation of minorities in medicine.

Pursuant to that commitment, the AAMC was also troubled by both under-representation and maldistribution of blacks in the medical profession. The primary responsibility for the training of black students was and had been assumed for almost one hundred years by Meharry and Howard. As a consequence of racism, segregation and discrimination, by 1970 black students represented only a paltry 2.8 per cent of total enrollment in all U.S.

medical schools and Howard and Meharry still accounted for about 50 per cent of all blacks in medical colleges. To address this problem of opening up more of the historically white medical schools and, thereby, distributing new enrollees throughout medical colleges, the AAMC organized the first of two Task Forces on Minority Student Opportunites in Medicine in 1970. This task force was funded by the Alfred P. Sloan Foundation and was chaired by Dr. Bernard M. Nelson.

Principal objectives of the task force included a delineation of the barriers which militated against full participation of minority groups in medical education, the identification of programs and strategies that could facilitate increases in access of minority students to medical schools, and the recommendation of concrete proposals that would assist in that process.[10] In April 1970, the task force presented its findings and recommendations to the Inter-Association Committee on Expanding Educational Opportunities in Medicine. One of its more highly publicized recommendations was to increase minority group representation to twelve per cent of total enrollment in U.S. medical colleges by 1975-76. Implementation programs regarding this recommendation, coupled with a recommendation to establish regional career informational and tutorial centers, drew widespread attention under the National Medical Association's banner "Project 75."[11]

In view of immediate action taken upon them, at least four additional recommendations from the 1970 Task Force had long-term implications. It was recommended that: (1) a central national organization be identified that would have the primary responsibility for coordinating the solicitation and distribution of all financial aid to minority group medical students; (2) an educational opportunity bank be established as a more permanent solution to the problem of financial aid; (3) regional centers be set up throughout the United States which could more effectively disseminate essential information about opportunities in the health professions as a career; and (4) the AAMC be urged to seek funds from external sources that would enable it to expand the Office of Minority Student Affairs to broaden the scope of its services.[12]

Support for these recommendations was immediate and broad. The medical colleges endorsed the recommendations as did the leadership structure of the AAMC. The proposals were also accepted by the American Medical Association, the National Medical Association (which is the leading historically black organization of physicians), the American Hospital Association, the federal government, and several major foundations.

The organization selected to coordinate the solicitation and distribution of scholarship monies to minority group medical students was the National Medical Fellowships, Inc., headquartered in New York City. This strategy proved highly successful, especially in the early years of the decade. Its success is evidenced in the National Medical Fellowships, Inc. (NMF) giving awards in 1970-71 to 588 minority group medical students totalling some $923,750. In 1973, some 1,760 students received awards that totaled $2,293,800. Unfortunately, 1973-74 was the peak year for both the number of awards and for the total amount of monies available for distribution by the NMF. Since that year, both declined in each succeeding year. According to the latest available data from the NMF, in 1977-78, it awarded a total of $1,314,380 to 1,345 minority group students.

In 1977-78, recipients were primarily from the States of New York (142); California (141); Pennsylvania (108); the District of Columbia (75); Illinois (74); and Massachusetts (74). This distribution is in part attributed to the presence of such a large number of medical colleges in these jurisdictions compared to others which have only one or two institutions. Comparatively, the nine southern "Adams States," with fewer institutions, have not been as successful in obtaining funds from this source. For instance, in 1977-78, the combined total of recipients for the States of Maryland (15), Arkansas (17), Florida (2), Georgia (9), Louisiana (17), Mississippi (11), North Carolina (2l), Oklahoma (3), and Virginia (17) was less than the total number of awards granted to either California or New York and was only four more than the total number of awards received in the tenth "Adams State", Pennsylvania (108).

A major contributor of scholarship monies through the AAMC was the Robert Wood Johnson Foundation. This organization gave some ten million dollars to American medical schools between 1971 and 1975 as loans or scholarships for minority, female, and rural medical students. More financial support for scholarships and loans came from the federal government with the establishment of Health Professions Scholarship and the Health Professions Loan Program.[13]

Expeditious action on these resolutions paved the way for significant recruitment programs to be established at the nation's medical schools and by professional organizations such as the National Medical Association. As a component of Project 75, in 1970, the NMA embarked on a program of establishing counseling and tutorial centers in various regional areas throughout the country. This project was supported through funds obtained from the Office of Economic Opportunity (OEO). A major initial purpose of these centers was to assist prospective applicants to improve their academic skills in preparation for possible medical school training and to achieve a better understanding of their own commitment to the medical and health care professions. Subsequent evaluation of Project 75 revealed extensive shortcomings and inadequacies in the academic preparation of many of the students who sought assistance especially in the basic sciences. It was soon evident that the kinds of services provided through Project 75 should, in fact, be incorporated in early training programs even in elementary school levels.[14] The Department of Health, Education and Welfare provided additional monies for recruitment activities and programs developed for increasing access to health care fields through its newly created Office of Health Professions Scholarship Program.

Participants were also exposed to a number of role models at these regional centers who provided information about medical careers and who often tutored them in the basic sciences. Any number of black physicians, scientists, and health care practitioners offered their services at these centers on a regular basis. Informed encounters also more effectively socialized potential applicants into the world of medical education.

The AAMC was able to obtain sufficient funds to permit expansion of its Office of Minority Affairs. As a result, three major activities of that Office have continued and broadened their scope. The first of these is Med-Mar. This is the Medical Minority Applicant Registry which is a mechanism through which minority student applicants to medical colleges may have pertinent biographical data circulated to the admissions offices of the various medical colleges and to other health service organizations without cost. Participation in this program is voluntary after a request for minority identification is provided on the Med-Mar Questionnaire. This is completed at the time the MCAT or the New MCAT is taken. The Med-Mar lists are circulated in July and November of each year for the benefit of both the applicant and interested medical colleges. After medical colleges make assessments of the biographical data submitted by the applicant, interested institutions may contact the applicant or the institution may request additional information from Med-Mar.[15]

The second service provided by the AAMC Office of Minority Affairs is the Minority Student Information Clearinghouse which furnishes information of all types that is of special interest to minority group medical school applicants and students as well as to institutions. This service includes information about the application process, programs of specific medical institutions, fee waivers, and so forth. The third related service to minority affairs is the continued publication of *Minority Student Opportunities in U.S. Medical Schools*. This book has been published almost annually since 1969. Its functions were described in Chapter II. However, it is a comprehensive document containing an impressive collection of information of value to anyone interested in medical education. The AAMC also participates in a centralized application service (which will be explained in a subsequent section) and provides pertinent data on "fee-waiver" programs of participating medical colleges.

Despite these activities, and for reasons to be amplified in a subsequent section, the goals of Project 75 and of the AAMC of achieving a twelve per cent minority student enrollment by 1975-76 were not attained. In fact, the enrollment of minority groups peaked in 1974 and for some groups there has been a noticeable continuing decline in their proportionate representation in medical colleges. This trend is especially true of black student enrollment despite increases in their absolute numbers.

Given this situation, the Executive Council of the AAMC established a second Task Force on Minority Student Opportunities in Medicine in February 1976. The essential charges to the Task Force were to obtain information, to analyze and assess problems that impede efforts to enroll more minority students in medicine, to explain the difficulties encountered by applicants from these groups in attempting to gain access, and to investigate how both minority group students within medical institutions themselves have responded to the challenges posed by a changing racial composition, and finally, to shed light on the perceptions that minority group students themselves may have of the desirability of medical careers when other alternative choices are available to them.[16]

The Second Task Force identified seven major goals in response to its charges. In the order of importance as a guideline to the AAMC's interpretation of problems encountered in providing greater access to minority groups students, they are:

Goal 1: Increase the pool of qualified racial minority applicants to levels equivalent to their proportion in the U.S. population with progress toward that goal reviewed on a biennial basis.

Goal 2: Enlarge the number of qualified racial minority students admitted to medical school through improvement of the selection process.

Goal 3: Emphasize the importance of financial assistance for racial minority students pursuing careers in medicine.

Goal 4: Strengthen programs which support the normal progress and successful graduation of racial minority students enrolled in medical school.

Goal 5: Increase the representation of racial minority persons among basic science and clinical faculty.

Goal 6: Encourage the establishment of faculty development programs aimed at fostering an understanding of the history

and culture of racial minority groups and at improving the
quality of medical school instruction.

Goal 7: Ensure that graduate and medical education needs and
opportunities for racial minority students are met.[17]

After more than two years of investigation, analysis and evaluation of
all of the issues related to its charges and its seven goals, the Task Force
made some thirty-four recommendations in its Report of June 1978. These
recommendations, which influenced many recommendations made in this
study, reveal the immense difficulties encountered in sustaining overall
commitment to increase equality of opportunity in medical education.
They also point to the fluctuations in patterns of enrollment and the retrenchment in financial as well as public support for programmatic efforts to
expand the racial composition of medical schools beyond mere tokenism.
By 1978, the mood of the nation had shifted dramatically from a manifest
concern for improving the status of minority groups in the direction of self-
centered dominant group prerogatives. Individual members of the majority
group had challenged the legitimacy and constitutionality of special programs designed to attain the goals articulated by both the first and second
Task Force Reports. Coalitions forged during the civil rights period of the
sixties were now dissolved as former partners took opposing and oftentimes
acrimonious positions during the debates over admissions policies, practices and procedures in graduate and professional schools.

The Cases of Defunis and Bakke

Although names such as DeFunis and Bakke emerged as symbols of
resistance to efforts to achieve equality of educational opportunity, other
court cases which challenged affirmative efforts in the professional schools
arose with less public notice and fanfare during the seventies. It should be
borne in mind that about fifty per cent of U.S. medical schools initiated
special recruitment programs in 1969. By 1975, eighty-nine of the 112
medical institutions which responded to a special questionnarie in a study
conducted by John S. Wellington and Pilar O. Gyorfy indicated that they
had established special recruitment programs for minority students.[18]
Frank Atelsek and Irene L. Gomberg reported in 1978 that of the institutions they studied, forty-five per cent who awarded graduate and professional degrees had at least one program designed specifically for minorities
and women; thirty per cent had special recruitment programs; thirty-five
per cent offered academic assistance, and schools of medicine and law were

considerably more active in these endeavors than were other professions.[19]

This study shows that for 1975, 66 per cent of the medical schools had special recruitment programs; 55 per cent had special admissions programs of some type; 85 per cent had minority financial aid programs; and 50 per cent had special academic assistance or retention programs. Without questions, these programs played a major role in whatever increases in the enrollment of black and other minority group students occurred during the seventies. But the legal challenges to modest successes boldly asserted that minority group students were entering medical schools at the expense of white students.

Between 1970 and 1976, aside from the challenge mounted by Marco DeFunis, Jr., the constitutionality of these special efforts and preferential admissions, in particular, was raised in several lower courts. These reached the Court of Appeals of New York for instance, which raised the question of the constitutionality of the minority programs at the Downstate Medical Center. Also, a challenge to the admissions policy of the Sophie Davis Center for Biomedical Education of the City University of New York reached a federal court in New York. Similarly, the financial aid program of the Law School of Georgetown University had to be defended in a federal court.[20] But, the two most important and far-reaching cases with national implications were the cases of *DeFunis v. Odegaard* (1974) and *Regents of the University of California* vs. *Allan Bakke* (1978).

Certain features of the *DeFunis* case were described in Chapter I. An excellent account of his family and educational history is presented in *Bakke, DeFunis, and Minority Admissions: The Quest for Equal Opportunity* by Allan P. Sindler.[21] This reference also provides a cogent analysis of the *DeFunis* case from its origins through the decision by the United States Supreme Court that the case was moot and without legal merit.[22] Just as this case was particularly significant as a precursor to the *Bakke* case, it had the immediate effect of identifying central issues germane to mainstreaming in professional school education. This case showed how many former partners in civil rights struggle against racial injustices during the sixties found themselves in an explosively contentious relationship.

For example, DeFunis gained support from such groups as the AFL-CIO and the National Association of Manufacturers. Strong support for his case came from the Anti-Defamation League of B'nai B'rith, the American Jewish Committee, the American Jewish Congress, and the Jewish Rights Council. Opponents to the position taken by DeFunis included the National Organization of Jewish Women, Harvard University, the Massachusetts Institute of Technology, Rutgers University, the Law School Association, the AAMC, the NAACP-LDF, the National Urban

League and the major civil rights organizations that represented the Black, Puerto Rican, Mexican American and Japanese American communities.[23] Essentially, then, battle lines were drawn and positions staked out in preparation for forthcoming cases in this issue.

In 1973, Allan Bakke, a white male of Norwegian ancestry and a resident of California, applied for the first time to the Medical College of the University of California at Davis. He was rejected and applied for the second time to UC/Davis in 1974 only to be rejected for a second time. Like DeFunis, Bakke also had strong academic credentials and had performed quite well on all objective measures of evaluation. The UC/Davis Medical School enrolled 100 students in its first year class but through its special admission program, sixteen of the 100 seats were specified exclusively for minority students. As a result of its special admissions program, the medical school had become one of the more successful non-black medical colleges in enrolling minority group students.

By 1974, as in the case of law schools and despite the expansion of the number of medical colleges since 1964, competition for positions in medical school classes had become keen and severe. Admissions standards were indeed raised and, even though there may have been an increasing tendency to assign greater weight to objective measures of evaluation, many medical schools required personal interviews to enable them to develop a more acceptable assessment of subjective or non-traditional factors employed in evaluation of criteria for admission. Allan Bakke was interviewed at UC/Davis and with mixed results. He considered options of suing Stanford University to test the constitutionality of its special admissions programs which set aside twelve seats for minority group students. He was assisted in his decision about options available to him by a person who was then a staff member of the UC/Davis Admissions Office. Ultimately, in 1974, following his rejection from the UC/Davis Medical School (as well as rejection from eleven other medical colleges), he sued and attacked the constitutionality of preferential admissions programs.

This case attracted more national attention than perhaps any other ethnic case that reached the Supreme Court of the United States since Brown vs. Board of Education. It was followed with exceptional interest as it made its way through the lower courts. The California Supreme Court supported Bakke but the Regents of the University of California decided to appeal the decision which led the case to the U.S. Supreme Court for resolution. Many persons questioned the motives of the University of California in pursuing what was described as a weak case supported by a weak legal defense at the State level. One persistent question was whether or not some form of collusion existed to eliminate special admission programs based upon racial or ethnic preference. Hence, the University was placed in a double bind of defending both its program and its legal course of action.

Some 57 amici briefs were filled in this case. Supporters of Bakke advocated an end to what they defined as "reverse discrimination," preferential treatment for minority groups and quotas. Opponents to Bakke argued against the legitimacy of the concept "reverse discrimination" and for rectification of past and present acts of discrimination that resulted in inequitable treatment between the races, secondary citizenship for minorities, and relegation of minority groups to the status of outsiders in the American society. They insisted on parity in all aspects of social, economic, political and educational life, social justice, and equality of opportunity. They took the position that race is a necessary variable in consideration of past patterns of discrimination and genuine efforts to eradicate every vestige of such behavior. They warned, and justifiably so, of the chilling effects of a decision totally in support of *Bakke* since the evidence indicated that institutions, federal, state, and local governments had already begun to return to the pre-affirmative era in their treatment of minority groups in any number of ways.

The printed and electronic media dutifully reported public opinion polls and the results of various studies on the attitudes of the American public regarding the critical issues in the *Bakke* case. In the opinion of many, the electronic media seemed to have demonstrated a definite pro-*Bakke* bias that was persuasive in shaping public sentiment toward his behalf.

Perceptions, no matter how unreal they were in terms of what objective data show, were pervasive that women and minorities received preferential treatment in jobs and income and that the white male was paying an unnecessary and unconstitutional price for their gains. Racism was displayed in an increasingly common assumption that all minorities were "less qualified" than whites for whatever position that was in question — whether in terms of admissions to graduate and professional schools, the acquisition of jobs, salary increases, and promotions. This was and is today the social milieu in which the *Bakke* decision was rendered and in which it presently operates.

On June 28, 1978, the United States Supreme Court rendered a two-part decision in the Bakke case. The first part of the decision, on a 5 - 4 vote, ordered Bakke to be admitted to the UC/Davis School of Medicine and rejected as unconstitutional "set aside" programs specifically for minority students. Since four Associate Justices abstained on the second part, the Court voted 5 - 0 in favor of the use of "race" as a positive factor in making decisions in certain circumstances. It should be stressed that the

Court *did* support the consideration of race or ethnicity in college and professional school admissions decisions; it *confirmed* the authority of colleges and universities to invoke discretionary powers in the formulation of admissions policies and in the implementation of admissions procedures; it *did* support the utilizations of flexible goals, (but *rejected* the employment of rigid, fixed quotas) so long as flexible goals are not in violation of Title VI of the Civil Rights Act of 1964.

The decision *does permit* the continuation of voluntary affirmative action practices designed to provide a diversified academic community which includes students, faculty, administrators and/or executives from different racial and ethnic groups. However, the Court warned that "race conscious" programs formulated to meet the social goal of diversity must satisfy compelling State interests. Therefore, one can read the decision as positive support for special admissions programs not involving arbitrary, fixed quotas and not deliberately racially exclusive. The question remains nonetheless as to whether or not the *Bakke* case has indeed produced a chilling effect on the admission and retention of black students in graduate and professional schools. The analysis of data collected for this study may be illuminating on this point. However, it may be argued that UC/Davis has already felt the impact of the decision since not a single black student enrolled in its 1980 first year medical school class.*

Traditional and Non-Traditional Admissions Criteria

Admissions personnel in medical colleges (as well as in other professional fields) agree that institutions have always drawn upon some combination of traditional (so-called "objective") criteria and non-traditional (or "subjective" factors) in making assessments of eligibility for entry into medical colleges. The relative weight assigned to each component varies over time, under specific circumstances, and from institution to institution in various jurisdictions.

Objective criteria include such variables as score or performance on the Medical College Aptitude Test (MCAT or new MCAT), college grade point average (GPA), the quality of the undergraduate college attended, and references. Many have argued that in some institutions ability and willingness to make a substantial financial contribution to the medical college is

---

*Vertis Thompson, M.D., quoted in *The Montclarion.* September 24, 1980, p. 12. He is 1980-81 President of NMA.

viewed as an objective factor. One characteristic which distinguishes objective measures from subjective measures is the ease by which the former are subject to quantification or the arbitrary assignment of numerical cut-off points.

Of the institutions participating in this study, all say that they relied on objective criteria in determining eligibility for admissions. However, all admitted the use of subjective measures in their overall evaluation of candidates. Between 1974 and 1978, as competition for space increased and with the uncertainty of Supreme Court action on the *Bakke* case, greater reliance on grade point average and new MCAT scores was discernable. It should also be pointed out that the proportion of first year medical school students with college grades point averages of "A" increased from 19.7 per cent in 1970 to 39.5 per cent in 1973. The proportion of such students reached 44.2 per cent in 1974 and a high of 50.4 per cent in 1977-78. It showed a slight decline just below the 50 per cent mark (48.6%) in 1978-79. The proportion of first year students who entered medical colleges with a grade point average of "B" dropped significantly from 73.3 per cent in 1970 to 50.8 per cent in 1974 and continued to decline the 46.9 per cent in 1978-79. Students with "C" averages declined from 7.0 per cent in 1970 to 1.8 per cent in 1978-79.[24]

Clearly the level of academic competition has sharpened dramatically during the past decade. Similarly, MCAT and new mean MCAT scores have climbed and become more competitive. In general, white students have scored considerably higher than black students on such tests. In some institutions, the performance of white students on such tests has been made as much as 2.5 points higher than the average mean score* presented by black students.

This disparity is not explainable by greater intellectual ability or reasoning capacity of the one group over the other, as some have indeed advocated, but more so by cultural factors and the socio-economic condition or milieu from which the majority of black students in medical colleges are drawn. The majority of white students have had long-standing experience with the kinds and quality of questions encountered in aptitude tests and are, therefore, psychologically prepared to answer them. Most are middle and upper-middle-class students who have also had continual reinforcement in home, school, and peer group environments of the kinds of verbal skills required for success on these tests. A distinct majority of white medical school students come from homes with family earnings equal to or

---

*This refers to mean computed on the new MCAT. Previously, the number of points often averaged 200 above mean scores for blacks.

above the median family income of white families in contrast to the over-representation of black students with low income origins. Some persons argue that white students are financially more able to attend "coaching schools" and thereby raise their scores on MCATs. The effectiveness of coaching is a matter of continuous and heated debates in educational communities.[25]

Subjective factors, on the other hand, are not easily amenable to quantification or precise statistical abstraction. One must rely often upon personal hunches, judgments informed by experience, and astute observations as well as good common sense in making subjective appraisals of a candidate's merit. The institutional participants in this study uniformly agreed that some form of subjective or non-traditional factors played an important role in their admissions decisions. These non-traditional factors included such things as: personal interviews; motivation for medicine as demonstrated, for instance, by work experience, membership in a premedical organization, community and volunteer work; personal characteristics evaluated in interviews by members of the admissions committee (faculty and student members); communication skills; working while carrying a full academic load as an undergraduate; maturity; sense of direction; promise for becoming leaders and contributors to the field of medicine; integrity; personal stability; commitment to service; ability to relate to people; awareness of world events; disadvantaged background; and race.

Of special importance to note is that literally all institutions utilize a combination of cognitive and non-cognitive factors. During the early seventies, a significant proportion of the medical colleges did over-weight the non-cognitive or non-tradition/subjective factors in ways that permitted compensation for poorer test scores among black students as a group in contrast to whites as a group. *However, it must also be reiterated that most black students did not enter medical colleges exclusively through special admissions programs. They had high MCAT scores and exceptionally high GPAs and often considerably higher than many of their white classmates.*

Medicine is a "people oriented" and people-contact profession. It requires individuals who not only have excellent mental and intellectual capacities but persons who are sensitive to the problems of others and who have the personality to work toward the accumulation of understandings of diverse personalities. It cannot and should not rely solely upon performance on objective measures of evaluation for to do so results in mistaken assumptions about the kind of person who can be the most effective physician or health care specialist. Consequently, non-cognitive factors have come to occupy a position of paramount importance in admissions decisions. This practice is so despite the tendency to use rigid cut-off points when confronted with difficult choices. All medical colleges insist that they wish to train the most well-rounded person to deliver effective health care services. Such a person is not necessarily the person with the highest MCAT score or with an "A" average from a prestigious undergraduate college. Nor does this mean that a "C" student with a "good personality and high motivation" is ipso facto such a person. A good balance must be made somewhere in between and that is undoubtedly why so many medical colleges suggest or insist upon personal interviews of all applicants.

The findings in this study showed that a substantial proportion of the medical institutions employed some form of special admission program both prior to and following the decision in the *Bakke* case. Special admissions programs obviously take a variety of forms. They include the set-aside types of programs established by the University of California at Davis and Stanford, as only two examples prior to 1978, in which minority students were guaranteed a specific number of seats in each entering class. They also take the form of separate admissions committees, in which all decisions on minority group applicants were rendered by this group. Some took the form of sub-committees of the Admission Committees in which the sub-committee considered minority applicants and reported their findings and recommendations to the parent committee. Often the parent committee merely rubber-stamped approval of sub-committee recommendations.

A few were labeled committees for "disadvantaged students" — often a euphemism for minority students. Frequently, unitary Admissions Committees attempted to assure that minority faculty and/or students were members of the Admissions Committee in order to guarantee the minority perspective in the decision-making processes. Less than 25 per cent of the institutions surveyed indicated that they never employed any type of special admissions committee, sub-committee, or minority membership on the parent committee during the 1970s. However, by 1978, most institutions had modified their "special character" of admissions to be in conformance with the *Bakke* decision. The evidence clearly supports the argument that those institutions that have employed aggressive recruitment and sensitive admissions programs have been the most successful institutions in enrolling black students in medical colleges.

Applications Activity in the 70s

Opponents to special admissions programs and affirmative efforts in higher education insist that black and other minority students are entering medical colleges at the expense of white students. The underlying assumption appears to be that so very few blacks are deserving of admissions and that all whites who are turned away are considerably more deserving than those black and other minority students who successfully enrolled in medical colleges. Common sense should ordinarily inform us of the erroneousness of such assumptions. But gaining admission to a professional school is such an emotionally charged phenomenon that "rejects" must identify scapegoats to rationalize their presumed failures. Nevertheless, this erroneous assumption is often conveyed, either wittingly or unwittingly, by the press and television. To establish truth or falsity of such assertions in a truly scientific way, it would be necessary to examine every applicant, case by case, total score by total score on both objective and subjective measures, at every institution and make comparisons over every time span. That is neither a practical nor a sensible undertaking.

The controversy over who gets into medical school may be viewed in the context of overall application and applicant activity. As the decade of the sixties began, some 14,397 persons filed 54,662 applications to the 86 existing medical colleges in the U.S. That equation represented a rate of 3.8 applications per person. At that time, the Medical College at Howard University and Meharry Medical College enrolled approximately 83 per cent of all black medical school students. This, of course, meant that the remaining 17 per cent of black students were scattered in 84 institutions and several of that 84 had no black students at all. As the sixties closed, there was a 59 per cent increase in the total number of applicants which almost tripled the number of applications from 54,662 to 133,822 by 1969-70* (See Table 10).

A pattern of escalation in the number of applicants and applications filed was well established by then at precisely the moment that the movement for promoting greater access of black students to non-black medical institutions achieved its greatest momentum. The two movements coincided: the escalation of white students interest in the field of medicine and the push for more black and other minority students in this field. The attraction of white students to medicine may be partially explainable by some students selecting medicine who might have chosen high technology fields like engineering had it not been for the temporary decline in engineering hiring at this time. It is also explainable by the influence of numerous publications which advertised the critical shortage of physicians who of course enjoyed exceptionally high social status, prestige, and power as well as excellent remuneration. For whatever reason, the number of applicants soared dur-

---

*Multiple applications simply mean that applicants file simultaneous applications at more than one institution to enhance opportunities for acceptance.

ing the seventies.

It can be assumed that the establishment of the American Medical Colleges Application Service benefitted black students who applied to medical colleges. This service sends all pertinent application data to a central site for computerization and distribution to member institutions. An applicant pays a single processing fee and this manages to reduce the overall costs normally involved in applying to multiple institutions. Interested institutions then contact the applicant for scheduling possible interviews as one component in the admissions process. As a result of participation in centralized application services, one may assume that the number of applications filed by black students closely approximates the overall number of black applicants. This is in contrast to the white situation in which applications exceed applicants.

In 1970-71, some 24,987 persons filed an average of 6.0 applications per person resulting in a total of 148,797 applications. Only 1,250 of those applications or 0.8 per cent were made by black applicants. The following year, applications exceeded the two-hundred thousand level and the ratio of applications reached 7.2 per person. However, applications from black students increased by only 302. Black students only obtained a 33 per cent rate of acceptance compared to 38.1 per cent for the applicant population as a whole. In the same year that *Bakke* filed his suit against the University of California (1974), some 42,624 persons filed 362,376 applications. By that time, the number of applications from black Americans had climbed by a mere 1,173 over the 1970 figure compared with an overall increase of approximately 16,000 in the four year period.

By 1974, eleven new medical colleges had opened or an increase from 103 to 114 institutions. Not only were more applications made but the space available in medical schools had also increased, although not sufficient to accommodate all applicants, deserving or not. In 1974, 43 per cent of all black applicants were accepted to medical school in contrast to 38.3 per cent of the total number of applicants. This disparity suggests that a special effort was being made to accommodate more qualified black applicants than black students received preferential treatment. On the other hand, if one examines the applicant characteristics very carefully, it is quite easy to discern that comparatively few black persons were then applying to medical colleges. Moreover, Howard and Meharry still accounted for a substantial proportion of all blacks who entered medical colleges. Their dispersion in the historically white institutions was so minimal as to occupy only about five seats per entering class per 112 medical colleges.

In the fall of 1978, the year in which the U.S. Supreme Court handed down its controversial decision in the *Bakke* case, there was a decline by

Table 10. Applications and Applicants to U.S. Medical Schools by Race and Year, 1968-79: All Institutions Combined

Year	Applicants	Applications	Applicants Per Person	Applications From Black Students
1968-69	21,118	112,195	5.3	—
1969-70	24,465	133,822	5.5	540
1970-71	24,987	148,797	6.0	1,250
1971-72	29,172	210,943	7.2	1,552
1972-73	36,135	267,306	7.4	2,186
1973-74	40,506	328,275	8.1	2,227
1974-75	42,624	362,376	8.5	2,423
1975-76	42,303	366,040	8.7	2,288
1976-77	42,155	372,282	8.8	2,523
1977-78	40,589	371,545	9.2	2,487
1978-79	36,636	335,982	9.2	2,564
1979-80	36,100	329,470	9.1	2,599

Sources: *Journal of American Medical Association* 243:9 (March 7, 1980), p. 852 and Mrs. Lillye Mae Johnson, Minority Affairs Office, AAMC Correspondence.

almost four thousand in the total of applicants from the preceding year (40,589 to 36,636) and a precipitious drop in the total number of applications filed (from 371,445 to 335,982). Yet the number of applications filed per person was precisely the same as in 1977-78 or at its highest level ever, 9.2 per person. Noticeably, there was only an increase of 77 more applications from black students over the 1977-78 figures.*

In essence, the application pool among black students for U.S. medical schools has remained virtually unchanged for the past eight years. During the same period, overall applications increased by approximately 100,000. *Who is getting in and at whose expense? Competition is keener, and cut-off points are higher. White students, in short, are knocking each other out, since there are so few minorities entering first year classes.* Even without the presence of blacks and other minorities in the applicant pool, the over-whelming majority of rejected whites would continue to be rejected for medical schools classes.

Between 1970 and 1978, approximately twenty-three thousand black

---

*More recent data show a further decline, however, in both applicants and applications. But in 1979-80, there was a modest increase in the number of applications filed by Blacks, and a further decline in total applicants.

students applied to medical schools in the United States (See Table 10). This represented less than .8 per cent of the total number of 2,531,238 applications filed. Despite initial higher acceptance rates for black students in the earlier part of the decade, today about the same proportion of white and black student who apply are enrolled. That is, approximately 46 per cent of the applicants are admitted to at least one of the 126 colleges of medicine.

Although the optimistic goals of Project 75 of enrolling minority students to equal twelve per cent of total enrollment by 1975 were not attained, one cannot justifiably categorize that effort as a failure. Nor can a similar characterization be made of the conscientious recruitment efforts and substantial financial outlays from the private sector simply because comparatively few generated more black students. The number of applications from black students during the seventies was about five times greater than the total number produced during the sixties. During the seventies, medicine had to compete with other career choices that were opening up to black students, some of which offered substantial financial rewards without incurring encumbrances or mounting indebtedness created by the rising costs of medical education. Some professions did not demand the continued deferring of gratifications as a prerequisite for goal attainment. Hence, the problem of underrepresentation does not imply a failure of Project 75.

Enrollment Trends in the 70s

The pattern of access of black students to medical colleges during the seventies can best be characterized by initial optimism and success, followed by declines and, possibly, a leveling off. Tables 11 and 12 depict enrollment fluctuations.

In 1968-69, approximately 60 per cent of the 266 first year black students enrolled in U.S. medical colleges were enrolled at Meharry and Howard Medical Colleges. About 40 per cent were scattered among 54 of the remaining 97 medical institutions.* As the sixties ended and the seventies opened, the 440 black medical school students, who represented 4.2 per cent of all first year classes combined, were still concentrated primarily at the two historically black medical colleges. Table 11 and Figures 1 and 2, reveal a systematic increase in the overall enrollment of first year students as well as a climb in the proportion of first year black enrollees. Although the proportions of black students reached a peak of 7.5 per cent in 1973 and in

---

*This distribution means that there were approximately two blacks in the first year classes at white institutions and that 43 colleges enrolled no blacks in the first year class.

Table 11. First Year Medical School Enrollment by Race and Year, 1968-79: All Institutions Combined

Year	First Year Enrollment All Students	Black Student First Year Enrollment	Per cent Black of Total
1968-69	9,740	266	2.7
1969-70	10,269	440	4.2
1970-71	11,169	697	6.1
1971-72	12,088	882	7.1
1972-73	13,352	957	7.0
1973-74	13,876	1,027	7.5
1974-75	14,579	1,106	7.5
1975-76	14,910	1,036	6.8
1976-77	15,282	1,040	6.7
1977-78	16,136	1,085	6.7
1978-79	16,530	1,064*	6.4
1979-80	16,930	1,108*	6.5

*Includes new students and first year repeaters. If these figures were not included, the number of first year - first-time Black enrollees falls to 922 in 1978-79 or 5.8 per cent of all first year, first-time enrollees.

Source: The Association of American Medical Colleges.

1974, the same year Bakke filed his suit in California, the absolute numerical change from year to year is not that significant except in the years, 1969, 1970 and 1971. For instance, the increase in first year enrollees for black students between 1973 and 1974 was by 70 students compared to an overall increase of 524 first year enrollees. The following year increase for black students was 79 students compared to an overall increase of 703. Each succeeding year thereafter shows a noticeable decline in proportionate enrollment of black students and significant reductions in absolute numbers are apparent. Hence, the overall first year enrollment increased continually; only four more black students enrolled in first year classes in 1976 over the 1975 figure and a dramatic drop occurs thereafter because the reported first year figures for black students also include first year repeaters.

When one removes the first year repeaters from first year enrollment in

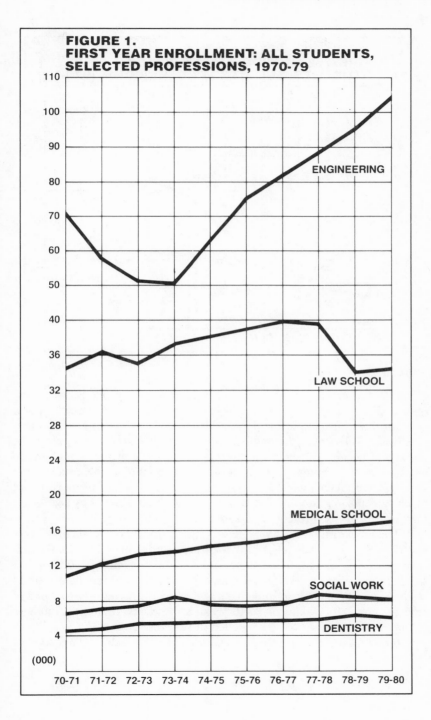

FIGURE 1.
FIRST YEAR ENROLLMENT: ALL STUDENTS,
SELECTED PROFESSIONS, 1970-79

ENGINEERING

LAW SCHOOL

MEDICAL SCHOOL

SOCIAL WORK

DENTISTRY

(000)

70-71 71-72 72-73 73-74 74-75 75-76 76-77 77-78 78-79 79-80

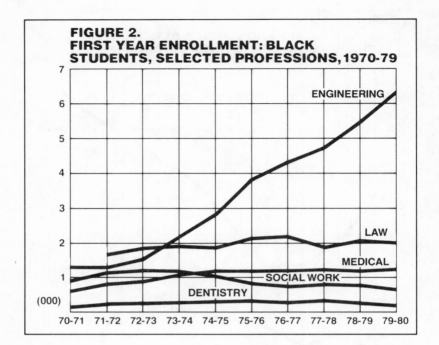

FIGURE 2.
FIRST YEAR ENROLLMENT: BLACK
STUDENTS, SELECTED PROFESSIONS, 1970-79

1978-79, for instance, the number of black students enrolled for the first time in all U.S. medical colleges falls sharply to 922 or 5.8 per cent of first year enrollees throughout the nation. Hence, the number of blacks in first year classes is less than the number enrolled in 1972, two years before the Bakke suit began, and the proportion is less than the proportion of blacks observed at the beginning of the decade.[26] If equally precise data on first year repeaters were available for all preceeding years, the proportions of black students in first year classes would undoubtedly be less by significant margins than those presented in Table 11.

The repeat rate in first year classes for all students in medical colleges in 1978-79 was 1.7 per cent of the total first year class of 14,342. By contrast, the repeat rate for black students was 14.4 per cent of 1,077 or 155 students. For Mexican American students, it was 9.2 per cent of 280 or 24 students. For Puerto Rican students, it was 4.6 per cent or twelve out of 262 students. And, for Asians or Pacific Islander students, the repeat rate was 2.7 per cent or 13 out of 483 enrolled in first year classes.[27]

Explanations for repeating a class are not always clear. They may be found in inadequate undergraduate preparation in the basic sciences, weak self-discipline and poor study habits, insufficient time to study because of job commitments, family problems which interfere with conscientious studying, a hostile learning environment, prejudiced behavior by professors who are not inclined to be fair to minority students, and a whole range of adjustment problems experienced by black students in a substantially new, often overpowering, and intimidating environment.

According to AAMC data, aside from Howard and Meharry Medical Colleges, the most successful medical institutions in enrolling black students in first year classes during the 1970's were: Temple University (282), The University of Illinois (263), Harvard University (180), Wayne State University (179), The University of North Carolina (169), Case Western Reserve University (163), State University of New York/Buffalo (157), State University of New York/Downstate (146), The University of Cincinnati (130), The University of Maryland (128), Washington University/St. Louis (125), The University of California/San Francisco (124), Indiana University (120), The University of Pennsylvania (118), Hahneman (117), The University of Florida (117), Michigan State University (111), CMD/NJ - Rutgers (111), Ohio State University (108), Jefferson Medical College (107), and Tufts University (107).

Seven of the twenty-three institutions listed above are located in the "Adams States": (1) Temple, (2) the University of North Carolina, (3) the University of Maryland, (4) the University of Pennsylvania, (5) the University of Florida, (6) Hahneman Medical College, and (7) Jefferson. Four of these are in the State of Pennsylvania.

Forty-four medical colleges that were in existence throughout the seventies enrolled fifty or fewer black students in that decade. Although about one-third of these institutions are located in States with exceptionally small black populations (e.g., Vermont, North Dakota, South Dakota, Nevada and Utah), a clear majority with less than average black enrollments included such institutions as the University of California/Davis; Pritzker Medical College (University of Chicago), the University of Kentucky, the University of Louisville, the University of Oklahoma, the University of Tennessee, and Vanderbilt.

Total Enrollments

The proportion of black students in the total enrollment in U.S. medical colleges more than doubled between 1969-70 and 1978-79. The increases in total enrollment can be attributed to a combination of factors that encompass institutional commitment, internal and external pressures on medical colleges, aggressive recruitment, special admissions, and federal mandates to desegregate all components of post-secondary education as dictated by the "Adams" decisions. Unquestionably, the upward climb in total enrollment of black students can also be attributed to a growing interest in medicine as an attainable career by black students from all socio-economic statuses. This change in and of itself reflected a perceptual change from the viewpoint that medicine was a career choice open only to the economically affluent and socially influential.

Increasing availability of financial aid programs was a highly significant determinant of enrollment patterns during the seventies. Recruitment itself was and continues to be affected to a large degree by the presence of adequate role models, black faculty and administrators in the university; the location of the institutions, and the type of institution — whether private, public, or state assisted. However, it also appears that the declines in total enrollment and the possible leveling off of black students around 5.8 per cent of total enrollment may reflect an actual loss of commitment to mainstream blacks through medical careers. That loss of commitment is evidenced by reductions in funds available for the recruitment and enrollment of black students. It is suggested by the eagerness with which the Bakke decision was received and the ensuing rapid abandonment of special admissions programs without commensurately restructuring them in ways that would assure both conformity to legal expectations and the continued commitment to racial and ethnic diversification. It is manifested in the increasingly heavier weight assigned to objective measures of evaluation and the presumed downgrading of subjective measures.

An examination of Table 12 and figures 3 and 4 show that total enrollment in the nation's medical colleges has moved steadily upward over the past decade. The yearly total enrollment has risen from 35,833 in 1968-69 to almost 64,000 in 1979-80. There was an increase of more than ten thousand students in total enrollment between 1970-71 and 1973-74. In the same period, the increase in total enrollment of black students rose by approximately 1,500 persons. The percentage increase may be somewhat phenomenal but, considering the low base from which the change is measured, the absolute numerical change is disappointing.

Similarly, it is observed that the percentage of black students enrolled in medical colleges peaked at 6.3 per cent in the same year that the Bakke suit was filed in California. The proportional decline each year thereafter is highly suspicious and suggests that a number of institutions deliberately embraced a policy of calculated cautiousness in anticipation of a pro-Bakke decision. Since 1974, the total yearly enrollment of all students combined

Table 12. Total Medical School Enrollment by Race and Year, 1968-79:
All Institutions Combined

Year	Total Enrollment	Black Student Enrollment	Percent Black Of Total
1968-69	35,833	783	2.1
1969-70	37,690	1,042	2.8
1970-71	40,238	1,509	3.8
1971-72	43,650	2,055	4.7
1972-73	47,366	2,582	5.5
1973-74	50,751	3,059	6.0
1974-75	53,554	3,353	6.3
1975-76	55,848	3,456	6.2
1976-77	57,760	3,517	6.1
1977-78	60,099	3,587	6.0
1978-79	62,242	3,540	5.8
1979-80	63,800	3,627	5.6

Source: The Association of American Medical Colleges.

climbed by more than 10,000; that is from 53,554 in 1974-75 to 63,800 in 1979-80. During the same period the annual total enrollment for black students has risen by a mere 300 students or from 3,353 in 74-75 to 3,627 in 1979-80. In fact, the total enrollment of black students dropped by some 47 students between 1977-78 and 1978-79 (See Table 12 and figures 3 and 4).

Clearly, that we now have forty more medical colleges in 1980 than we had in 1968-69 has not proven a boon to the total enrollment of black students in American medical colleges. Although it is true that black students are now more widely distributed in U.S. medical colleges than ever before, when students enrolled at Howard and Meharry are deleted from the total enrollment figures, the problems of dispersion and concentration are readily apparent.

Meharry and Howard account for 20 per cent of all black medical students. This means that these two institutions enroll as many black students as 74 medical institutions combined. According to the preliminary data collected by the AAMC for Fall 1979-80 on individual medical schools, slightly more than 50 per cent of all mainland U.S. medical schools enroll twenty or fewer black students. More than one-quarter of the 122

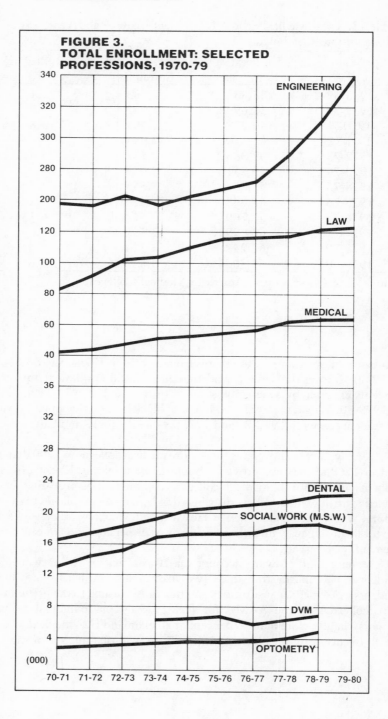

FIGURE 3.
TOTAL ENROLLMENT: SELECTED PROFESSIONS, 1970-79

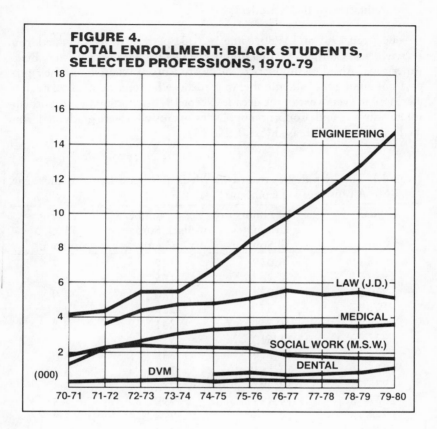

FIGURE 4.
TOTAL ENROLLMENT: BLACK STUDENTS,
SELECTED PROFESSIONS, 1970-79

mainland U.S. institutions enroll ten or fewer black students. However, some 35 per cent (43 institutions) have a total black student enrollment of between 21 and 50 students.

Only 9.8 per cent of the medical colleges enroll between 51 and 100 black students. This group consists of the following institutions: Case Western Reserve University, The University of Cincinatti, Hahneman Medical College, Harvard University, the University of Illinois, Michigan State University, CMDNJ/Jew Jersey Medical, CMDNJ - Rutgers, the University of North Carolina, Temple University, and Wayne State University. Of this group, only Hahneman, Temple and The University of North Carolina are in the "Adams States." These institutions are about evenly divided in terms of classification as a private or public institution.[29]

Graduation of Black Students

Between 1969 and 1980, some 141,331 persons were graduated from U.S. Medical Colleges. Of that number, 6,041 (or 4.2 per cent) were Black Americans. Absolute numbers of black students graduated and their proportions among the total number of students who received medical degrees increased in every year of the decade except 1978-79. In that year, both absolute numbers and proportion of black students who were graduated from medical colleges decreased (See Table 13).

Table 13. Total Graduates from U.S. Medical Schools by Race and Year, 1969-80: All Institutions Combined

Year	Total Number of Graduates — All Students	Total Number of Black Student Graduates	Per cent Black of Total
1968-69	8,059	142	1.7
1969-70	8,367	165	1.9
1970-71	8,974	180	2.0
1971-72	9,551	229	2.4
1972-73	10,391	341	3.2
1973-74	11,613	511	4.4
1974-75	12,714	638	5.0
1975-76	13,561	743	5.4
1976-77	13,607	752	5.5
1977-78	14,393	793	5.5
1978-79	14,966	774	5.2
1979-80	15,135	768	5.1
Total	140,331	6,041	4.2

Source: Association of American Medical Colleges.

The data provided in Table 13 attest to the successes experienced in the early seventies in enrollment of black students as a consequence of institutional commitment, federal assistance, aggressive recruitment, and, perhaps, flexible admissions policies. Without this combination of factors, it is highly probable that Howard University and Meharry Medical College would have continued to graduate more than four-fifths of all the black

physicians in the United States. As a direct result of changes in the recruitment and enrollment patterns, larger and larger numbers of the nation's
medical colleges are participating in the effort to expand equality of opportunity in medical education. The apparent loss of a substantial number of
these students between the first and fourth years, however, suggests problems of serious proportions.

Problem Areas in the Medical Education of Black Students

One of the most serious problems is the comparatively high attrition
rate of black students. Many of the institutions who responded to our
survey referred to the lack of adequate preparation in the basic sciences as a
serious weakness of most of the students who are either dismissed for
academic reasons or who repeat a year of study. Indeed, one of the cautions
about Table 11 is the fact that we are not able to identify the precise number
of "repeaters" in each first year class. In fact, repeaters are probably included in each of the preceding years. Hence, it may be somewhat presumptious
to conclude that the fall-out apparent between first year enrollment (Table
11) and the number of black students who should have graduated four
years later is as great as it appears to be (Table 13). Further, some of the loss
may be from withdrawals for non-academic reasons. Albeit, most institutions do report a serious problem of retention among black students when
compared to the student population as a whole.

According to one study, black Americans have a retention rate of approximately 87 per cent through four years of medical education compared
to 97 per cent for the medical population as a whole. The retention rates for
Mexican Americans (88 per cent) and American Indians (89 per cent) in
only slightly better. By contrast, Mainland Puerto Rican Students have a
retention rate of 98 per cent which is slightly better than the national
average.[29] According to Johnson and Sedlacek, however, minority group
retention rates approximate those rates observed for white students about a
decade or so ago.[30] A primary goal of medical institutions is, of course, to
raise the retention level of black and other minority group students to approximate national retention averages.

To address this problem, most institutions provide some form of
academic and non-academic support services. Several institutions obtained
special grants through either the Health Careers Opportunity Program or
Special Projects of the Department of Health Education and Welfare, as
well as from other sources, to help finance programs designed to aid those
minority students who needed academic assistance to enhance their chances

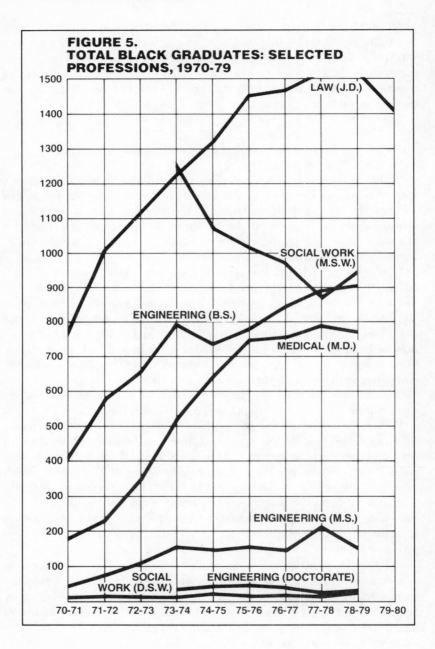

FIGURE 5.
TOTAL BLACK GRADUATES: SELECTED PROFESSIONS, 1970-79

of success.

For instance, Summer Institutes and Pre-Enrollment Enrichment Programs were instituted at a number of institutions (e.g., Rutgers Medical School, the University of California/Davis, University of Colorado Medical School, the University of Kansas, and Harvard University). These summer programs provide students an overview of first year medical school courses (e.g., gross anatomy) and conduct sessions that concentrate on understanding the basic sciences, the development of effective study skills, and a general orientation to the medical school.

In many instances, these initial services are followed up through organized academic support services made available to students who encounter academic difficulties or who wish to avoid them (e.g., Michigan State University, Cornell University, and the University of Texas/Houston, and others). Supportive services, where provided, include such components as paid tutorial assistance, voluntary tutorial assistance by some faculty members, courses on study skills, time management and test-taking strategies, assistance in selecting academic advisers, alternative curriculum planning and flexible curricula or self-paced studies, and assistance in financial aid planning. Personal counseling is frequently a component of comprehensive support services. In addition, as previously mentioned, mutual psychological support is gained through participation in organized student medical associations, Black Student Unions, recognition of the existence of a critical mass of students, and the presence of black faculty members.

Black Faculty

Although the absolute number of black faculty at U.S. medical schools increased during the seventies, about forty per cent of total black faculty are employed at the medical colleges of Howard University, Meharry Medical College, and Morehouse College. (When the number of black faculty of the Charles Drew Post Graduate Medical Center in Los Angeles is added to this group, that proportion is raised above forty per cent).[31] When all institutions are combined, black faculty at U.S. medical schools comprise only about 1.8 per cent of the total number of medical school faculty.[32]

Most black faculty teach in the clinical programs in contrast to the basic sciences programs of medical schools. This underrepresentation of black faculty in the basic sciences (e.g., Anatomy, Biochemistry, Immunology, Physiology, and Microbiology) is attributable in part to the paucity of black Americans who hold the doctoral degree in these subjects.

As discussed in Chapter Twelve, there is a critical shortage of doctoral degree recipients in the physical and natural sciences among black Americans. It is also explainable in part by the preference of black physicians, who can also teach basic sciences, for other medical careers.

Without black faculty who teach in the basic sciences curriculum, many medical students must await the clinical aspects of their training before being exposed to role models. Several participating institutions in this study recognized the need for more black faculty to serve as role models in a variety of academic and non-academic functions but few seemed imaginative as to how to attack the problem. Yet, the data presented in Chapter III suggest that the presence of black faculty may be the most important contributor to successful recruitment, enrollment, and graduation of black students.

Other problems cited include: a struggle for financial support which interferes with medical studies, lack of a critical mass of black students, emotional difficulties (at home and within the institution), and failure to adjust to the new institutional environment. Clearly, the problem of retention is multi-dimensional. It can only be solved through a coordinated, systematized program which adequately addresses each of the areas specified in the preceding discussion.

According to some participants in this study, the Bakke decision has already had some subtle influences on practices in medical colleges. These influences were described in terms of "diluting forces" which arise from the tendency to lump together a variety of groups (e.g., women, and handicapped) under the fabric of "disadvantaged." While the needs of these groups are compelling and legitimate, these officials argue that this arrangement places them in competition with other underrepresented minorities and dilutes the efforts to combat the nation's historic and lingering problem of racism and racial discrimination.

Finally, it would be a tragic mistake to assume that the societal problem of a shortage of black physicians is no longer a significant issue simply because 5,899 blacks were graduated from medical colleges during the seventies. The nation does not have a sufficient number of black physicians to meet the health needs of either black communities or the demands for medical researchers, and requirements for medical school faculty positions. Parity has not been reached in the production of black physicians in the United States. Hence, when the federal government warns of an oversupply of physicians by 1990, it is grossly misleading to assume that such a prediction assumes that the country will have an excess of black physicians.[33] On the contrary, should the matriculation of blacks in medical colleges remain at or below its present rate, parity will not be reached in this century.

Footnotes

1. Karen Albarnel, "Is A Relapse Ahead for Minority Medical Education?," *Foundation News* (March, 1977), 24-27.
2. *Minorities in Medicine: Report of A Conference*. New York: Josiah Macy, Jr. Foundation, 1977. p. 5 and James E. Blackwell, Philip Hart, and Robert Sharpley, *Alientation Among Metropolitan Blacks*. New Brunswick, N.J.: Transaction Books, Forthcoming, Chapter I.
3. *Minorities in Medicine. Supra,* p. 3.
4. *Report of the AAMC Task Force on Minority Student Opportunities in Medicine.* Washington, D.C.: The Association of American Medical Colleges, June 1978, p. 1.; *Report of The American Association of Medical Colleges Task Force,* April, 1970, pp. 1-18, and James L. Curtis, *Blacks, Medical Schools and Society.* Ann Arbor: University of Michigan Press, 1971, p. 57-58.
5. Clinical appointments are distinguished from appointments as Basic Sciences Faculty. The former are usually adjunct, non-tenure-track appointees who work with medical students in clinical programs. A decided majority of black faculty members listed in medical colleges have these appointments. Basic Sciences faculty are more likely to teach basic courses such as Biochemistry and Endocrinology. These faculty are likely to be on the tenure-trank, have both the M.D. and Ph.D. or one of the two degrees and are likely to be involved in medical training from the first year onward. Black faculty are less likely to teach Basic Sciences curriculum, except at Howard, Meharry and Morehouse. It should also be pointed out that, because of the nature of clinical training, the majority of medical faculty, in general, are likely to have such appointments and they are ordinarily physicians in practice, with needed specialities, and are attached to hospitals in which internships and resident training occur.
6. *Op. Cit,* 1978 Report of AAMC.
7. *Ibid.,* p. 4.
8. Ralph C. Kuhli, "Education For Health Careers," *Journal of School Health.* 41 (January 1971), p. 17.
9. Davis C. Johnson, *et. al.,* "Recruitment and Progress of Minority Medical Schools Entrants, 1970-72," *Journal of Medical Education* 50: 721, 1975.
10. *Report of the AAMC Task Force on the Inter Association Committee on Expanding Educational Opportunities in Medicine for Blacks and Other Minority Students.* Washington, D.C., April 22, 1970.
11. *Ibid.*
12. *Ibid.*
13. *Report of the AAMC Tast Force,* June 1978.
14. James E. Blackwell, *Access of Black Students to Graduate and Professional Schools.* Atlanta: The Southern Education Foundation, 1975, p. 16.
15. "Information for Minority Group Students," *Medical School Admission Requirements, 1976-77.* p. 51.
16. *Report, Op. Cit.,* p. 19.
17. *Ibid., p. 3*
18. *Cf. Allan P. Sindler, Bakke, DeFunis and Minority Admissions.* New York: Longman, 1978, p 18.
19. Frank J. Atelsek and Irene L. Gomberg, *Special Programs for Female and Minority*

Graduate Students. Washington, D.C.: American Council on Education, HEP Report, Number 41, 1978.

20. *Minorities in Medicine, Op. Cit.,* p. 34.

21. Sindler, *Op. Cit.,* p. 36.

22. *Ibid.,* p. 205.

23. *Ibid.*

24. *Journal of the American Medical Association* 243-9 (March 7, 1980) p. 853.

25. A major controversy has arisen with the past few years on the value of coaching schools in raising scores on standardized/aptitude tests. For information, cf., Cheryl M. Fields, "SAT Scores Don't Measure Scholastic Aptitude, Can Be Influenced by 'Coaching', Report Says," *The Chronicle of Higher Education,* (May 27, 1980), p. 1.

26. *Op. Cit.,* p. 855.

27. *Ibid.*

28. Cf. Table 7-4 in *MSAR, Minority Student Information by Individual Medical Schools.* Washington, D.C.: Association of American Medical Colleges, 1980 (Unofficial Results).

29. *Report of the Association of American Medical Colleges Task Force on Minority Student Opportunities in Medicine* Washington, D.C.: Association of American Medical Medical Colleges, 1978, p. 40.

30. Davis F. Johnson and William E. Sedlacek, "Retention by Sex and Race of 1968-1972 U.S. Medical School Entrants," *Journal of Medical Education.* 50: 932, 1975.

31. Cf. Report, Supra., p. 45: *Participation of Women Minorities in U.S. Medical Faculties.* Washington, D.C.: DHEW Publication No, (HRA) 76-91, Prepared by H.P. Jolly and Thomas Larson of the AAMC), 1976; and personal interview with Dr. Daniel Wooten of the Charles Drew Post Graduate Medical Center, August 1980).

32. Report, *Op. Cit.*

33. *A Report to the President and Congress on the Status of Health Professions Personnel in the United States (1980).*

CHAPTER 5

Mainstreaming Black Americans in Dentistry

This chapter focuses on the processes involved in the mainstreaming of black Americans in the field of dentistry during the 1970s. Dentistry has never attracted as many persons to it as have other scientific fields such as medicine. Compared to medicine, fewer black Americans have become dentists during the 140 years since dentistry became recognized as a major profession in the United States. However, the attractiveness of the field increased dramatically after WW II. Many were drawn to dentistry during the 1960s when entrance into medical schools became so restrictive. As a result, many gifted persons interested in the health care fields selected other professions such as dentistry. Nevertheless, there is a major crisis in the dental education of black Americans. This is not a crisis of the quality of dental education, rather of the shortage of black dentists in the United States.

In this chapter, specific attention is called to the serious underrepresentation of black Americans in dentistry and to concrete efforts to increase their participation in this profession. The chapter focuses especially on the historic roles performed by the College of Dentistry of Howard University and of Meharry Medical College in the training of black dentists. It describes the influence of policy changes articulated by the American Dental Association and by the Association of American Dental Schools in stimulating access and in reducing structural discrimination in dental education. However, the primary and most essential emphasis of this chapter is on enrollment and graduation trends of black dental students during the past decade and the salience of certain variables for understanding why these fluctuations or patterns occurred.

Historical Context

Prior to the opening of the Baltimore College of Dental Surgery in 1840, there was no college of Dentistry in the United States. Dentistry was not regarded as a major profession and there was no formalized institution for the training of individuals in the art of dental surgery. Persons who "extracted teeth" were viewed as technicians or artisans who learned their trade exclusively through apprenticeships under practitioners of this skill. The first college of dentistry in America discriminated on the basis of race. Although blacks could not enroll at Baltimore College of Dental Surgery, it is estimated that when that instituition opened in 1840, some 120 blacks were practicing the art of dentistry.[1]

When Howard University opened in 1867, it did not establish either a department or a school of dentistry. Not until 1881 was such a "department" organized at Howard. Even then, the dentistry department did not have the same status as the School of Medicine. This low priority to which dentistry was assigned reflected its second-class status in the larger society as well.

Dentistry did not become a matter of special priority until Dr. Mordecai W. Johnson became the first black to serve as President of Howard University. In his inaugural address, in 1926, Dr. Johnson committed himself to the expansion of the dental education program at Howard University. Three years later, Dentistry was re-organized to become an independent, autonomous unit, separate and apart from the School of Medicine and headed by its own College Dean. Structural improvements continued to be made at Meharry also and both institutions were able to broaden the scope of their programs in the training of dentists.[2]

Dentistry fared little better at Meharry Medical College in the early years. Since Meharry was founded in 1886, some five years following the establishment of the Department of Dentistry at Howard University, the early training of blacks in this profession lagged somewhat behind that at Howard. Even today, Meharry Medical College remains somewhat smaller than federally supported Howard University.

For more than seventy-five years, these two historically black Colleges of Dentistry trained more than 90 per cent of all the black dentists in the United States. With the advent of *Plessy* vs. *Ferguson* a pattern had already been established of developing strong parallel institutions in the black community to meet the felt needs of the population. Blacks were excluded from historically white colleges of dentistry either by de jure or de facto systems of discrimination. Many white dentists would not provide dental care to blacks who needed such health services. Hence, partially in response to recognized needs and partially as a reaction to systematic segregation, institutional racism, and persistent discrimination, the dental education programs at Meharry and Howard expanded and attracted increasing numbers of black students. Without these two institutions, the crisis in dental education and in the delivery of dental health care services to the black community would be considerably more disastrous than it is in the 1980's.

By 1910, there were some 478 black dentists in the United States. Within a span of some two decades, that is by 1930, there was a 400 per cent increase in the number of black dentists. More specifically, according to the 1930 Bureau of Census Report, the number of black dentists reached some 1,774.[3]

That progress was slowed to a trickle by the Great Depression. Since black Americans are so frequently treated as outsiders with limited access to highly prized services and the economic reward system, it is not surprising that periods of economic stress create devastating problems for black Americans. Hence, the Great Depression of the 1930s resulted in widespread economic disabilities for Americans, in general, but it had its most severe impact upon black Americans who were at the bottom of the economic ladder.

One consequence of this situation was a shocking decline in the enrollment of black students in dentistry. Such a precipitous fall in enrollment is illustrated by the enrollment of only 34 students in the College of Dentistry at Howard University and by only three students receiving the D.D.S. from Howard in 1934. The effects of the Depression — high unemployment and absence of money income — prevented all but a very few black families from attempting to support higher education for their offspring. Many young men and women found it almost impossible to obtain even the most menial jobs that could generate income sufficient to support dental education. This situation only improved after the enactment of federal relief and work programs to combat economic stagnation. It was alleviated, too, by a wartime economy necessitated by World War II which expanded economic opportunity. Thereafter, both the enrollment of blacks in dental schools and their graduation from them increased. But the major responsibility for the training of blacks in dentistry was still assumed by Howard University and Meharry Medical College.

Earlier on, after graduation, licensing, and entry into the profession, black dentists continued to encounter discrimination within the profession. The American Dental Association denied them membertship or even attendance at workshops, lectures on scientific advances in dental specializations, and in-service training programs. Most of its component or constituent societies also excluded blacks. Because of these policies of exclusion, black dentists took steps to provide their own mechanisms for assuring their exposure to new knowledge, new technologies in dental education and advances in the profession. Dr. David A. Ferguson, an 1899 graduate of the Dental School at Howard University, was instrumental in organizing a group of black dentists in 1901 who became the nucleus for what subsequently became the National Dental Association.[4]

Professional Associations and the Training of Black Dentists

Inasmuch as discrimination and segregation were widespread in dentistry, it was inevitable that this field would be forced to make radical changes. The professional associations played a profound role in inducing reforms which, ultimately, expanded access to dental schools. For instance, the National Dental Association joined with the major formal organizations of the Civil Rights Movement in demanding equality of access of black students to colleges of dentistry throughout the nation. It also insisted upon the abandonment of policies of discrimination followed so rigidly by the American Dental Association and its affiliated societies. The American Association of Dental Schools and the American Dental Association also deliberated on the importance of structural reforms, policy changes, and the need to respond in aggressive ways to the dictates of the Civil Rights Movement for broader participation of all Americans in professional school education. The American Dental Association issued a policy change in 1965 in which it urged all its constituent and component societies to drop all barriers to membership based upon race, religion, ethnicity, or creed. Some of these groups had already taken steps in this direction; the problem was not with groups themselves but with those components who were most resistant to social change and unwilling to share rewards, power, influence, and status with members of other racial groups.

Ultimately, all of the societies complied, at least in principle, but many black dentists refused to join either the constituent societies or the American Dental Association. In 1980, many of them divert that $500 or $1500, which would be paid to the ADA in membership fees, to support the all black National Dental Association's activities and other civic or social action programs in the black community.

The Association of American Dental Schools took further steps to dramatize its commitment to increasing enrollment of black students in the schools of dentistry. It urged the re-examination of recruitment, selection, and admissions policies to end the previous discriminatory practices. Black students could apply with greater assurance of being admitted to dental schools. Forty-four of its membership institutions became early participants in the centralized admission program which enabled black students, especially those from impoverished backgrounds, to reduce their applications costs.

A variety of special programs was initiated with the encouragement of the American Association of Dental Schools in order to stimulate greater equality of opportunity. Finally, the American Dental Association and the Association of Dental Schools began to collect systematic data on minorities in dental schools. They also encouraged private and federal governmental financial support to assist in dental education in ways that would ultimately benefit impoverished students educationally qualified for dental schools. Concerted action of this type, a reflection of an early commitment to changing the racial composition of dental schools, had immediate and far-reaching results.

Admissions Criteria

As in professional education in general, dental schools employ a mixture of objective or cognitive criteria and subjective or non-cognitive, nontraditional criteria in reaching admissions decisions. As a response to both internal and external pressures for change, some institutions modified selection criteria in ways that facilitated greater access for black and other minority students. Special programs for minority students have taken diverse forms. They have varied from a kind of open admissions format to a more rigid, highly quantified system. However, litigation against special admissions programs, as in the Bakke case, shows a high correspondence to the growing number of dental schools which now claim that they have "no special programs for women and minority students." Approximately nineteen of the 60 schools of dentistry in 1979-80, publicly stressed the absence of specific programs for special groups of students. This is in contrast to the relative paucity of these statements in the early seventies.

Objective admissions criteria include a Dental Admissions Test Score (DAT) of an average of 4 or above on a 5 point scale. All U.S. schools of dentistry required the DAT but all do not make the same use of test performance in overall evaluations of the candidate. All do not assign the same relative weight to these scores in determining eligibility. The DAT is designed to test the applicant's knowledge in the following four areas:

(1) knowledge of natural sciences (biology and inorganic and organic chemistry), (2) reading comprehension (natural and basic sciences), (3) verbal and quantitative ability, and (4) perceptual-motor ability (two-and-three-dimensional problem solving).[5]

These components purport to permit the student to demonstrate his/her manual and academic aptitude as well as aptitude for science. However, performance on these tests is conditioned by the same set of factors which affect performance on any objective test. These may include previous experience and demonstrated skill in taking tests, the nature of the test setting, the test administrator, sense of security and confidence, the student's emotional condition at the time of the test, coaching experience, and other factors.[6] Other traditional factors utilized in admissions and selection decisions include college grade point average (GPA), grade point average in Science, and letters of recommendation.

Since there are eighteen States without schools or colleges of dentistry, these states have made cooperative arrangments often with the nearest college of dentistry for the training of students from their States. In effect, some colleges of dentistry not only give preference to residents of their own State but to students from their contracting institutions or State. This policy may raise the level of competition for admission to particular institutions. In addition, some institutions assign a value to the quality of the applicant's undergraduate college which, in turn, may affect acceptance irrespective of residence.

Forty of the U.S. dental schools now require a personal interview of the candidate by admissions committee members or its designees prior to formal acceptance. Another eight of these institutions leave the interview requirement entirely to the discretion of the Admissions Committee. Only three schools of dentistry state categorically that no interview is required of their applicants. No information was collected on the remainder. This pattern was consistent throughout the seventies. One explanation for the interview is to provide admissions committees members with opportunities to fathom more information about the applicant's personal characteristics not sufficiently detailed in the autobioghraphical data, to find further manifestations of the person's commitment to the fields of dentistry, and to obtain some impression of the person as a whole.

Essentially, non-cognitive criteria are examined through personal interviews, although some opportunities may be provided for the candidate to demonstrate manual dexterity and his/her facility at artistry during time spent on campus. But a major effort is made here to discern motivation, integrity, commitment, maturity, promise as a practicing dentist, and, as one institution says "who are winners across the board." Several institutions, such as the University of Kentucky, specifically state the DAT scores are considered in conjunction with other important factors which include the applicant's background and overall record of personal achievements. Thus, rigid cutoff points are not employed. Several institutions maintain that they seek a good balance between cognitive and non-cognitive criteria in selection and admissions. However, it is often difficult to unravel precisely how this balance is achieved.

Enrollment Trends in the 1970s

Although schools of dentistry are currently experiencing a decline in applicants and applications, first year enrollment continues to increase. For instance, between 1960 and 1977, first year enrollment in U.S. schools of dentistry increased by some 64.1 per cent.[7] Some of that change may be explained by the opening of fourteen additional schools of dentistry since 1960. When that fact is taken into consideration, the net result is that those institutions then in existence in 1960 realized a forty per cent growth in enrollment during an eighteen year period.[8]

In 1960, total enrollment in U.S. schools of dentistry was 13,580 students. By 1970, total enrollment had climbed to 16,551 in 51 institutions located on the mainland of the U.S. plus one institution in Puerto Rico. Total black student enrollment, not including Puerto Rico, was 451 students. About 62 per cent of all those students were enrolled at either Howard University (169) or at Meharry Medical College (110). Only the University of Illinois (17 black students), the University of California at San Francisco (13), the University of Maryland (13), and the University of Michigan (10) had an enrollment of ten or more black students.

Eight of the fifty-one mainland USA dental schools did not have a single black student enrolled in 1970. These were the University of Louisville, Louisiana State University, Loyola University of New Orleans (since closed), the University of Oregon Dental School, Baylor University College of Dentistry, the University of Texas/San Antonio, Virginia Commonwealth University, and West Virginia University. Seventeen dental schools enrolled only a single black student and there were only ten institutions who had black students in the fourth year of dental school training. As a result, 47 of the 55 black seniors were enrolled at the two historically black dental schools in the nation. Ultimately, they graduated about 80 per cent of black graduates in 1971 (See Table 14).

First Year Enrollment:

Of special interest is the first-year enrollment profile of black students in dental schools in 1970. According to Table 14, 185 black students were

Table 14. Black Student Dental School Enrollment October 1970

SCHOOLS	1st Year	2nd Year	3rd Year	4th Year	Total
University of Alabama	3	—	—	—	3
University of the Pacific	1	—	—	—	1
University of California, San Francisco	6	6	1	—	13
University of California, Los Angeles	6	—	—	—	6
University of Southern California	—	1	1	—	2
Loma Linda University	—	—	1	1	2
University of Connecticut	—	1	—	—	1
Georgetown University	1	—	—	—	1
Howard University	59	52	35	23	169
Emory University	1	1	—	—	2
Medical College of Georgia	3	—	—	—	3
Loyola (Chicago)	—	—	1	—	1
Northwestern University Dental School	1	—	1	—	2
University of Illinois	9	7	1	—	17
Indiana University-Purdue University	1	3	3	2	9
University of Iowa	1	—	—	—	1
University of Kentucky	1	—	—	—	1
University of Louisville	—	—	—	—	—
Louisiana State University	—	—	—	—	—
Loyola University - New Orleans	—	—	—	—	—
University of Maryland	9	3	1	—	13
Harvard School of Dental Medicine	3	3	—	—	6
Tufts University School of Dental Medicine	2	4	—	1	7
University of Detroit	7	1	2	—	10
University of Michigan	5	2	2	—	9
University of Minnesota	1	—	—	—	1
University of Missouri-Kansas City	2	1	—	—	3
Washington University School of Dentistry	1	—	—	—	1
Creighton University School of Dentistry	1	—	—	—	1
University of Nebraska	—	—	1	1	2
Fairleigh Dickinson University	—	—	1	—	1
New Jersey College of Medicine & Dentistry	—	—	—	1	1
Columbia University	2	—	—	—	2
New York University	2	3	1	—	6
State University of New York at Buffalo	4	3	—	—	7
University of North Carolina	1	—	—	—	1
The Ohio State University	2	—	—	1	3
Case Western Reserve University	7	1	—	—	8
University of Oregon Dental School	—	—	—	—	—
Temple University	1	1	1	—	3
University of Pennsylvania	—	1	—	—	1
University of Pittsburgh	4	2	1	—	7
Medical University of South Carolina	1	—	—	—	1
Meharry Medical College	27	31	28	24	110
University of Tennessee	5	—	—	—	5
Baylor University College of Dentistry	—	—	—	—	—
The University of Texas Dental Branch	1	1	—	1	3
University of Texas at San Antonio	—	—	—	—	—
Virginia Commonwealth University	—	—	—	—	—
University of Washington	2	—	1	—	3
West Virginia University	—	—	—	—	—
Marquette University	1	1	—	—	2
University of Puerto Rico	1	—	—	1	2
TOTAL	185	129	83	56	453
Total Dental School Enrollment	4563	4216	3979	3793	16,551
Percentage	4.1%	3.1%	2.1%	1.5%	2.7%

Source: American Dental Association," Analysis of Black Applicants To Dental Schools," 1971.

enrolled in first year classes as the decade began. Again, Howard and Meharry accounted for a substantial portion of those students. Eighty-six of the 185 students were enrolled at these two institutions. Only five additional institutions enrolled more than five black students in their first year classes. In fact, sixteen of the fifty-one institutions or almost one-third of them did not enroll a single black student. Approximately 30 per cent of them enrolled only one black student and another quarter of them enrolled from two to five black students in their first year classes. What was missing from a significant majority of these institutions, inter alia, was a critical mass of black students sufficient to create a sense of mutual support and psychological well-being in an alien environment.

Less than 10 per cent of the black students (19) were enrolled in those states that became known as the "Adams States". Of that number, four were enrolled in Georgia, nine in Maryland, and one in North Carolina and five were enrolled in the three dental schools of Pennsylvania. Howard's first year enrollment was more than twice the combined enrollment of black students in the "Adams States" and Meharry's first year enrollment was significantly greater than their combined total enrollments of new students. This inequality of access of black students to schools of dentistry underscores the importance of external pressure exerted by the NAACP — Legal Defense Fund on the Department of Health, Education and Welfare and its Office of Civil Rights to force those states which had historically operated dual systems of education to expand opportunities for black students. The various court decisions during the 1970s that ordered such expansion of educational opportunity are among the factors which account for those growth patterns observed in the participation of black students in the "Adams States" (See Tables 15, 16, and 17).

The first year enrollment of black students in dental schools increased steadily in every year since 1970 with the exception of 1978-79, when an absolute numerical decline of sixteen students occurred in relation to the preceding year. The downward trend continued in 1979-80 when 274 black students enrolled in first year classes. As in medicine, the percentage of black students in first year dental school classes peaked in 1974 and in 1975 but declined thereafter. In effect, first year enrollment of black students are not keeping pace with overall first year enrollments which are steadily climbing. Between 1974 and 1979-80, some 35,702 students enrolled in first year classes in dental schools. Of that number only 1,717 or 4.8 per cent were black students. As Table 15 and figure 2 show, there has been some progress in the participation of blacks in dental education but that progress can only be described as limited access.

The combined enrollment of black students in the thirteen schools of

Table 15. Total First Year Enrollment in Dentistry (DDS) Programs All Institutions Combined, By Year and Race: 1968-1979

Year	Total First Year Enrollment	Black Student Enrollment	Per cent Black of Total
1968 - 69	4,203	—	—
1969 - 70	4,355	—	—
1970 - 71	4,565	185	4.0
1971 - 72	4,745	245	5.1
1972 - 73	5,337	244	4.5
1973 - 74	5,445	273	5.0
1974 - 75	5,617	279	4.9
1975 - 76	5,763	298	5.1
1976 - 77	5,835	290	4.9
1977 - 78	5,954	296	4.9
1978 - 79	6,301	280	4.4
1979 - 80	6,132	274	4.4

Source: American Dental Association, *Minority Reports.*

Table 16. First Year Black Student Enrollment in Schools of Dentistry in the "Adams States" by School, Sex and Year: 1974 - 79.

Dental School	Fall 1974 M	F	Fall 1975 M	F	Fall 1976 M	F	Fall 1977 M	F	Fall 1978 M	F	Fall 1979 M	F	Total
Univ. of Florida	—	—	4	0	0	0	2	0	3	1	0	1	11
Emory University	—	—	1	0	0	0	0	1	0	1	1	2	6
Med. Col. of Ga.	5	1	6	2	5	3	1	1	4	4	7	3	42
Louisiana State U.	1	1	0	2	2	1	0	3	2	0	0	0	12
Loyola Univ. (La.)	—	—	—	—	—	—	—	—	—	—	—	—	—
Univ. of Maryland	5	2	5	4	10	5	9	6	8	1	4	2	61
Univ. of Miss.	—	—	1	1	1	0	4	3	4	1	3	2	20
Univ. of N.C.	2	2	4	1	1	3	2	2	2	2	6	3	30
Univ. of Okla	1	—	2	0	2	1	0	1	—	—	2	0	9
Oral Roberts Univ.	—	—	—	—	—	—	—	—	—	—	0	0	0
Temple Univ.	2	2	2	0	3	1	2	1	0	3	1	2	19
Univ. of Penna.	1	0	6	3	1	2	2	0	1	2	1	1	20
Univ. of Pittsburgh	4	3	3	5	2	2	2	2	3	2	3	0	31
Va. Commonwealth	2	1	2	0	1	4	2	0	0	0	2	0	14
Totals: Adams States	23	12	36	18	28	22	26	20	27	17	30	16	
Total:	(35)		(54)		(50)		(46)		(44)		(46)		275

Source: American Dental Association, *Minority Reports.*

Table 17. First Year Black Student Enrollment in Schools of Dentistry in Non- "Adams States" By School, Sex and Year: 1974 - 79

Dental School	1974 M	1974 F	1975 M	1975 F	1976 M	1976 F	1977 M	1977 F	1978 M	1978 F	1979 M	1979 F
Univ. of Alabama	2	1	4	0	2	2	3	2	3	2	0	4
U. of Pacific	2	1	0	0	0	0	0	0	0	0	0	0
Univ. of Calif./S.F.	6	0	1	1	3	4	6	1	0	2	4	1
UCLA	4	0	4	6	6	5	6	9	4	9	5	3
U. of So. Calif.	6	0	4	1	0	2	3	0	2	0	0	0
Loma Linda Univ.	3	1	0	0	2	1	2	1	0	0	6	0
Univ. of Colorado	0	0	0	0	1	1	0	0	0	0	1	0
Univ. of Conn.	0	0	2	0	1	0	0	0	0	0	0	0
Georgetown Univ.	4	0	3	2	3	1	0	0	2	6	3	2
Howard Univ.	68	19	60	20	49	22	48	24	45	22	45	28
Loyola Univ. (Ill.)	2	0	2	2	0	1	1	1	0	4	1	1
Northwestern Univ.	0	0	0	1	0	1	0	0	1	0	0	0
U. of So. Ill.	0	0	0	0	1	0	0	0	1	0	0	1
U. of Illinois	3	4	4	2	3	3	2	4	2	2	2	3
Indiana Univ.	1	0	2	0	0	3	1	0	2	0	0	0
Univ. of Iowa	2	0	0	1	2	1	0	0	2	0	1	0
Univ. of Kentucky	2	0	2	0	0	2	1	3	0	2	3	1
U. of Louisville	1	0	1	0	0	1	0	1	1	0	0	0
Harvard Univ.	2	0	2	1	1	0	2	3	0	2	1	0
Boston Univ.	0	0	1	0	0	0	1	0	2	1	0	1
Tufts Univ.	4	3	1	2	4	1	1	1	0	1	1	1
Univ. of Detroit	2	3	5	3	3	0	3	0	1	1	1	1
Univ of Michigan	10	2	6	4	7	7	6	7	5	5	8	5
Univ. of Minnesota	1	0	1	0	2	0	1	0	1	1	1	0
Univ. of MO/KC	5	1	1	1	4	0	5	1	0	1	2	1
St. Louis Univ.	0	0	0	0	0	0	0	0	0	0	0	0
Washington Univ.	2	1	1	1	2	1	2	1	2	1	2	0
Creighton Univ.	2	0	2	1	0	1	0	1	2	0	0	0
Univ. of Nebraska	0	0	0	0	0	0	0	0	0	0	0	0
Fairleigh Dickinson	4	0	0	1	0	1	1	0	2	0	4	2
N.J. Dental School	3	0	8	2	9	2	8	5	8	4	4	1
Columbia Univ.	0	0	0	0	0	1	0	0	1	0	0	1
New York Univ.	0	3	4	0	1	0	4	1	2	4	0	1
SUNY/Stony Brook	1	2	1	0	0	0	0	0	0	0	0	0
SUNY/Buffalo	2	1	2	2	2	3	1	0	0	0	1	2
Ohio State Univ.	1	1	1	1	2	0	1	0	6	3	6	2
Case Western Reserve	2	3	4	1	3	0	0	2	1	3	3	2
Univ. of Oregon	0	0	3	1	2	0	2	0	0	0	0	0
Med. Col. of S.C.	0	0	4	1	2	0	2	0	2	0	0	1
Meharry Medical Col.	32	8	25	16	34	10	41	11	30	18	32	19
Univ. of Tenn.	4	0	0	1	1	1	0	1	1	1	0	1
Baylor College	2	1	0	1	0	0	1	0	1	1	1	0
U. Of Texas/Houston	1	1	3	0	4	1	4	5	3	4	0	2
U. of Texas/San Anton.	0	0	1	1	1	0	3	1	0	0	2	0
U. of Washington	2	0	2	0	2	1	2	0	0	1	0	0
West Va. Univ.	2	0	0	0	0	0	1	0	0	0	0	0
Marquette Univ.	0	0	0	0	0	0	0	0	1	0	1	0
Total:	190	54	106	77	160	80	165	80	136	100	141	87
Totals: (Non-Adams States)	(244)		(183)		(240)		(245)		(236)		(228)	

Source: American Dental Association, *Minority Reports.*

dentistry in the "Adams States" was 35 in 1974; 54 in 1975; 50 in 1976; 46 in 1977; 44 in 1978-79, and 46 in 1979-80. (See Table 16). In terms of absolute numbers, the most successful institutions in the "Adams States," regarding first year enrollment of black students, are the University of Maryland, the University of Georgia, and the University of Pittsburgh. Not only are sizeable black colleges located in these states but they also have a substantial black population. With the exception of Oklahoma, blacks comprise a substantial proportion of the total population of all of the "Adams States". Yet, they are grossly underrepresented in their dental schools.

As shown in Table 15, black students comprised 4.9 per cent of first year enrollees in dental schools in 1971. By 1974, dental schools enrolled almost 100 more black students in first year classes than they did in 1970-71. The 279 black students enrolled in that year represented 4.9 per cent of the 5,617 students of all races enrolled in first year classes. That the percentage of black students enrolled during the following year was almost identical to the 1973-74 percentage reflects the commensurate increase in first year students in general and indicates that black students were not necessarily gaining ground at the expense of any particular group of students.

The decline in the black first year enrollment thereafter may be explained by a combination of diverse factors whose cumulative impact resulted in an enrollment drop. For example, the Bakke suit was proceeding through the lower courts. That reality may have fostered a state of uncertainty and dimmed optimism among recruitment and admissions committees concerning the constitutionality of some of their affirmative actions and institutional policies. Simultaneously, the nation experienced a serious recession which impacted with far graver consequences on the black population than perhaps on any other segment of the population.

The economic crisis, combined with severe reductions in the available financial assistance, heightened suspiciousness among gifted black students about the practicality of deferring their gratifications or of saddling themselves with debts of $20,000 or more in order to complete the requirements for a degree in dentistry. Consequently, some of these students, who may have normally enrolled in schools of dentistry, opted for lucrative job opportunities then available to them. Further, actual institutional behavior relative to increasing weights assigned to objective measures of evaluation was of special importance in curtailing an expansion of black student enrollment in first year dental school classes.

In "non-Adams States" and excluding Howard University and Meharry Medical College from the analysis, Table 17 shows that four institutions were comparatively successful in the recruitment and admission of black students in their first year classes during the seventies. These dental schools are the University of Michigan, the University of California at Los Angeles, New Jersey Dental School, and the University of Illinois. That each one of these institutions is a public institution suggests that public institutions were and are considerably more responsive to pressures for increasing access to all students than were the private dental schools. This suggestion appears to be borne out by the regression analysis, described in Chapter III, which depicts a trend in that direction during the late seventies.

In earlier years, private institutions had far more latitude in opening their doors to black students. The current enrollment profile suggests that this may no longer be the case. A plausible assumption is that either their commitment deteriorated; their focus was on token admission, or that they may have come to depend more heavily on quantitative factors in making admissions decisions or on a combination of these factors. As a result private institutions have not done as well as previously anticipated. Exceptions to this generalization are such institutions as Georgetown University, the University of Detroit, Case Western Reserve University, New York University, Tufts University, and the University of Southern California.

The success of this latter group of institutions is relative to what other institutions in their State or jurisdiction accomplished during the five year period between 1974 and 79 (See Table 17). For instance, Case Western Reserve University did somewhat better than Ohio State University in enrolling black students in first year classes. But the University of Detroit, much smaller in size, was not as successful in terms of absolute numbers as was the University of Michigan. Nor was the University of Southern California as successful as UCLA. Tufts did better than either Harvard University or Boston University but all are private institutions located in the same city. Similarly Georgetown and Howard Universities are located in the same city and both draw students from throughout the nation. Howard is historically black and federally supported but Georgetown is not.

Whether in "Adams States" or elsewhere, the most important dependent variable in this analysis is of course first year enrollment of black students. Obviously, if there is no, or limited success in enrollment of black students, there will be little or no success in the production of black dentists. This is a problem of institutional behavior—recruitment selection, admissions, quality of the learning environment, resources— and of the available student pool. Institutional behavior is not uniform throughout the nation.

Actually, all dental schools enrolled at least one black student during the 1970s. In view of the limited access of black students in several of them, however, it would seem that several are confronted in the 1980s with the task of whether or not *to move* beyond mere tokenism. Presence of more serious instances of tokenism in the "Adams States" adds further credence to the hypothesis that pressures engendered by such litigation as the *Adams* vs. *Califano* case are of inestimable importance in promoting equality of opportunity beyond tokenism. If that is the case, the question remains as to what precise type of litigation is appropriate in States not presently under litigation to desegregate but contain institutions whose policies help to create widespread underrepresentation of blacks in the field of dentistry.

Total Enrollment of Black Students

In 1970, the 451 black students enrolled in the nation's schools of dentistry represented a mere 2.7 per cent of all students enrolled at that time. It is evident from Table 18 and figure 1 that total enrollment in dental schools increased by approximately five or six hundred students annually. Even the absolute numbers of black students enrolled in dental schools almost doubled between 1970-71 and 1974-75. Since that year, black students total enrollment has been conspicuously inconsistent. Total enrollment of black students actually peaked in 1975-76, followed by an actual decline in absolute numbers or a loss of some twenty-two students in 1976-77. Thereafter, total enrollment increased each succeeding year and reached its previous high of 977 black students in 1978-79. (See Table 18).

Only in 1979-80 did total black student enrollment exceed 1,000. Even then, the 4.4 per cent of total enrollment which they represented did not approach parity — a percentage equivalent to the percentage of black people in the total U.S. population. This incontrovertible fact speaks dramatically to the need for major reforms in dental education policies, procedures and activities that will significantly transform the racial composition of the dental profession. With such paucity of black students enrolled, coupled with retention problems, and despite apparent successes, it is understandable why in 1980 there are only 3,000 black dentists in the United States.[9]

Table 19 presents total black student enrollment in dental schools located in the "Adams States". One illustration of activity in the "Adams States" may be drawn from the University of Florida in 1974 which had two black students out of a total enrollment of eighty-eight students in its School of Dentistry. This number represented about two per cent of total enrollment. In 1979-80, seven black students were enrolled there out of a total of 249 students. These seven black students represented less than three per cent (2.8) of the total enrollment in that institution. By contrast, the Univer-

Table 18. Total Enrollment - All Classes and Institutions in DDS (Dental School) Program By Year and Race: 1968 - 79

Year	Total Enrollment	Total Black Student Enrollment	Per Cent Black of Total
1968-69	15,408	—	—
1969-70	16,008	—	—
1970-71	16,553	451	2.7
1971-72	17,505	—	—
1972-73	18,376	—	—
1973-74	19,369	—	—
1974-75	20,146	945	4.6
1975-76	20,767	977	4.7
1976-77	21,013	955	4.5
1977-78	21,510	968	4.5
1978-79	22,179	977	4.4
1979-80	22,482	1009	4.4

—: data unavailable
Source: The American Dental Association, *Minority Reports.*

sity of Maryland had a total black student enrollment in its School of Dentistry in 1974 of thirty-eight out of 527 students. This represented slightly more than seven per cent of the total enrollment. By 1979-80, the proportion of black students in the dental school in Maryland had declined but only by an insignificant fraction. (See Table 19).

In North Carolina in 1974, black students comprised slightly more than two per cent of the enrollment in the University of North Carolina's College of Dentistry. By 1979-80, the eighteen black students enrolled there represented 5.7 per cent of the total enrollment of 312 students. One could argue that their proportionate enrollment had more than doubled during a five year period. However, that assertion would be grossly misleading since, in the first instance, the initial base for comparison was so minute. Similar observations could be made about other institutions in the "Adams States."

Some of the "Adams States" fare quite well when compared to "non-Adams States" in terms of the percentage distribution of black students in schools of dentistry. A detailed analysis of such findings is unwarranted since most of the States never enrolled anymore than from two to seven per

Table 19. Total Black Student Enrollment in Schools of Dentistry in "Adams States" By School, Sex and Year: 1974 - 79

Dental School	1974		1975		1976		1977		1978		1979	
	M	F	M	F	M	F	M	F	M	F	M	F
Univ. of Florida	1	1	5	1	4	1	5	0	8	1	5	2
Emory Univ.	0	0	1	0	1	0	1	1	1	2	1	4
Med. Col. of Ga.	18	4	16	4	14	6	9	4	9	6	16	7
Louisiana State Univ.	2	1	2	2	3	1	2	4	4	1	3	1
Univ. of Maryland	33	5	26	9	23	14	25	17	24	13	25	11
Univ. of Mississippi	—	—	1	1	2	1	5	4	9	4	10	5
Univ. of North Carolina	5	2	8	2	9	5	9	6	10	6	11	7
Univ. of Oklahoma	3	0	5	0	5	1	4	2	3	2	4	2
Oral Roberts Univ.	—	—	—	—	—	—	—	—	—	—	—	—
Temple Univ.	8	2	9	2	10	3	7	4	5	6	4	7
Univ. of Penn.	7	0	10	3	9	5	7	4	5	5	5	4
Univ. of Pittsburgh	14	4	13	9	9	9	12	10	12	10	11	5
Va. Commonwealth U.	9	2	8	2	6	6	6	3	4	2	6	2
Totals	100	21	104	35	95	52	92	59	94	58	101	57
	(121)		(139)		(147)		(151)		(152)		(158)	

Source: American Dental Association, *Minority Reports.*

— = data not available

cent of black students of total enrollment. Even a cursory glance at these data strikes the astute observer that a crisis of underrepresentation of black students in the field of dentistry remains in 1980 despite corrective action taken by some institutions, States, and representatives in the public and private sectors. (See Table 20).

According to the latest available data, the 1979-80 statistics provided by the Dental Education Section of the American Dental Association, only the University of Nebraska does not have a single black student enrolled in its School of Dentistry. According to the institutional information for special applicants cited in the *Admission Requirements of U.S. and Canadian Dental Schools: 1979-80,* the University of Nebraska reported that, at the present time, its dental college is without necessary funds to "support special programs for minority recruitment and retention." This situation also pertains to disadvantaged students, women and aid to pre-dental students.[10] The absence of a strong recruitment program may very well explain both the absence of black students and the underrepresentation of minority students in general in that institution.

Table 20. Total Black Student Enrollment in Schools of Dentistry in Non- "Adams States" By School, Sex and Year: 1974 - 79

Dental School	1974 M	1974 F	1975 M	1975 F	1976 M	1976 F	1977 M	1977 F	1978 M	1978 F	1979 M	1979 F
Univ. of Alabama	8	3	9	3	9	2	9	5	10	6	8	9
Univ. of Pacific	4	0	1	0	1	0	0	0	0	0	0	0
Univ. of Calif./S.F.	26	4	19	5	18	8	14	6	11	8	13	9
UCLA	20	3	17	7	18	12	19	20	19	27	26	19
U. of So. Calif.	22	2	23	3	15	2	10	3	6	2	6	2
Loma Linda Univ.	8	2	6	2	5	2	5	4	3	4	10	3
Univ. of Colorado	0	0	0	0	1	1	1	1	1	1	2	1
Univ. of Conn.	1	2	3	2	2	2	2	0	1	0	1	0
Georgetown Univ.	7	1	9	3	10	4	9	2	7	8	5	6
Howard Univ.	242	60	233	69	193	66	191	80	171	79	179	94
Loyola Univ. (Ill.)	5	0	5	2	4	2	5	2	4	5	2	5
Northwestern Univ.	0	0	0	1	1	1	0	2	1	2	1	2
U. of So. Illinois	1	1	1	0	1	0	1	0	2	0	0	1
U. of Illinois	12	5	14	5	13	7	12	11	12	11	7	11
Indiana Univ.	5	1	4	2	4	4	4	3	4	3	3	3
Univ. of Iowa	8	0	5	1	5	2	3	1	3	1	4	0
U. of Kentucky	11	2	11	1	7	3	5	5	3	7	6	7
U. of Louisville	3	1	3	0	3	2	1	2	2	2	1	1
Harvard Univ.	8	1	8	1	8	1	7	4	4	6	4	5
Boston Univ.	0	0	1	0	1	0	2	0	3	1	1	2
Tufts Univ.	14	5	11	7	8	6	4	4	4	3	2	3
Univ. of Detroit	9	4	10	5	6	2	7	2	7	0	5	1
Univ. of Michigan	37	8	36	10	38	15	26	18	21	21	21	22
Univ. of Minnesota	3	2	3	2	4	0	3	0	4	1	5	1
Univ. of Mo./KC	10	2	7	3	10	3	15	4	9	2	11	3
St. Louis Univ.	0	0	0	0	0	0	0	0	0	0	0	0
Washington Univ.	4	0	5	1	5	2	4	3	5	2	5	1
Creighton Univ.	12	0	13	1	8	2	4	4	4	4	2	2
Univ. of Nebraska	0	0	0	0	0	0	0	0	0	0	0	0
Fairleigh Dickinson	10	0	9	1	8	1	7	1	3	1	6	3
N.J. Dental School	12	0	14	2	17	4	20	8	22	10	17	8
Columbia Univ.	4	2	2	1	2	1	0	1	1	1	0	2
New York Univ.	10	4	9	1	6	1	8	1	5	4	3	5
SUNY/Stony Brook	1	2	2	2	2	2	2	2	1	0	0	0
SUNY/Buffalo	8	2	9	4	7	6	5	5	4	4	4	4
Ohio State Univ.	6	1	6	2	3	1	2	1	8	3	12	5
Case Western Reserve	7	4	9	4	8	4	6	5	5	5	6	6
Univ. of Oregon	2	0	4	1	6	1	5	1	4	2	3	1
Med. Col. of S.C.	0	0	5	1	6	0	7	0	7	0	4	1
Meharry Medical College	119	21	103	28	107	35	107	37	117	52	128	55
Univ. of Tenn.	6	0	5	2	4	2	3	3	2	3	0	2
Baylor College	2	1	2	2	2	2	1	1	2	1	3	1
U. of Texas/Houston	2	1	5	1	8	2	9	7	10	9	8	10
U. of Texas/San Anton.	0	0	1	1	1	0	4	0	4	0	5	0
U. of Washington	4	0	5	0	6	1	5	2	4	2	2	2
West Va. Univ.	3	0	2	0	2	0	3	0	1	0	1	0
Marquette Univ.	1	0	0	0	0	0	0	0	1	0	1	0
Totals												
Non-"Adams States"	669	147	649	203	593	215	556	261	522	303	533	318
(All Colleges)	777	168	753	224	688	267	648	320	616	361	634	375

Source: American Dental Association, *Minority Reports.*

Factors Affecting Enrollment Trends

An explicit assumption of this study is that first year enrollment is dependent upon one or more of twelve independent variables, the presence of which may prove to be powerful predictors of enrollment patterns or trends. Previous analyses have demonstrated that among the most powerful predictors of first year enrollment are the presence of role models (black faculty), quality of financial aid programs, being under litigation to desegregate higher education, proportion of blacks in the state population, type of institution-whether public, private, or state-assisted, and the special recruitment and admissions programs. Two of these variables are addressed below:

(1) The availability of adequate financial aid programs fluctuated during the 1970s despite increases in possible sources of financial assistance for students in the health fields and for institutions that operated such programs. One of the critical factors to bear in mind here is that the cost of dental education exacts a heavy burden upon disadvantaged students of all races and upon students who are not able to draw upon family resources to help defray these costs. Table 21 depicts both current first year and total estimated costs of dental education for resident students in each of the fifty-nine mainland U.S. dental schools. These costs are not stable and have shown major increases in recent years as a result of uncontrolled inflation and the escalating expenses of dental school training in general. If students are relatively disadvantaged and must seek their financial support outside of family resources, this situation presents staggering problems. (See Table 21).

Hence, outside or external assistance is required for acquiring a degree in dentistry. According to Table 22, the proportion of students receiving either loans or scholarships or both to finance dental education ranges from a low of fourteen per cent of students enrolled at the University of the Pacific to a high of 95 per cent at Temple University in Pennsylvania. First year costs at the University of Pacific are estimated at $12,310 while total costs for the 36 consecutive month-program are estimated at $30,425. The first year costs at the School of Dentistry at Temple University are estimated to be $6,220 for resident students and $8,520 for non-resident students. Total costs for all four years run from $17,810 for residents to $27,000 for non-residents. (See Table 22).

Among institutions in which at least three-quarters of its students receive loans, scholarships, or both to finance their dental education are Loma Linda University, the University of California - Los Angeles, the University of Southern California, Harvard University, Creighton Univer-

Table 21. Estimated Student Expenses, Excluding Living Costs, At Mainland U.S. Dental Schools for 1979-80 (Arranged by State)

Institution	First Year Resident $	Non-Resident $	Total Expenses (all years) Resident $	Non-Resident $
Univ. of Alabama	3,845	5,045	10,625	15,425
UCLA	2,840	4,745	9,092	16,712
Univ. of Calif./S.F.	4,053.50	5,958.50	10,244	17,864
Loma Linda Univ.	10,365*		33,745*	
Univ. of the Pacific	12,310*		30,425*	
Univ. of So. Calif.	8,527*		29,038*	
Univ. of Colorado	16,899**	N/A	63,142	N/A
Univ. of Conn.	2,662	4,062	10,787	16,387
Georgetown Univ.	11,063*		40,058*	
Howard Univ.	3,835.50*		11,117*	
Univ. of Florida	2,690	4,958	12,630	21,135
Emory Univ.	7,730*		26,095*	
Univ. of Georgia	3,790	4,945	9,650	14,930
Univ. of Ill./Chicago	3,397	5,335	9,928	17,680
Loyola Univ./Chicago	6,692*			22,570*
Northwestern Univ.	7,573*		29,634*	
So. Ill. U./Edwardsville	3,135	N/A	7,485	N/A
Indiana Univ.	2,480	4,100	8,130	14,610
Univ. of Iowa	2,950	4,330	8,415	13,935
Univ. of Kentucky	3,104	4,404	8,501	13,701
Univ. of Louisville	3,090	4,390	8,900	14,100
Louisiana State Univ	3,755	5,255	8,765	14,765
Univ. of MD/Baltimore	4,618	6,418	13,592	20,792
Boston Univ.	6,750*		28,950*	
Harvard Sc. of Den. Med.	6,475*		33,835	
Tufts Univ.	12,678*		31,154*	
Univ. of Detroit	6,825*		21,505*	
Univ. of Michigan	5,085	7,285	16,020	24,820
Univ. of Minnesota	3,105	6,276	11,163	23,847
Univ. of Mississippi	2,202	3,085	7,908	11,240
Univ. of Missouri/K.C.	4,329	5,529	12,041	16,841
Washington Univ.	7,370*			
Creighton Univ.	6,575*		24,101*	
Univ. of Nebraska	2,443	3,843	9,982	15,870
Fairleigh Dickinson Univ.	7,425*		26,500*	
N.J. Dental School	5,655	6,655	18,935	22,935
Columbia Univ.	6,885*		26,715*	
New York Univ.	10,544*		33,711*	
SUNY/Buffalo	7,137.50	8,137.50	17,000	21,000
SUNY/Stony Brook	3,730	5,130	14,920	20,520
Univ. of North Carolina	4,689.30	6,939.80	8,962.80	17,027.80
Case Western Reserve U.	7,035*		25,540*	
Ohio State Univ.	5,488	7,408	10,254	16,014
Univ. of Oklahoma	3,870	5,464	10,429	16,815
Oral Roberts Univ.	8,945*		31,140*	
Univ. of Oregon	3,727	7,102	9,490.50	22,990.50
Univ. of Pennsylvania	10,365*		35,080*	
Univ. of Pittsburgh	6,856	8,856	20,817	28,817
Temple Univ.	6,220	8,520	17,810	27,010
Med. C. of So. Carolina	3,335	4,735	9,685	15,285
Meharry Medical College	5,111.91*		20,710.07*	
Univ. of Tennessee	3,329		11,132	
Baylor C. of Dentistry +	4,492	6,812	8,291	17,251
Univ. of Texas/Houston	2,943	3,643	6,020	9,220
Univ. of Texas/S. Anton	2,964	3,564	6,712	9,112
Va. Commonwealth U.	4,496	5,901	14,375	19,995
Univ. of Washington	4,224	6,954	10,801	21,721
W. Va. U. Sch. of Dent.	4,370.90	5,660.90	7,259.85	11,477.45
Marquette Univ.	7,070*		23,350*	

*Resident and Non-Resident.
**Tuition may be reduced for service to the state.
+ Three year program
Source: American Association of Dental Schools, *Admission Requirements of U.S. Canadian Dental Schools: 1979-80.* Washington, D.C., 1978.

sity, and Meharry Medical College. Three of these institutions enroll a substantial number of black students: UCLA, USC, and Meharry.

Ninety per cent of the first year enrollees at both Loma Linda University and Temple University require loans. Eighty-five per cent of Harvard University's first year students and 80 per cent of Georgetown's students take out loans to assist in the financing of the first year. Although slightly more than half of Meharry's first year students rely upon loans, the proportion climbs to 85 per cent in the sophomore year, drops to 80 per cent for juniors and 70 per cent for seniors.

Except at Georgetown University, in which 80 per cent of its first year students receive scholarships, assistance from scholarship programs is considerably below the proportion of students receiving loans. This lack of scholarships and the dollar value of the scholarships in relation to costs of dental education also help to explain the underrepresentation of black students in dental schools. Sixty-five per cent of Meharry's first year students, eighty-five per cent of its sophomores and juniors and seventy-five per cent of its seniors receive scholarship assistance.

Minimum scholarships awarded, according to the most recent data, vary from $2,900 to a maximum of $7,500. However, the mean amounts awarded through scholarships range from $63.00 to $3,777 (See Table 23). The mean of loans taken by students from university-controlled funds ranges from a low of $773.00 at the University of California/San Francisco to a high of $4,800.00 at the University of Washington. It is interesting to note that the University of California at San Francisco enrolls three times more black students than does the University of Washington.[11] Precisely why this is so is not clear.

Even though all dental schools do not participate in the twelve national sources of financial aid listed below, all of them do assist students in obtaining some form of needed financial assistance either before or following enrollment. Institutions most effective in enrolling black students are also those with the best financial aid packages and who, during the initial recruitment process, inform these students of their willingness to assist them.

Sources of Financial Aid

Under the Armed Forces Health Professions Scholarship Program, a student may receive coverage for educational expenses plus a $400.00 per month stipend, provided they sign a contract by which they agree to serve a minimum of two years in a branch of the armed services. With escalating

Table 22. Percent of Students Receiving Scholarships or Loans from University-Controlled Funds, by Year in School: 1976-77

State	Dental school	Percent receiving either loan or scholarship or both	Percent receiving scholarships				Percent receiving loans			
			1st Year	2nd Year	3rd Year	4th Year	1st Year	2nd Year	3rd Year	4th Year
AL	Univ. of Alabama	21	-	-	-	2	26	22	16	17
CA	Univ. of California at LA	75	39	42	37	38	45	55	50	42
	Univ. of California at SF	54	50	52	53	52	49	51	50	48
	Loma Linda Univ.	82	10	30	22	10	90	70	78	90
	Univ. of the Pacific	14	1	1	-	4	7	13	-	16
	Univ. of So. California	80	22	22	26	26	75	80	70	70
CO	Univ. of Colorado	40	56	28	50	24	52	16	50	24
CT	Univ. of Connecticut	42	40	24	37	41	40	22	39	24
DC	Georgetown Univ.	65	80	65	60	40	80	70	73	52
	Howard Univ.	40	1	4	7	28	20	29	39	36
FL	Univ. of Florida	29	-	-	3	13	8	48	33	13
GA	Emory Univ.	30	7	9	11	10	21	29	36	33
	Med. College of Georgia	37	3	5	20	-	27	30	40	-
IL	Univ. of Illinois, Chicago	27	1	1	1	4	29	25	20	32
	Loyola Univ.	30	°	°	°	°	30	26	29	33
	Northwestern Univ.	46	4	15	6	12	34	51	47	53
	So. Ill. U. at Edwardsville	31	39	20	-	7	41	22	-	17
IN	Indiana Univ.	36	8	5	10	17	33	34	29	34
IA	Univ. of Iowa	69	°	°	°	°	46	67	82	62
KY	Univ. of Kentucky	38	7	15	24	15	22	42	34	35
	Univ. of Louisville	41	15	7	4	5	31	38	37	50
LA	Louisiana State Univ.	56	6	12	10	21	49	47	46	39
MD	Univ. of Maryland	40	26	30	30	23	35	35	35	27
MA	Boston Univ.	31	-	-	-	-	27	35	44	12
	Harvard Sc. of Dental Med.	77	60	71	69	50	85	81	69	54
	Tufts Univ.	41	10	13	-	15	30	31	-	35
MI	Univ. of Detroit	**	**	**	**	**	**	**	**	**
	Univ. of Michigan	30	24	25	28	30	32	31	32	24
MN	Univ. of Minnesota	38	25	30	37	22	35	37	44	26
MS	Univ. of Mississippi	65	4	-	-	-	50	79	-	-
MO	U. of Missouri at Kansas C.	37	5	13	13	19	35	34	30	33
	Washington Univ.	39	2	9	-	11	-	46	-	54
NB	Creighton Univ.	75	57	39	34	32	62	65	38	44
	Univ. of Nebraska-Lincoln	23	25	30	25	20	30	25	25	10
NJ	Fairleigh Dickinson Univ.	21	2	4	8	7	7	26	22	23
	N.J. Dental School	40	30	16	-	12	55	40	-	30
NY	Columbia Univ.	44	33	41	43	60	40	28	18	24
	New York Univ.	42	2	4	-	31	44	37	-	6
	SUNY at Buffalo	68	-	22	22	22	53	47	44	39
	SUNY at Stony Brook	24	-	25	40	14	-	33	40	33
NC	Univ. of North Carolina	22	6	6	-	3	28	21	21	18
OH	Case Western Reserve U.	23	-	2	6	5	15	27	24	24
	Ohio State Univ.	43	4	6	-	6	47	45	-	36
OK	Univ. of Oklahoma	53	23	15	15	-	40	49	46	42
OR	Univ. of Oregon	52	-	-	-	8	56	70	27	53
PA	Univ. of Pennsylvania	55	47	42	35	36	50	55	49	55
	Univ. of Pittsburgh	35	8	12	10	7	39	39	34	18
	Temple Univ.	95	50	40	40	40	90	90	80	75
SC	Med. U. of South Carolina	50	26	23	29	-	76	35	39	-

Table 22. Percent of Students Receiving Scholarships or Loans from University-Controlled Funds, by Year in School: 1976-77 *(Continued)*

State	Dental school	Percent receiving either loan or scholarship or both	Percent receiving scholarships				Percent receiving loans			
			1st Year	2nd Year	3rd Year	4th Year	1st Year	2nd Year	3rd Year	4th Year
TN	Meharry Medical College	75	65	85	85	75	55	85	80	70
	Univ. of Tennessee	54	16	22	21	15	63	59	51	46
TX	Baylor College of Dentistry	40	-	-	-	-	45	50	50	25
	Univ. of Texas at Houston	39	12	12	21	18	32	30	31	19
	U. of Texas at San Antonio	19	-	-	*	*	23	18	14	*
VA	Virginia Commonwealth U.	46	23	32	24	15	47	51	48	34
WA	Univ. of Washington	43	8	32	46	28	39	43	57	31
WV	West Virginia Univ.	32	6	11	11	10	20	31	41	36
WI	Marquette Univ.	72	-	-	-	11	74	78	75	65
PR	Univ. of Puerto Rico	47	23	29	29	33	11	25	17	25

*Less than one per cent.
**Information not available.
Source: American Dental Association Council on Dental Education *Anual Report on Dental Education, 1977/78.*

educational costs, a sizeable portion of dental school students are taking advantage of this offering. Under the National Health Service Corps Scholarship Program, similar financial assistance is available for those students who agree to serve one year for every year in which they participated in the program in an underserved area or in a public service. This service may include Indian Health Services, federal prison medical facilities, the National Health Service, Public Service Hospitals and clinics, and similar public service structures.[12]

A dental student may borrow up to a total of $15,000.00 or as much as $5,000.00 per year under the Federal Guaranteed Loan Program and up to $10,000 for the National Direct Student Loan Program (NDSL).* Other important sources of financial assistance include the Health Professions Insured Loan Program, Scholarships for Health Profession Students of Exceptional Need, the College Work/Study Program, the Southern Regional Education Board, (SREB), the Western Interstate Commission for Higher Education (WICHE), the United Student Aid Funds, and the Robert Wood Johnson Student Loan Guarantee Program. In addition, there are numerous small foundations and other private sources of scholarship monies.[13]

One of the major problems with these programs is their inconsistency. There is no assurance of how much money will be available for student aid

*This amount was raised to $12,000.00 under the Higher Education Amendments of 1980.

Table 23. Amounts Awarded to Recipients of Scholarship and Loans from University-Controlled Funds: 1976-77

State	Dental school	Scholarship amounts			Loan amounts		
		Mini-mum	Maxi-mum	Mean	Mini-mum	Maxi-mum	Mean
AL	Univ. of Alabama	$2,000	$2,000	$2,000	$ 100	$ 3,805	$1,923
CA	Univ. of California at LA	100	5,800	1,100	100	3,500	900
	Univ. of California at SF	29	5,450	1,821	50	3,810	773
	Loma Linda Univ.	500	3,000	1,500	500	3,500	1,500
	Univ. of the Pacific	2,000	3,250	2,290	500	4,334	2,400
	Univ. of So. California	150	3,000	1,500	150	15,000	4,000
CO	Univ. of Colorado	500	3,000	1,340	300	3,500	2,275
CT	Univ. of Connecticut	100	3,500	1,240	100	2,000	1,035
DC	Georgetown Univ.	500	1,500	1,000	500	1,500	1,000
	Howard Univ.	300	1,650	975	25	2,500	1,263
FL	Univ. of Florida	400	800	677	100	2,000	1,099
GA	Emory Univ.	300	2,200	1,250	300	3,500	2,000
	Medical College of Georgia	400	4,000	1,406	325	7,000	2,028
IL	Univ. of Illinois, Chicago	**	**	**	**	**	**
	Loyola Univ.	3,300	3,500	3,400	150	3,500	1,450
	Northwestern Univ.	210	5,500	890	235	5,605	1,620
	So. Ill. U. at Edwardsville	25	1,344	63	350	2,500	1,332
IN	Indiana Univ.	100	3,470	1,008	165	8,000	2,291
IA	Univ. of Iowa	500	500	500	50	11,010	1,200
KY	Univ. of Kentucky	61	4,500	1,287	75	2,500	647
	Univ. of Louisville	346	1,080	618	270	2,500	1,135
LA	Louisiana State Univ.	125	2,000	542	200	3,500	2,190
MD	Univ. of Maryland	250	4,000	1,470	150	4,500	1,916
MA	Boston Univ.	—	—	—	500	6,000	2,976
	Harvard Sc. of Dental Med.	500	3,300	1,500	2,500	5,500	4,000
	Tufts Univ.	400	4,989	3,777	300	4,989	2,138
MI	Univ. of Detroit	100	3,000	**	200	3,000	**
	Univ. of Michigan	225	6,900	1,877	150	3,750	1,816
MN	Univ. of Minnesota	100	4,200	535	210	5,000	1,520
MS	Univ. of Mississippi	700	700	700	300	2,500	1,459
MO	Univ. of Missouri at KC	100	1,500	1,345	100	3,500	2,550
	Washington Univ.	500	4,925	2,738	238	1,708	1,039
NB	Creighton Univ.	375	3,000	1,564	100	6,500	3,206
	U. of Nebraska at Lincoln	400	2,000	1,000	200	3,500	1,000
NJ	Fairleigh Dickinson Univ.	730	5,000	3,525	100	3,500	1,912
	New Jersey Dental School	400	4,090	1,045	200	4,665	1,162
NY	Columbia Univ.	200	3,800	1,745	400	3,500	1,480
	New York Univ.	55	5,000	1,889	100	3,500	2,693
	SUNY at Buffalo	200	800	400	100	3,500	1,800
	SUNY at Stony Brook	650	1,900	1,500	200	1,800	1,120
NC	Univ. of North Carolina	200	1,975	810	250	2,850	1,097
OH	Case Western Reserve Univ.	500	2,500	1,485	500	7,500	2,135
	Ohio State Univ.	150	5,000	2,575	300	4,650	2,475
OK	Univ. of Oklahoma	794	2,500	1,647	105	6,450	4,116
OR	Univ. of Oregon	500	962	625	500	3,500	1,900
PA	Univ. of Pennsylvania	50	5,500	1,100	50	9,200	3,700
	Univ. of Pittsburgh	100	6,672	1,971	200	3,500	1,781
	Temple University	500	3,500	2,000	500	6,000	3,500
SC	Med. U. of South Carolina	650	2,400	1,525	350	3,500	1,925
TN	Meharry Medical Colllege	500	7,500	2,000	500	3,000	2,000
	Univ. of Tennessee	500	2,900	500	400	4,000	2,500

Table 23. Amounts Awarded to Recipients of Scholarships and Loans from University-Controlled Funds: 1976-77 *(Continued)*.

State	Dental school	Scholarship amounts			Loan amounts		
		Mini-mum	Maxi-mum	Mean	Mini-mum	Maxi-mum	Mean
TX	Baylor College of Dentistry	—	—	—	1,000	4,600	2,000
	Univ. of Texas at Houston	50	1,500	400	150	3,500	1,269
	U. of Texas at San Antonio	840	1,620	1,310	400	2,000	967
VA	Virginia Commonwealth U.	100	1,500	550	200	4,700	2,258
WA	Univ. of Washington	300	1,500	350	120	7,800	4,800
WV	West Virginia Univ.	200	1,950	730	200	4,500	1,271
WI	Marquette Univ.	900	3,500	1,550	200	5,000	4,200
PR	Univ. of Puerto Rico	1,000	5,500	3,205	1,000	3,500	1,373

** Information not available

Source: American Dental Association Council on Dental Education *Annual Report on Dental Education, 1977/78.*

or institutional assistance from year to year. Indeed, declines in total monies available during the mid-and-second half of the seventies are associated with declines in student enrollment. This observation is borne out by this study which suggests that quality of financial aid is conducive to attracting larger number of black students into professional schools.

A 1978 interim report on student education finances from the American Association of Dental Schools addressed this problem of sources to support the cost of dental education. Of the 3,337 participants in this study, 42.3 per cent identified the federal health professions loan programs as the source of their loans; 17.9 per cent borrowed from the National Direct Student Loan or National Defense Education Loan source; 52.9 per cent used State or Federally Insured Loan services through a bank or lending institution; 13.4 per cent made personal loans from a bank or lending institution (not federally or state insured), and 39.4 per cent obtained loans from other sources.[14] Only 16.7 per cent reported "own earnings or savings as the principal source of funds to meet expenses during dental schools." Spouses earnings were important in 20 per cent of the cases, and loans or gifts from the family were reported in 36.9 per cent of the cases. Almost 15 per cent of these students relied upon scholarships to meet expenses. The largest single source of scholarship monies was the dental school or its parent university.[15] According to this study, the mean debt reported upon entry into dental school was $1,516 but the mean graduating debt was $12,022.[16] Indebtedness has direct bearing upon post-dental graduation employment (e.g., private or group practice, public service, etc.).

(2) Strong recruitment programs are also essential for successful enrollment of significant numbers of black students. Without question, many in-

stitutions with low enrollment of black students do recruit with exceptional diligence. Yet, they are unsuccessful in attracting black students. For them, the situation is particularly frustrating since poor results do not adequately reflect the magnitude of their effort nor their overall financial investment in the recruitment process. These recruitment programs include the utilization of minority recruiters, extensive travel to regional centers, visits to historically black colleges and to career days sponsored by other institutions, participation in the Health Professions Summer Program, waiver of application fees, participation in the DDS-Op program, and the availability of a self-paced flexible curriculum. Parenthetically, a few of the institutions have abolished the flexible curriculum in recent years.

Several dental schools maintain that the recruitment and admissions problem is complicated by the small pool of "qualified black students" and the increasing competition for that group. "Qualified", in so many instances, refers exclusively to performance on the DAT. The fact of the matter is that black students, as a group, fall from 1.5 to as much as 2 points below the mean DAT scores of the acceptees at several institutions. That is one of the reasons why some institutions, such as the University of Alabama, insist that it is necessary to scrutinize the *entire* profile of the applicant in making admissions decisions rather than rely exclusively on one or two criteria.

Other institutions candidly admit that their institutional behavior leaves a great deal to be desired regarding black applicants. Institutional behavior may include an intimidating environment within which the personal interview is conducted, and the lack of standardized concept of expectations for the personal interview which often means that different interviewers are searching for entirely different qualities in the applicant. Frequently, as in any subjective encounter, "people see what they look for and look for precisely what they see."

There may be considerable fall-out from the inability of many black students to afford the cost of travel to an institution which expresses an interest in them. Those institutions frequently state that no travel money is available to bring prospective students to the campus. They also fail to draw upon alumni in nearby cities or even to send their faculty to cities closest to the candidate to help eliminate travel problems. When they do follow-up, several institutions do not provide sufficient scholarship money that enables them to woo capable black students to their institutions.

Further, there is some evidence that institutional commitment is waning. This situation coincides with a rise of the new conservatism in the American society. One dimension of this shift is a disregard for anything more than limited access or tokenism and some institutions do not seem committed to that. A few institutions claim that black students are not interested in dentistry and that fact alone accounts for their underrepresentation. This position is unacceptable. While it may be more to the point to assert that the interest in dentistry is not as great as is the interest in medicine, which seems to be true of American students in general, the profession of dentistry has to become far more aggressive and less rigid in its practices in order *to enroll significant numbers of black students. If it does not enroll black dental students, clearly it cannot produce dentists.*

Graduation Profiles

We have only fragmented data on graduation rates of black students from schools of dentistry prior to the 1971-72 academic year. It is evident from historical information that, prior to the seventies, Howard University and Meharry Medical College produced more black dentists than all other schools of dentistry combined. Between 1971-72 and 1979-80, a total of 1436 black students received the Doctor of Dental Surgery degree from U.S. schools of dentistry. During the same period, dental colleges produced a total of 42,711 dentists. Black dental college graduates represented a mere 3.3 per cent of the total number of dentists produced in that period. This comparatively low productivity, although resulting in three times the number of black dentists in the work force, accentuates the persistent problem of underrepresentation of blacks in the dental profession.

An examination of Table 24 shows that the number of black dental schools graduates almost tripled between 1971-72 and 1973-74. In that two year period the absolute numbers rose from 55 black graduates in 1972 to 154 in 1974. The percentage change was from 1.1 per cent of total graduated in 1972 to 3.6 per cent in 1974. The absolute numerical increases in the number of graduates from dental schools among black students continued through the 1976-77 academic year. However, the proportion of black students did not reach its peak of 4.1 per cent of total number graduated until the following year when a slight decrease in absolute numbers began. That decline was followed by an extremely sharp drop in the number of black students who were graduated in 1978-79 as well as a return to the 1975-76 level in terms of proportion in the total graduating population. The pattern changed somewhat in 1979-80 when the 182 black students graduated from U.S. schools of dentistry represented 3.3 of the total graduating population of 5,424 students.

This pattern reflects the success and failure of retention programs in dental schools. It may also be instructive on institutional behavior and may

Table 24. Total Dentistry Degrees Conferred By Race and Year:
All Institutions Combined, 1969-80

Year	Total Degrees Conferred	Total Number of Blacks Receiving DDS Degrees	Per Cent Black of Total
1968-69	3,457	—	—
1969-70	3,433	—	—
1970-71	3,749	—	—
1971-72	3,775	55	1.4
1972-73	3,961	74	1.8
1973-74	4,230	154	3.6
1974-75	4,515	187	3.4
1975-76	4,969	213	3.8
1976-77	5,336	209	4.0
1977-78	5,177	203	4.1
1978-79	5,324	159	3.8
1979-80	5,424	182	3.3
Total:	53,350		
Total: 1972-80	(42,711)	1,436	3.3

Blank space indicates data unavailable
Source: American Dental Association, *Minority Reports.*

be suggestive of the varying quality of students recruited for dental education. Relative to attrition, one would assume that, if attrition were low, well over 90 per cent of the students who entered in 1970-71, for instance, would have graduated in 1974. One-hundred and eighty-four black students entered dental school in 1970-71 but only 154 of them were graduated in 1974. That is a loss of thirty students or almost 16.3 per cent of these students. By contrast, total enrollment in the 1970-71 first year class was 4,565; four years later, 4,515 of them were graduated resulting in a net loss of fifty students or less than one per cent. Thirty of them were black. Similarly, students who enrolled in 1975-76, would normally be graduated in 1978-79. In 1975-76, 298 black students were enrolled in first year classes. Only 203 received dental degrees in 1978-79. This was a fall-out of some 95 students or 32 per cent of those who entered. By contrast, in 75-76, some 5,763 students enrolled at all schools of dentistry and 5,324 received degrees in 1978-79.

However, these findings are highly suspicious due to failure to control for institutions with flexible or self-paced curricula. It is not known precisely how many of these students have in fact delayed graduation by one year by taking a reduced load. If these data were known, graduation rates by year may reflect substantial changes in the overall retention or attrition rates. Irrespective of these considerations, there is an apparent problem of the retention of black students. It is of primary importance in the nature of the outcome four years following entry into dental schools.

The first year for which systematic, State-by-State data are available on the race of dental school graduates is 1974. These are students who entered dental colleges in 1970-71. In 1974, the eleven schools of dentistry in the ten "Adams States" conferred a total of seventeen degrees on black students or 12 per cent of all dental degrees received by black students throughout the nation. Almost half of these degrees were conferred by one institution, the University of Maryland, which has consistently led the "Adams States" in the number of black dental graduates. Only the University of Georgia approaches the University of Maryland in total number of black students graduated in the period between 1974 and 1979. (See Table 25).

In 1974, the "Adams States" enrolled thirty-five black students in their colleges or schools of dentistry. Normally, these students would be expected to receive their degrees in 1978. In that year, these institutions awarded twenty-eight degrees to black students or 17.6 per cent of the national total of 203 degrees received by black dental students. Once again, an examination of Table 25 shows that the University of Maryland and the University of Georgia performed more effectively and were significantly more successful in retaining black students than were other institutions in the "Adams States."

A disproportionate number of black dental graduates continue to come from the two historically black dental schools. In 1974, the Dental College at Howard University and Meharry Medical College accounted for slightly more than fifty per cent of all black students who received degrees in Dentistry. Their proportions declined slightly to about 50 per cent again in 1975; then to slightly more than 49 per cent in 1976; to 43 per cent in 1977 and dropped to only 40 per cent in 1979. These changes suggest that an increasingly large number of historically white institutions are retaining greater numbers of black students through the year of graduation. Nevertheless, paucity in enrollment and in graduation is still troublesome. Only one institution in the United States, the University of Nebraska, failed to enroll a single black student in the entire decade of the seventies.* Only about half dozen historically white institutions are even approaching the

Table 25. Total Black Dental School Graduates from Institutions in the "Adams States" by School, Sex and Year, 1974-79

Dental School	1974 M	1974 F	1975 M	1975 F	1976 M	1976 F	1977 M	1977 F	1978 M	1978 F	1979 M	1979 F	Total
Univ. of Florida	0	0	0	0	0	0	1	0	0	0	2	0	3
Emory University	1	0	0	0	0	0	0	0	0	0	1	0	2
Med. Col of Ga.	2	1	5	1	5	2	5	1	5	1	0	0	28
Louisiana State Univ.	0	0	0	0	1	0	0	0	1	0	0	0	2
Univ. of Maryland	9	0	6	0	9	0	5	2	5	2	1	3	42
U. of Mississippi*	—	—	0	0	0	0	0	0	0	0	0	0	0
U. of North Carolina	0	0	1	0	1	0	2	0	2	1	4	1	12
U. of Oklahoma	0	0	0	0	1	0	1	0	1	0	1	0	4
Oral Roberts Univ.*	—	—	—	—	—	—	—	—	0	0	0	0	0
Temple Univ.	0	0	0	0	2	0	3	0	1	1	2	0	9
Univ. of Penn.	0	0	5	0	5	0	2	0	1	0	3	1	17
Univ. of Pittsburgh	3	1	3	0	2	0	1	1	2	2	2	5	22
Va. Commonwealth U.	0	0	2	0	2	0	1	1	2	1	1	0	10
Totals	15	2	22	1	28	2	21	5	20	8	18	10	152
	17		23		30		26		28		28		

*New School — data unavailable
Source: American Dental Association, *Minority Reports*.

level of responsibility assumed by Howard and Meharry. Once again, an explanation for their failure to attract and graduate sufficient numbers of black students can be found in institutional behavior as much as it can in undesirable societal conditions.

Although the University of Nebraska neither enrolled nor graduated a black student, partially because "they did not meet resident requirements," Creighton University, the private institution in that state, managed to graduate a dozen black dentists between 1974 and 1978. The University of Colorado graduated no black students between 1974 and 1978. Only one black student received dental degrees during that period from Northwestern University and West Virginia University. Table 26 shows an incredibly high number of institutions who graduated fewer than five and less than ten black students during that period.

The most successful historically white institutions outside the "Adams States" during this period were the University of Michigan, the University of California/San Francisco, the University of California/Los Angeles, Tufts University, the University of Illinois, and the University of Southern California. Some institutions, such as New Jersey Dental School rank

*This statement is based on data reported to the ADA.

Table 26. Total Black Graduates from Dental Schools in "Non-Adams States" by Schools, Sex and Year, 1974-79.

Dental School	1974 M	1974 F	1975 M	1975 F	1976 M	1976 F	1977 M	1977 F	1978 M	1978 F	1979 M	1979 F	Total
Univ. of Alabama	2	0	3	0	0	1	3	1	2	1	2	0	15
Univ. of Pacific	3	0	2	0	0	0	1	0	0	0	0	0	6
Univ. of Calif./S.F.	6	0	5	0	5	0	3	3	6	0	2	1	31
UCLA	3	1	7	1	4	1	2	0	6	0	2	5	30
U. of So. Calif.	0	0	2	0	6	1	7	0	5	1	2	0	24
Loma Linda Univ.	1	0	1	0	0	1	2	0	3	0	2	0	10
Univ. of Colorado	-	-	0	0	0	0	0	0	0	0	0	0	0
Univ. of Conn.	-	-	0	0	1	0	0	1	0	0	0	0	2
Georgetown Univ.	-	-	0	0	1	0	1	1	1	0	3	2	9
Howard Univ.	42	6	53	8	62	18	49	15	47	13	32	13	358
Loyola Univ.	-	-	0	0	2	0	0	0	1	0	3	1	7
Northwestern Univ.	1	0	0	0	0	0	0	0	0	0	0	0	1
U. of So. Illinois	-	-	0	0	1	0	0	0	0	0	1	0	2
U. of Illinois	7	0	2	1	3	0	3	0	2	1	6	3	28
Indiana Univ.	2	0	2	0	0	1	1	1	2	0	1	0	10
Univ. of Iowa	0	0	3	0	2	1	1	0	2	0	0	1	10
U. of Kentucky	0	1	2	1	4	0	3	1	1	0	1	0	14
U. of Louisville	0	0	0	1	0	0	2	0	0	0	1	1	5
Harvard Univ.	3	1	2	1	1	0	3	0	2	0	2	1	13
Boston Univ.	-	-	0	0	0	0	0	0	0	0	1	0	1
Tufts Univ.	1	0	5	1	5	1	4	3	0	2	2	1	25
Univ. of Detroit	8	0	2	0	4	1	0	0	1	1	1	0	18
Univ. of Michigan	3	0	5	1	5	1	17	3	8	2	3	2	50
Univ. of Minnesota	-	-	0	0	1	2	2	0	0	0	0	0	5
Univ. of Mo./KC	1	0	0	0	1	0	1	0	5	1	2	1	12
Washington Univ.	0	0	0	0	2	0	2	0	0	0	2	1	7
Creighton Univ.	1	0	1	0	4	0	4	0	2	0	2	1	15
Univ. of Nebraska	0	0	0	0	0	0	0	0	0	0	0	0	0
Fairleigh Dickinson	0	0	0	0	1	0	2	0	6	0	0	0	9
N.J. Dental School	0	0	5	0	2	0	2	0	3	1	3	2	18
Columbia Univ.	1	0	2	1	0	0	2	0	0	0	0	0	6
New York Univ.	2	0	3	1	4	0	1	1	3	0	1	0	16
SUNY/Stony Brook	0	0	0	0	0	0	0	0	0	2	1	0	3
SUNY/Buffalo	3	0	1	0	2	0	2	1	1	1	1	1	13
Ohio State Univ.	2	0	1	0	4	0	1	0	0	1	1	0	10
Case West. Reserve	4	0	1	1	3	0	1	0	1	3	2	0	16
Univ. of Oregon	0	0	1	0	0	0	1	0	0	0	1	0	3
Med. Col. of S.C.	1	0	0	0	0	0	0	0	3	0	1	0	5
Meharry Medical C.	24	1	26	6	23	2	23	6	20	3	16	1	161
Univ. of Tenn.	4	0	1	0	1	0	1	0	1	1	1	1	11
Baylor College	0	0	0	0	0	0	2	1	0	1	0	0	4
U. of Texas/Hous.	1	0	0	0	0	0	1	0	1	1	2	0	6
U. of Texas/San A.	0	0	0	0	0	0	0	0	0	0	0	0	0
U. of Washington	1	0	0	0	0	0	0	0	2	0	2	0	5
West Va. Univ.	0	0	1	0	0	0	0	0	2	0	0	0	3
Marquette Univ.	0	0	1	0	0	0	0	0	0	0	0	0	1
Totals	125	10	140	24	153	30	152	36	139	36	103	51	
Totals (All Schools Combined)	(154)		(187)		(213)		(215)		(203)		(182)		

Source: American Dental Association, *Minority Reports.*

relatively high in enrollment but have not been as successful in retaining black students.

The Sex Variable in Dentistry

Dentistry is primarily a male preserve. However, the number of women enrolled in schools of dentistry and who are graduating from them has increased significantly during the recent past. In 1974, for instance, there were 66 black women enrolled in dental schools. Their absolute numbers and their proportions of all black enrollees both increased significantly in each year thereafter. By 1978, the 117 black women matriculated in dental schools almost doubled their 1974 enrollment. In 1974, they represented 7.0 per cent of black enrollees. In 1978, they comprised 12.8 per cent of black enrollment in U.S. dental schools. These increases are observed throughout the nation, in "Adams States" as elsewhere.

Similarly, black women comprise an increasingly larger proportion of the graduates. In 1974, there were only a dozen black women who received dental degrees. By 1979, there were 61 black women dental graduates. This number was five times larger than the number who were dental degree recipients in 1974. Although the numbers are extremely small in the "Adams States", the data on black women graduates in dentistry show the combined total increasing from two graduates in 1974 to ten graduates in 1979. In all these states as a group, a special effort has to be mounted to improve the ratio of black women graduating from dental colleges.

With considerably more aggressive recruitment, appropriate modifications in selection and admissions practices and alterations in institutional behavior in general, the hiring of more black faculty and substantial increases in direct scholarships to students, dental school can improve upon their success rates in assuring greater equality of educational opportunity.

Footnotes

1. Clifton O. Dummett and Lois D. Dummett, "Afro-Americans in Dentistry: A Synopsis," *Crisis* (November 1979), p. 398.
2. *Ibid.*
3. *Ibid.,* p. 379.
4. *Ibid.*
5. *Admission Requirements of U.S. and Canadian Dental Schools, 1979-80.* Washington, D.C.: American Association of Dental Schools, 1979, p. 3.

6. These issues are a matter of continuing debate between advocates of standardized tests such as the Educational Testing Service and the College Board and by those who have recently challenged the validity of tests, including Ralph Nader, the National Association for the Advancement of Colored People (NAACP, and others.) The position of the Educational Testing Service and the College Board is that these factors have minimal effect on the overall performance of the test-taker.

7. American Dental Association, *Trend Analysis, 1977-78*. Chicago: American Dental Association, 1978, p. 1.

8. *Ibid.*

9. This figure was provided by the American Dental Association. It is based upon an address given by Dr. Elijah Richardson, President of the National Dental Association.

10. *Admission Requirements of U.S. and Canadian Dental Schools, 1978-79*, p. 110.

11. *Ibid.*

12. *Ibid.*, p.38

13. *Ibid.*

14. *Ibid.*

15. *Ibid.*

16. *Ibid.*

CHAPTER 6

Mainstreaming Black Americans in Optometry

Optometric educators are constantly warning of the imminent manpower crisis in the supply of optometrists necessary to meet vision care needs of the American population.[1] According to Henry B. Peters, a recent study conducted for the American Optometric Association placed optometric manpower issues at the lowest priority level. In his view, this situation is a result of misinformation about the numbers of optometrists produced annually by the thirteen schools and colleges of Optometry. Although these institutions are filled to capacity at the present time, Peters suggests that they are still not producing a sufficient number of optometrists to meet vision care needs of an aging population. One reason for this disparity is that a significantly large bulk of practicing optometrists are themselves nearing the age of retirement. At the present rate of O.D. degree productivity, the gap between optometrists produced and vision care needs will continue to widen in the foreseeable future.[2] Hence, there is presently an overall shortage of persons trained in the vision care areas for which optometrists are specialists.

Today, there are approximately 21,500 optometrists in the United States. Only about 225 or one per cent are black Americans and another twenty-five or thirty are members of other minority groups such as Hispanics and Native Americans. The manpower crisis described by Peters for the population as a whole is considerably more acute for black Americans. Since black health care specialists are likely to serve a largely black clientele, it is safe to assume that there are 225 black optometrists for about 30 million black Americans.

This chapter describes the trends in enrollment of black students in schools and colleges of optometry during the 1970s. It provides substantial support for the hypothesis that the presence of role models is directly correlated with growth in access of black students to graduate and professional schools. It points to success models for the recruitment and enrollment of black students in optometry and some of the myriad problems confronted by institutions in increasing access to outsider groups.

141

The Historical Context

Optometry has always stood in the shadow of Medicine and Dentistry as a health care profession. While these fields have enjoyed immense prestige, social status, and financial rewards associated with privilege, that has not always been the experience of optometry. Inducements for selecting optometry as a profession and as a field of scientific study have never been as powerful as have those factors which persuade individuals to enter medicine and dentistry. Further, recognition of optometry as a respectable occupation came considerably later that it did for dentistry, for instance, and the prestige granted to the medical profession can be traced back into antiquity.

Programs in optometry undoubtedly began in special colleges during the 19th century. In fact, the New England College of Optometry claims to operate the oldest continuous program in optometry in the United States. It traces its origins to the Massachusetts College of Optometry, founded in 1894. However, no formally recognized degree was conferred in its earlier program. The first institution to offer the Doctor of Optometry (O.D.) degree was Ohio State University in 1914. Since that time, twelve additional institutions were organized with degree-granting programs in optometry. Seven of these institutions are private and six are public. The six public colleges are either components of major universities or they are physically and administratively separate state-controlled schools of optometry.

Nine of the institutions were in existence prior to 1960 and four opened during the sixties and seventies. Ferris State College of Optometry (established in 1974), is the youngest of all schools and colleges of Optometry. Since there are only thirteen of these institutions in the United States, all of them are essentially regional schools of optometry. That is to say, they have contractual arrangements with other States in their regions for the training of out-of-state residents who wish to pursue a degree in optometry. There are no schools and colleges of optometry located at an historically black medical college. That absence may help to account for the low representation of blacks in this profession.

The 21,500 optometrists in the United States represent a ratio of about 9 optometrists for every 100,000 persons in the population.[3] In 1970, according to Edwin C. Marshall, the ration between optometrists to population was then one for 9,797 persons. However, the 109 black optometrists then practicing in the United States represented a ratio of one black optometrist for every 208,005 black Americans.[4] Because of this immense disparity between the total population and the optimal number of optometrists desired to provide effective and adequate vision care, the American Optometric Association committed itself to the goal of training a sufficient number of optometrists of all races to attain the optimal level of one optometrist for every 7,000 Americans.[5] That goal has not been attained despite significant increases in the production of optometrists during the 1970s.

Recruitment and Enrollment of Black Students in Optometry

As the decade of the sixties closed, it was apparent to optometric educators that the nation had a serious shortage of black optometrists. Few of the colleges and schools of optometry had ever given special attention to the recruitment and enrollment of black Americans. One notable exception to this generalization was the Illinois College of Optometry (ICO). After World War II, capitalizing on educational opportunities provided by the G.I. Bill of Rights, the ICO wooed black students into this field. Consequently, a substantial number of blacks trained in optometry in the post-WW II era and into the sixties were graduates of the ICO. As a result, Chicago in 1980 has the largest single concentration of black optometrists in the United States. Approximately thirty (30) black optometrists are currently practicing in Chicago today.[6]

The National Optometric Association (NOA) was established in 1969. Immediately, the NOA exerted pressure on the predominantly white American Optometric Association (AOA) to assume active leadership and take aggressive action toward the recruitment and enrollment of black students in optometry. The AOA responded to this urgency in three essential ways. First, for the first time in its history, it appointed a black optometrist, Dr. Charles Comer, to one of its major committees. Dr. Comer was appointed and served for three years (1970-73) as a member of the AOA Career Guidance Committee. Second, a minority recruitment officer was also appointed, and third, primarily due to the diligence and enthusiastic work of Dr. Henry Hofstetter of the College of Optometry at Indiana University, the AOA allocated some $15,000 for each of two years specifically for the recruitment of minority students. The Minority Recruitment Officer was supported primarily through an Urban Coalition Grant of $35,000.00 to the AOA.[7] This program was jointly sponsored by the AOA and the NOA.

In the early seventies, there was an initial upsurge of interest in the recruitment and enrollment of black students. Inducements for this action came largely from the federal government through capitation money awarded to institutions on the basis of the number of minority students actually enrolled. In addition, "Federal giving" to institutions in the health care fields, including optometry, signalled the availability of federal funds to support affirmative action programs and education in the health fields in general. Institutional behavior suggested an initial commitment to rectifying past injustices and the elimination of discriminatory actions which had resulted in the critical underrepresentation of black students in this field. In addition to a positive stance assumed by many schools and colleges of optometry, several of the State Associations also took an active role in the recruitment of black students. For instance, in January, 1972, delegates to the AOA Manpower Conference in HEW Region V States of Ohio, Michigan, Indiana, Illinois, Wisconsin, and Minnesota declared the recruitment of minority students as a major priority.[8]

The delegates urged their member associations to make optimal utilization of recruitment materials developed by the AOA and the NOA. They also recommeded a publicity campaign in the State Association's newsletter to call attention to minority recruitment, and to establish scholarships specifically for minority students as inducement for them to enter the profession of optometry.[9]

In the early seventies, in response to the NOA initiatives and AOA recommendations, several schools and colleges became far more aggressive in the recruitment of black students than at any time in their history. As early as 1970, the Division of Optometry of Indiana University developed a major recruitment program which was coordinated and administered by the black students enrolled in the O.D. program at that time. This Committee visited historically black colleges and other universities in Kentucky, Tennessee, North Carolina and Georgia to inform black students of opportunities at Indiana and to interest them in the optometry program. It sponsored lectures to senior high school College Upward Bound students; participated in Career Day programs, and distributed recruitment materials it had developed. It organized visits to the campus of Indiana University during which prospective students were exposed to the type of training that an optometry degree entails.[10] Importantly, Indiana University's School of Optometry hired black faculty who were visible role models to potential students and who could provide instant evidence of the possibilities of success in this profession.

Other institutions, such as the University of California/Berkeley, the University of Alabama/Birmingham, the Pennsylvania College of Optometry, Southern College of Optometry, and the University of Houston, similarly mounted aggressive recruitment programs to attract blacks and other minority group students. Invariably, whatever success they experienced can by attributed to some combination of the following factors: (1) the presence of role models (e.g., black faculty in either a full-time university position or a part-time clinical position), (2) the availability and utilization of federal grant money and/or State and institutional funds to provide financial assistance to needy students, (3) the use of flexible admissions criteria, (4) the availability of academic and psychological support services, and (5) evidence that recruitment efforts moved beyond tokenism. Several institutions established an office of minority affairs. Now all thirteen colleges and schools of optometry publicly acknowledge a commitment to affirmative action. However, recent declines in the enrollment of black students might suggest a disjunction between articulated policy and actual institutional behavior. Obviously, stating a policy is not necessarily synonymous to its implementation and enforcement.

Institutional efforts in the recruitment of black students were assisted immeasurably through activities sponsored by the National Optometric Association. The NOA not only exerted pressure on institutions resistant to changing the racial composition of colleges of optometry but organized special recruitment programs of its own. For example, the NOA appointed regional directors who were responsible for maintaining direct contacts with selected colleges of optometry and for monitoring their recruitment activities. Whenever they resisted or lagged in their efforts, NOA members reminded them of their responsibilities and commitments consistent with AOA policies. The NOA used some of its members as liaison officers to students and minority affairs at specific institutions. The purpose of this activity was to prevent academic and psychological problems from occurring and to find immediate solutions to those problems encountered.

The NOA designed retention programs, offered tutoring, counselling, and psychological support, and its members served as "advocates" before academic boards for students on certain occasions. Local chapters of the National Optometric Association were established on every college campus. These chapters were not only formal associations to which black students could belong but they also served a variety of important social and academic functions. For example, white students often studied together and formed working groups to discuss problems encountered in class and in the clinical setting but black students were often excluded. The black students were also frequently not invited into the cliques, social groups, and informal social activities which contribute to a sense of well-being and which may help to create a positive learning milieu. The local chapters of the NOA served these functions and presumably helped to reduce attrition and increase retention among black students in optometry.

In the early seventies, student financial aid consisted of more direct aid and scholarships to students instead of repayable loans. That pattern contributed in substantial ways to recruitment success. As these funds dried up or were curtailed and increasing reliance on repayable loans occurred, enrollment declines followed. This loss of black students is also attributable to the impressions communicated that in the Bakke era it was no longer popular to be black. In this period, the federal government was less than enthusiastic in supporting affirmative action in higher education. Consequently, for several institutions, recruiting, enrolling, and graduating black optometrists became a matter of low priority. Declines in enrollment may also be a direct response to a decreasing value attached to being "a doctor of optometry" when some of these students could very well become doctors of medicine or doctors of dentistry which offer considerably more prestige than the O.D. carries in the black community.

It is important to note, however, that some schools and colleges of optometry have magnificant recruitment programs on paper but they fail to implement them. Some have made strong efforts to recruit a critical mass of black students into their programs but without significant success. Their failure may be attributed in part to inadequate funds contributed from institutional sources. Institutional financial support was and is not sufficient to achieve recruitment and enrollment objectives. It may also be explained by the inability to attract federal support for their programs despite the high quality of the recruitment proposals submitted to federal agencies. The lack of commitment by higher level administrators may also help to explain the failure of these programs. In some instances, too, black students simply do not wish to be what is tantamount to a "first," or a guinea pig in an institution that has either enrolled no minorities in the past or is without a significant proportion in its total institutional population.

Admissions Criteria

The thirteen schools and colleges of optometry employ a combination of cognitive and non-cognitive criteria in making admissions decisions. The relative weight assigned to each of the categories by admissions personnel in optometric institutions varied markedly during the seventies. Shifts in weights plus the flexibility apparent in the utilization of discretionary powers by the admissions committees in assigning differential values to one or both sets of criteria had an unquestionable impact on the enrollment of black students during the 70s. In some instances, discretionary authority in favoring one set of criteria over the other may have worked to the benefit of both black and white students whom admissions committees deemed worthy of selection.

Cognitive criteria employed in admissions decisions include the scores on the Optometric College Admissions Test (OCAT), grade point average in college, and, in some instances, scores on Personality Profiles. Noncognitive factors weighed in admissions decisions include personal interviews, the candidate's autobiography, and the letters of evaluation. Several institutional representatives maintain that the underrepresentation of blacks in optometry degree programs and enrollment difficulties may be partially explained by the lack of competitiveness by black students in performance on the OCAT and by their lower college grade point average. An examination of the *Annual Survey of Optometric Educational Institutions* for 1974-75, 1977-78 and 1978-79 shows that only one institution, the Southern College of Optometry, reported that the cognitive portion of the OCAT had no influence on admissions decisions. The number of institutions that reported a "significant" influence exerted by scores on the cognitive portion of the OCAT went from three institutions in 1974-75 to six institutions in both 1977-78 and 1978-79. The number of colleges of optometry which reported a "moderate" influence exerted by OCAT scores dropped from eight institutions in 1974-75 to five institutions in 77-78 and in 78-79.

Similarly, a noticeable change is observed in the relative importance assigned by schools and colleges of optometry to the non-cognitive portion of the OCAT in reaching admissions decisions. In 1974-75, seven of the twelve schools and colleges of optometry then in existence reported that the non-cognitive portion of the OCAT had no influence on admissions decisions. No significant change in that number was observed for the later years. Between 1974 and 1977, the number of institutions that assigned a "slight" degree of importance to non-cognitive performance on the OCAT was reduced by fifty per cent, or from four institutions to two. By comparison, in 1974, one institution, Pacific University, stated that non-cognitive aspects were "significant". That institution remained the only one which continued to rate that portion as significant in 1978-79.

Regardless of the weights attached to OCAT scores, it appears that institutions do utilize their prerogatives and discretionary powers in making decisions regarding the relative importance of this test. Hence, one may justifiably speculate that OCAT scores may be employed as an instrument of exclusion in much the same manner as they can be utilized to include those persons an institution desires or deems otherwise worthy of admission. However, there, is an apparent trend in heavier weighting of the OCAT in 1980 than was the case in the early seventies.

In Edwin Marshall's 1972 study, four of the eleven responding institutions indicated that they made special concessions to minority students.[12] In this study, some institutions reported that they had lowered OCAT minimum requirements for minority or disadvantaged students. However, "relaxing standards", as some referred to this process, resulted in retention problems for these students in later years. It also appears that those institutions which claimed to be unyielding in their demand that universalistic criteria be equally applied to all students and who are without black faculty and major recruitment programs are unsuccessful in terms of black students. They enroll fewer black students and graduate even fewer black students than those who are willing to aggressively seek black students and make reasonable adjustments in formal admissions criteria.

The grade point average is sometimes cited as another impediment to successful enrollment of black students. Implicit in this argument is that a significant proportion of black students do not qualify for admissions because of poor academic performance in college. A review of the *Annual Survey* reveals that the mean Grade Point Average (GPA) for enrolled students in all schools and colleges of optometry combined was 2.67 on a 4.0 scale or C + in 1970-71. In that year, specific institutional mean GPAs ranged from a low of 2.46 for all entering students at Illinois College of Optometry to a high of 2.96 on a 4.0 scale at Ohio State University. The mean GPA for all institutions combined did not reach the level of B or 3.0 until 1973-74. It attained its highest value of 3.2 in 1976-77. However, since 1973-74, the mean GPA has hovered about the level of 3.1 or 3.2. About one-half of the institutions currently report mean GPAs of less than 3.0 for their entering students in the O.D. degree program. Although the GPA means have risen, it is difficult to accept the notion that a GPA of C, C + or B − has been a significant impediment to the enrollment of black students.[13] Yet, the claim continues to be made that blacks are deficient in pre-optometric studies, academic preparation, and in GPAs.

Eleven of the thirteen institutions interview from ten to one-hundred per cent of students who apply for admissions as a prerequisite for enrollment. Only the University of Houston reports that it interviews 100 per cent of its applicants. Five institutions interview from ten to twenty-five per cent. Five other institutions interview from twenty-six to fifty per cent of their applicants. The interview and the student autobiography are essential elements in assessment since they enable selection officers to make some determination of the strengths and weaknesses of the applicant's non-cognitive qualifications. The interview may also be utilized to test the applicant's overall ability to relate to other persons, and to ascertain the presence of desirable personality qualities deemed to be salient for successful optometrists.[14]

Since there were only three full-time known black faculty in the thirteen schools and colleges of optometry in 1979-80, it is difficult to assure representation from the black academic community on admissions committees. In those institutions in which blacks are faculty members, blacks do serve on these committees from time to time. Several institutions without black faculty claim a sincere and "good faith effort" to recruit black faculty but they are unsuccessful. They rationalize their failure by assuming that black optometrists are not interested in faculty positions since the financial rewards of private practice are substantially higher than the actual incomes received by professors, especially those who do not have a simultaneous opportunity to engage in private practice. In some instances, black optometrists in private practice are recruited for an institution's clinical program. Several of these persons occasionally serve on the institution's recruitment and admissions committee for these institutions. Nevertheless, both recruitment and admissions of black students, as important processes in the mainstreaming of outsiders in optometry, are impaired by the underrepresentation of black faculty in optometric education programs.

Enrollment Trends

On one level of analysis, an examination of Table 27 leads to the incontrovertible conclusion that significant progress was made during the 1970s in the recruitment and enrollment of black students in schools and colleges of optometry. That conclusion is supported by positive trends in absolute numbers of black students enrolled in O.D. degree programs. In 1970-71, for instance, a total of fifteen black students enrolled in all of the schools and colleges of optometry. Enrollment of black students increased significantly thereafter. This change was consonant with the recruitment programs mounted by the AOA and the NOA. As a result by 1972-73, total black student enrollment more than doubled to thirty-seven. The success of the recruitment is apparent for each succeeding year following 1976-77 when a total of eighty-nine black students were enrolled in optometry degree programs. However, the sharp increases in enrollment during the first half of the seventies are presently being off-set by disturbing enrollment declines since the 1976-77 peak. Total black student enrollment in optometry in 1980 is 22 per cent less than it was in 1976.

On another level, Table 27 also shows that at no time during the 1970s did black student enrollment reach 100 students. It is striking that every increase in total black student enrollment is matched by a significant incline in total student enrollment. This suggests that colleges of optometry, in toto, expanded both first year and total enrollment capacities during that period. Further, at no time during the 1970s did total black student enrollment reach a proportion of three per cent of total enrollment. In fact, the proportion of total black student enrollment ranged from an abysmally low of five-tenths of one per cent in 1970-71 to 2.9 per cent in 1976-77. Nevertheless, the statistical claim can be made that the proportion of black students enrolled in schools and colleges of Optometry increased approximately six-fold during the 1970s. Such assertions border on the ludicrous when one examines both the original numerical base and the absolute numbers represented by such claims. There is no concrete evidence whatsoever to support any argument that black students were enrolled at the expense of white students. The sparsity of their numbers alone defies that contention. Further, optometry colleges actually enroll from 75 to 84 per cent of all students admitted. There are no reasons to speculate significant departures from this trend in the case of black admittees.

Another observation that should be made is that no school or college of optometry enrolled a total of 100 black students during all of the 1970s.* Total enrollment of black students during that period ranged from a high of 99 black students at Indiana University to a low of seven at Pacific University among all institutions in existence for a substantial portion of the decade. (Ferris State College enrolled a single black student but that institution is the newest of all optometric colleges). Indiana University enrolled one-third more black students than did the New England College of Optometry (60 black students). It also enrolled forty students more than did both the University of California/Berkeley and the Illinois College of Optometry, each with a total of 59 black students during the entire decade. Indiana enrolled almost twice as many black students as did the Pennsylvania College of Optometry (52 black students) and more than doubled the 49 black students enrolled by the University of Alabama/Birmingham. Indiana, Alabama, and Pennsylvania were the three institutions with full-time black faculty during most of the seventies.

The Southern College of Optometry managed to enroll a total of 36 black students in the seventies; the University of Houston enrolled thirty-one. The State University of New York College of Optometry enrolled twenty-three while the oldest doctor of optometry degree-granting institution on the nation, Ohio State University, and Southern California College of Optometry (and its predecessor, Los Angeles College of Optometry) each enrolled a total of fifteen black students during the decade of the seventies.

*This statement does not include the 1979-80 data. When those numbers are included, Indiana exceeds 100 black students enrolled in the 1970s.

Table 27. Total Enrollment in Colleges of Optometry by School and Race, 1970 - 79/80
(T = Total Enrollment   B = Black Student Total Enrollment)

Name of Institution	1970-71 T	B	1971-72 T	B	1972-73 T	B	1973-74 T	B	1974-75 T	B	1975-76 T	B	1976-77 T	B	1977-78 T	B	1978-79 T	B	1979-80 T	B
Univ. of Alabama/Bham	28	0	46	1	70	1	85	3	96	4	105	8	112	11	127	11	146	10	151	3
UC/Berkeley	193	1	212	2	215	6	220	5	229	10	242	10	249	9	254	9	261	7	266	2
Ferris State (Mich.)	—	—	—	—	—	—	—	—	—	—	—	—	—	—	—	—	99	1	110	1
Univ. of Houston	238	2	243	2	250	5	265	4	257	4	258	4	290	3	334	3	374	4	405	3
Illinois College of Optometry	412	3	455	4	489	4	516	8	529	11	568	7	571	9	584	8	587	5	596	4
Indiana University	223	4	246	3	253	4	258	5	266	11	276	17	269	22	268	19	258	15	262	10
New England College of Optometry	197	0	226	1	256	6	281	11	281	12	292	12	313	8	346	6	344	4	362	3
Ohio State University College of Optometry	190	0	197	1	204	1	210	3	215	3	216	2	220	1	226	2	229	2	233	2
Pacific University	265	0	273	0	283	0	280	1	289	1	298	1	316	2	329	1	328	1	327	0
Pennsylvania College of Optometry	429	3	452	4	489	3	513	5	529	6	551	10	530	8	550	8	572	5	586	8
Southern California College of Optometry and LACO	245	1	247	1	260	3	280	0	305	1	344	3	391	3	398	2	387	2	371	4
Southern College of Optometry	406	1	479	2	514	4	566	4	559	6	568	5	573	5	580	5	588	4	582	5
S.U.N.Y. New York College of Optometry	0	0	21	0	45	1	65	3	83	4	105	4	149	4	185	5	218	2	249	1
Totals	2826	15	3097	22	3328	37	3638	52	3824	73	3824	83	3983	89	4177	79	4626	64	4500	48
Percent Black of Total	.5		.7		1.1		1.4		1.9		2.1		2.2		1.89		1.3		1.0	

Source: Association of Schools and Colleges of Optometry: Annual Survey of Optometric Educational Institutions.

These enrollment trends underscore the primacy and importance of role models for successful recruitment and enrollment of black and other minority students in professional schools. A major portion of the success experienced by such institutions as Indiana University, throughout the decade, and the University of Alabama/Birmingham, at certain points in the decade, can be attributed to the presence of black faculty who were aggressive in the recruitment and enrollment process. Although the utilization of black faculty in part-time clinical or adjunct appointments may also facilitate matriculation, that practice, especially when done sporadically, does not seem to have the same quality or as enduring impact on the recruitment and enrollment process as does the employment of full-time black faculty in tenure-track positions who are available to students throughout the O.D. program. If students are recruited by black faculty, there is good reasons to believe that their retention through graduation is also positively affected by their interaction with those black faculty members at critical points in their training. However, sincere, interested non-black faculty who exact high standards and who evaluate students fairly also play an important role in the retention of black students.

Many of these institutions insist that they are actively seeking "qualified" black students. They insist that finding them is next to an impossibility. It is also argued that most of the black students who are qualified for optometry either are not particularly interested in the profession or use it as a safety valve while they await acceptance into a medical or dental college. The argument posed relative to the lack of "qualified" black applicants raises a number of questions about the meaning of "qualifications". Most of these questions were addressed in the section on Admissions; however, one unanswered but essential question that remains is: To what degree are some institutions using the OCAT scores selectively as a means of excluding black students compared to their assignment of low priority to these scores for students they desire to admit?

Male - Female Ratios

Another significant pattern observed in the total enrollment of black students in optometry is a shift in male - female ratios. The number of black females enrolled in O.D. degree programs, compared to black males, showed striking increases in every year of the decade. In 1970-71, for example, black males comprised almost 87 per cent of black student enrollment and black females accounted for only about 13 per cent. By 1976-77, black males represented approximately fifty-two per cent of all black students in

Optometry and black females had climbed to 48 per cent. However, in 1977-78 and thereafter, black female enrollment not only equalled but surpassed black male enrollment so that in 1980 black females represent slightly more than 54 per cent of total black student enrollment.

How long this trend will continue is an uncertainty. It may be a function of the degree to which the more qualified black males are drawn to medicine, dentistry and engineering or the degree to which black females count as a double minority in affirmative action considerations. It may mean that some black females are able to secure sustained parental contributions, compared to black males, or that more of them are better prepared academically for the O.D. degree program than are black male applicants. It could be a manifestation of an actual change in career focus among black females in contrast to black males in comparable age cohorts.

Retention

Retention of black students is facilitated by the maintenance of a variety of academic and non-academic support services made available to black and other minority students. Mention was made earlier of the retention activities sponsored by the National Optometric Association. Several institutions offer tutorial programs, counseling services, and advising through the minority affairs office. Some institutions frequently provide special assistance in the most difficult first year courses such as geometric optics and human anatomy since the highest attrition rates occur during the first year. However, this service is apparently more often provided by interested professors who are frequently over-loaded with other responsibilities.

Hence, there is an urgent need for an organized tutorial program to assist all students with weak academic preparation in most schools and colleges of optometry. An organized program includes not only adequately paid tutorial staff in sufficient numbers, but also strong counseling services, a reasonable supply of audio-visual materials that may heighten retention of substantive material in subject matter, the presence of role models with whom some students may feel more at ease, and sufficient financial aid to reduce attrition and accelerate the graduation rate of black students from schools and colleges of optometry.

Black Graduates in Optometry

It is especially difficult to obtain data on the precise number of Doctor of Optometry degrees awared to blacks. One reason for this problem lies in the method of reporting graduation data to the Association of Schools and Colleges of Optometry. The graduation data are not disaggregated by race.[15]

According to HEGIS reports, it appears that the thirteen schools and colleges of optometry are producing about sixteen or seventeen black recipients of the Doctor of Optometry degree each year.[16] This is approximately 1.4 per cent of all O.D. degrees conferred in the United States. These students are primarily the products of such institutions as the University of California at Berkeley, the Illinois College of Optometry, Indiana University, New England College of Optometry, the University of Houston, and the State University of New York's College of Optometry. During the seventies, the University of California/Berkeley graduated eight blacks with O.D. degrees. By contrast, Pacific University produced one black male in 1970 and another black male in 1974 with the O.D. degree. Retention through graduation is a problem that deserves special attention particularly in institutions without solid, comprehensive retention programs. All available evidence suggests that successful graduation is not only a function of the academic strengths that a student brings into the first year class but also of the overall quality of retention programs and the learning environment of the institution itself.

Major Problems

These institutions do not claim to be affected by the 1978 Bakke decisions because they insist that they never employed admissions quotas in the first place. Indeed, most of them seemed to have struggled to enroll even two or three black students per year. However, the lack of a critical mass might be interpreted as a negative quota which impeded the enrollment of a sizeable number of black students each year. In fact, the total enrollment of black students in some colleges of optometry in 1979 had reverted to the proportions reported in 1972-73. In five institutions, the proportion of black students in total enrollment in 1978-79 was less than one percent: (1) The University of California/Berkeley, (2) Ohio State University, (3) Pacific University (Oregon), (4) Pennsylvania College of Optometry, and (5) Southern College of Optometry. Only in Indiana University and the University of Alabama/Birmingham did black students comprise five per cent or more of the total student enrollment in 1978-79. These figures were 5.6 per cent and 6.6 per cent, respectively.

Nevertheless, institutions cite as major problems confronting them in their efforts to recruit and enroll black students successfully such factors as lack of minimal qualifications of black applicants; the need for better competitive academic preparation among black applicants; well organized and better supported recruitment programs with a clear institutional commitment to recruit more minority students; more scholarship money, and minority or black faculty to serve as role models. Some individuals maintain that the field of optometry itself must become more competitive and somehow assure greater salaries or wage-earning capacities for practicing optometrists. This is, indeed, a major problem since it is estimated that the average practicing optometrist in the United States earns approximately $35,000 per year compared to about $50,000 for a physican.* That a newly minted doctor of optometry is likely to have incurred from $25,000 to $50,000 in debt by the time he acquires the degree plus the expectation of earning no more than approximately $15,000 per annum inevitably deters entry into this profession, especially when other professional options are available.

More effective recruitment may also be aided by a more equitable distribution of practicing optometrists. As previously noted, it is estimated that about thirty of the 225 practicing black optometrists are located in the city of Chicago. There are five in the State of Ohio; eight in North Carolina, and none in many States and large cities. For example, there is one black optometrist in the entire city of New York and he is located in Brooklyn. Gary, Indiana has one but Indianapolis none. One practices in New Orleans but no one in Shreveport. There is one in Cleveland and there are four practicing black optometrists in Detroit. Not only do cities, such as Birmingham, Houston, Dallas, Jackson, Baltimore and Greensboro need more black optometrists, but rural areas are particularly underserved.[17] This shortage will never be fully alleviated until the rate of productivity approximates the proportion of blacks in the general population.

Financial assistance is especially critical to black students who come from low income families. Tuition for residents in 1979 ranged from none at the University of California/Berkeley and the Pennsylvania College of Optometry to $4,200 at the Southern College of Optometry. Non-resident tuition ranged from "none" at PCO to $4,200 at both the New England College of Optometry and Southern College of Optometry. PCO, however, has a $7,994 required fee for non-residents and $3,994 in required fees for residents. Add to tuition demands other fees and living expenses, the

---

*According to Farrell Aron of the American Optometric Association, the 1977 mean net income of optometrists was $37,403 and the mean was $35,000. It takes 8 years following graduation to reach an average annual income of $37,594.

estimated costs of optometric education rise to staggering sums over four years.

Finally, most institutions will have to devise concrete programs to halt the overall decline in enrollment of black students that began in the second half of the 1970s. These steps involve, among other things, an identification of the conditions responsible for them. No institution established for the training of optometrists can afford to conclude that it has already done its part or all that it can do to stimulate equality of access. It is apparent that what has been done has been insufficient to provide an adequate number of black optometrists for meeting the vision care needs of a major segment of the American population. Consequently, the production of black optometrists in even larger numbers must become a matter of high priority in the 1980s.

Footnotes

1. Henry A. Peters, O.D., "Critical Optometric Manpower Issues", *Journal of Optometric Education.* 4:4 (Spring 1979), p. 8.
2. *Ibid.* p. 9.
3. *Ad Loc.*
4. Edwin C. Marshall, "Social Indifference or Blatant Ignorance". *Journal of the American Optometric Association.* 43: 12 (November 1972), p. 1261-1266.
5. *Ibid.*
6. Charles Comer, O.D., in personal communication.
7. According to Dr. Edwin C. Marshall, the $35,000 covered recruitment activities for a period of 12 to 15 months. A consolidated recruitment program was to be mounted by the constituent societies of the AOA, the NOA, the American Optometric Student Association and the Association of Schools and Colleges of Optometry (ASCO). Cf. *Op. Cit.*
8. *Ibid.* p. 1265.
9. *Ibid.*
10. *Ibid.*
11. *Annual Survey of Optometric Education Institutions,* Washington, D.C.: Association of Schools and Colleges of Optometry. The surveys for each year from 1969 through 1978-79 were made available for this research by ASCO.
12. *Op. Cit.*
13. *Op. Cit.*
14. *Information for Applicants to Schools and Colleges of Optometry.* Washington, D.C.: Association of Schools and Colleges of Optometry, Fall, 1979.
15. Parenthetically, the same problem exists in determining the number of black students in first year classes. Black student enrollment is reported by total enrollment each year. Unless institutions provide this specific data set, obtaining such data remains problematic. Further, HEGIS data are reported biennially which precludes presentation of annual data from this source.
16. Cf. *Racial And Ethnic Enrollment Data From Institutions Of Higher Education.* Fall 1976, Washington, D.C.: U.S. H.E.W.: Office of Civil Rights, Table 9.
17. Charles Comer, O.D., in personal communication.

CHAPTER 7

Recruitment and Enrollment of
Black Students in Schools of Pharmacy

Pharmaceutical education in the United States is deeply rooted in the historic tradition of institutional segregation, neglect and controlled access of black students into the profession. That fact is a principal explanation for the current underrepresentation of black Americans in pharmacy. Today, it is estimated that the 4,275 black pharmacists in the United States comprise approximately three per cent of the total number of American pharmacists which is now set at 142,500.

A significant majority of the black pharmacists received their training at one of the four colleges or schools of pharmacy located at historically black institutions. These colleges are (1) Florida A & M University, (2) Howard University, (3) Texas Southern University, and (4) Xavier University of Louisiana. Although these institutions are historically black universities, founded when blacks were excluded by law from enrolling in traditionally white institutions, their schools and colleges of pharmacy in 1980 are considerably more desegregated than any of their white counterparts.

This chapter focuses primary attention on recruitment and enrollment of black students in schools and colleges of pharmacy during the 1970s. It describes enrollment trends in both graduate and undergraduate pharmaceutical training. Factors associated with enrollment fluctuations are delineated. A case history of the College of Pharmacy of Xavier University in New Orleans demonstrates the type of success that is attainable in the recruitment, enrollment, and graduation of black pharmacists when institutional commitment is positive.

Recruitment

Organized efforts to recruit more black students into the field of pharmacy were far from uniform during the seventies. Recruitment activities in the 72 schools and colleges of pharmacy in the United States may be characterized in one of three ways. The first type of institution may be described as "serious and committed" in the sense that its academic and administrative leaders facilitated the development of systematized plans for the recruitment of black students, financially supported them with staff and program funds, and encouraged their actual implementation. These institutions were more aggressive in the overall recruitment process. They did not abandon their commitment in the face of initial failures to achieve articulated goals. They were effective in communicating a sense of integrity and seriousness of purpose to those black students they wished to attract.

The second type of institution consisted of those who had "good programs on paper" but failed to operationalize them. They made public pronouncements, either in their literature or proposals to the various funding agencies, about their special concerns for blacks and other minority students underrepresented in pharmacy. However, they quickly abandoned the process when their initial results were not commensurate, in their view, with the time and energies expended. As one informant stated, many of these institutions were often interested in the "super-black" but the "super blacks" were not interested in them.

The third group consisted of those with "limited or no interest." They were committed to either the maintenance of the status quo or to calculated tokenism. They had no implementable plan for the recruitment of black and other minority students. Nor did they appear to be particularly concerned about changing the racial composition of their institutions. They displayed little concern about a moral responsibility for making pharmaceutical education more accessible to all Americans. It appears that the enrollment results reflect these varied degrees of institutional commitment during the seventies.

As recruitment is conditioned by institutional behavior and structural factors prevalent in the majority community and the larger society as a whole, so it is also a function of perceptions that individuals in the minority group have about the profession itself. Pharmacy to the professional person is a complex and important component of the allied health science fields. It is substantially more complicated than simplistic perceptions the public has of the profession. Its scope is much broader than what the public believes to be the major function of the pharmacists.

The rank and file person in the population is unaware of the six traditional and general fields of specialization: (1) pharmacy, (2) pharmaceutical chemistry, (3) pharmacology, (4) pharmacy administration, (5) pharmacognosy, and (6) hospital pharmacy. His/her perceptions are influenced by the *man* (because only in recent years have we observed an increasing number of female pharmacists) in the white jacket hidden behind a counter in the rear of the corner drug store or in the apothecaries of shopping centers. Even then, he/she may not possess a clear understanding of what the pharmacist is doing other than filling physicians' precriptions for drugs that are already compounded and safely stored in mysterious containers.

Community attitudes are reflected in an invidious occupational ranking system that places black pharmacists at a substantially lower level in social status, prestige, and influence scales than the status conferred on physicians and dentists. Their position in the social hierarchy has undoubtedly played a significant role in their ability to recruit students in general. Frequently, many students who applied for schools of pharmacy did so as a safety valve while they awaited the results of applications to other professional schools such as medicine and dentistry. Some went into pharmacy graduate programs, particularly, only after they were not accepted into medicine or dentistry. Other students took an undergraduate pharmacy degree and used it as a major path into dental or medical schools and never practiced pharmacy. Some in this group practiced both professions.

The ambiguous roles and attitudes about the profession of pharmacy are also prevalent in the black community. These perceptions are strongly shaped by the nature of interaction between community members and pharmacists, their visibility in the community, and leadership roles, and by their success in wearing the mantle of "doctor." Hence, community attitudes are far from uniform. Because there are so few pharmacists in the black community, black Americans may have an even more distorted view of the profession. With such a limited number of role models, recruiters often have to do an exceptionally fine "selling job" regarding pharmacy as a desirable profession for black Americans.

The contextual situation in which blacks see physicians and dentists accentuates their characterization as high status professionals and stimulates the desire among many black youths to emulate them as professionals. The recruiters must also be adept in convincing black students that pharmacy is not only a profession but offers avenues to other entrepreneural opportunities.

During the seventies, organized recruitment programs took a variety of forms. The American Association of Colleges of Pharmacy (AACP) encouraged its member institutions to become more aggressive in the recruitment of the underrepresented minorities in the profession. A series of policy statements issued by this association called attention to the escalating urgency of the problem of underrepresentation of minority groups in this field and to institutional as well as organizational responsibilities to rectify past inequities.[1]

In 1971 and 1972, at least seventeen pharmacy schools received special grants or funds through the Office of Health Manpower Opportunity to explicitly increase their recruitment activities for minorities into pharmacy. These institutions were: (1) the University of California School of Pharmacy, (2) Florida A & M University, (3) Mercer University School of Pharmacy (Georgia), (4) Purdue University School of Pharmacy and Pharmacal Sciences (Indiana), (5) University of Kentucky College of Pharmacy, (6) Xavier University of Louisiana School of Pharmacy, (7) University of Maryland School of Pharmacy, (8) Massachusetts College of Pharmacy, (9) University of Michigan College of Pharmacy, (10) University of Montana School of Pharmacy, (11) University of New Mexico College of Pharmacy, (12) State University of New York at Buffalo School of Pharmacy, (13) Ohio State University College of Pharmacy, (14) University of Oklahoma College of Pharmacy, (15) Temple University School of Pharmacy (Pennsylvania), (16) Medical University of South Carolina School of Pharmacy, and (17) Texas Southern University School of Pharmacy.[2]

Of these institutions, six were located in the "Adams States" of Florida, Georgia, Louisiana, Maryland, Oklahoma, and Pennsylvania. Four were located in historically black colleges; fourteen were located in States with substantial black populations.

In addition to these institutions, others also obtained federal funds and /or institutional financial support for pre-health careers, allied health professional training programs, and general science training. These funds aided the recruitment of blacks and other minorities into pharmacy.

The recruitment program at Temple University's school of pharmacy illustrates component activities taken by several of the institutions which received Health Professions Special Project Grants for Pharmacy Students. This program, funded initially at $200,000, ran from 1972-75. Its primary goal was the recruitment and enrollment of minority and low income students in pharmacy schools in order to alleviate shortages in underserved areas (e.g., minority communities and rural areas). The program also provided guidance and counseling services to high school students in urban and rural areas concerning careers in pharmacy. Minority counselors were trained to work with high school, community, or junior college and four year college advisors in disseminating information about the pharmacy program at Temple University. Numerous visits were made to high schools and colleges to inform individuals of opportunities in pharmacy programs.

During this period, some 258 teachers, administrators, and advisors were either visited or contacted through correspondence. Not only was specific contact by four counselors made with countless students but many students participated in the counseling program. Thirty-six black students and fifteen disadvantaged students from the Apppalachian region were admitted to the School of Pharmacy as a result of this organized recruitment program.[3]

In other "Adams States", recruitment activities for undergraduate pharmacy programs were reported for this study from Mercer University, a private institution in Georgia, the University of Georgia, the University of Florida, and the University of Maryland. Mercer University received a Special Project Grant for 1973-74. The school of pharmacy also operated, in connection with historically black Clark College in Atlanta, a special manpower program for Nigerian students.

Although the University did not provide a precise description of its recruitment program, one obtains some measure of the program's success by the following enrollment data. In 1972-73, Mercer enrolled fifteen black students (9 men and 6 women) in Pharmacy. In 1974-75, the enrollment consisted of a total of eleven black students (3 men and 8 women). The number remained at eleven black students in the Fall of 1975 but it then included five men and six women. By 1977, there were seven black males and three black females enrolled in the Mercer University School of Pharmacy. (These data refer to total enrollment figures. However, they suggest that first year enrollments were minimal following the grant period)[4] A total of twelve black students were enrolled in 1979-80.

Between 1974-78, the school of Pharmacy at the University of Georgia had a Minority Recruitment Grant funded by the Health Resources Administration. Previously, during the period of the grant, and into the present time, the recruitment program consisted of contact with the predominantly black schools in the State of Georgia. Under the grant, a "Minority Recruiter" was hired "to generate applications" to the school of pharmacy. However, no special academic support program was offered to black students. One black instructor taught in the school of pharmacy for a short period of time until she moved out-of-state. As of October, 1979, no other black faculty has either applied or been hired in the School of Pharmacy.[5]

Between 1968-69 and 1978-79, the University of Georgia School of Pharmacy enrolled 1,694 white students and 39 black students. Of those thirty-nine black students, twenty five were enrolled during and immediately following termination of the grant. About five black students per year were enrolled during this eleven year period. (In 1979-80, the total number of black students was twelve). When the program was in full operation, two recruiters were utilized. Financial aid was also provided from state funds.

Although the State of Georgia is under litigation to become more desegregated, and even though it is increasing minority presence in pharmacy, the critical variable in the recruitment of black students in general appears to have been both the use of a black recruiter and the presence of a black faculty member, even for a relatively short period of time.[6]

The University of Florida operated both a special admissions and a regular admissions/recruitment program in the period between 1968 and 1978-79. Recruitment activities centered on both undergraduate and graduate degree programs in pharmacy. However, based upon the data provided for this study, it is not possible to describe fully the precise characteristics of the recruitment program at either the undergraduate or graduate level of pharmacy education. The University of Florida operates a "2-3" program which involves two years of pre-pharmacy and three years of pharmacy training. In general, the University enrolled all black students who were "accepted". During the period between 1968 and 1973, when no form of special admissions operated and prior to the full impact of the enrollment mandates consequent to the *Adams v. Richardson* litigation, all black students who applied were accepted and enrolled. This *number varied* from three to a maximum of five in any given year.

The proportion of white applicants who were ultimately admitted and enrolled was about 95 per cent during the same period. The number of white students enrolled rose from 72 in 1968-69 to an all-time high of 175 in 1972-73. Since 1973, about one-third of all whites who applied were accepted and approximately 95 per cent of the white acceptees were enrolled. In the same period, when some form of special admissions programs operated for black students, the number of black applicants rose from three applicants in 1968 to a high of eight black applicants in 1977 and in 1978. However, the number of acceptances and enrollment for black students has remained fundamentally unchanged. An average of four black students are enrolled in the B.S. in pharmacy program each year. (The 1979-80 total black student enrollment was six). No black faculty were employed in the school of Pharmacy during this period.[7]

The University of Maryland operated a recruitment program under a "Recruitment, Retention and Replacement Grant" for a three-year period from 1972 to 1975. This effort was supported by approximately $80,000 and utilized two full-time staff members for all recruitment functions. The school of Pharmacy also received another $30,000 to support recruitment and retention on a two-year grant from 1975-77. One full-time person was employed under this grant. In 1978, the University received $125,000 for its recruitment and retention program in medicine, dentistry, and pharmacy. Six full-time staff persons were hired. At least one full-time black faculty member, who teaches pharmacology, was employed throughout the study period. The University of Maryland, like a significant proportion of institutions with colleges or schools of pharmacy, responded to federal capitation programs by expanding overall capacities in the health sciences in order to obtain increased federal government financial support.[8]

Hence, in any overall discussion of enrollment, the actual increases in the total enrollment of black students in pharmacy must be viewed within the context of capitation grants designed to stimulate overall enrollment of a subtantially larger number of student in the health sciences. In the specific case of the University of Maryland, during the recruitment and retention grant periods of 1972-78, the total number of applications increased from 169 in 1972 to a high of 303 applications to the school of pharmacy in 1974-75, and declined the next two years to an average of 278. It then rose again to 300 applications in 77-78 and dropped to 266 in 1978-79. In 1977-78, ten percent of the applicants were black but in 78-79 only 18 of the 266 applicants or about seven per cent were black.

Total enrollment, in the meantime, rose from 77 students in 1972 to 92 students in 1978-79. First year black student enrollment increased from four students in both 1972 and 1973 to twelve in 76-77 (when total, enrollment was 94 students) and reached a high of thirteen blacks in the first year class of 1977-78. However, the 1978-79 first year black student enrollment declined to eleven.* An important dimension of the Maryland program is that, like the program at Temple University, it combines continuing financial support, good staffing and the presence of at least one or more full-time black faculty which effectively foster greater success in recruitment of black students.[9]

The University of Wisconsin School of Pharmacy is one example of a recruitment program in a "non-Adams State." This school also initiated its special recruitment activities in 1972 for the purpose of attracting more minority students into the field of pharmacy. The program was based upon five principal goals. These were to increase the: (1) actual number of minority students enrolled in its two-year pre-pharmacy program; (2) number of minority students who completed requirements for admission to the school of pharmacy; (3) number of "qualified" applicants for the school of pharmacy; (4) number of minority students who successfully complete professional degree requirements; and (5) number of minority students who actually pursue pharmacy as a career and enter, therefore, into the labor force.[10]

Wisconsin employed one director-counselor to administer, coor-

*Total black student enrollment was 31 in 1979-80.

164 MAINSTREAMING OUTSIDERS

dinate, and implement its program. An allocation of approximately $27,000 was made for financial support of the program in each year. To augment recruitment efforts, the staff director was assisted by campus-wide recruiters from the Office of Undergraduate Orientation of the University. The program involved the development and distribution of special bulletins that described opportunities for minority students in pharmacy. An eight-week Summer orientation program was initiated. Pre-pharmacy advisors visited Wisconsin high schools with a high proportion of minority students. These advisors also participated in a variety of projects developed by the campus-wide recruiters. Financial support provided by the School of Pharmacy enabled some of the minority pharmacy students to participate in recruitment activities. The School granted some of the funds required for the establishment of the Office of Minority Programs in the Center for Health Sciences. A coordinated program was not implemented until July 1, 1976.[11]

As a result of these activities, the School of Pharmacy of the University of Wisconsin reports that pre-pharmacy enrollment increased by 100 per cent. The number of minority students who successfully completed admission requirements increased by 50 per cent. The number of minority applicants rose by 25 per cent. However, the original base was exceptionally low. This modest success is attributed to recruiting efforts at college fairs, the publicity generated from the MAPP brochures, and contact developed through the Mid-Western Pharmacy Consortium. However, the number of black graduates in pharmacy continues to be noticeably small: one in 1978 and possibly four in 1979.[12]

Nevertheless, it should be stated that the consolidated minority program at Wisconsin includes tutorial assistance, counseling and guidance, financial aid, and work-study opportunities. A possible measure of the program's success is that in 1970-71, prior to its recruitment activities, there was one black student enrolled in Pharmacy. In 1979-80, there were 15. However, there are no black faculty in the School of Pharmacy and it is this absence that may help to account for the limited gains made in the recruitment of black students.[13]

Admissions Requirements

One rationalization offered for the underrepresentation of blacks in health fields, such as pharmacy, is their inability to meet rigorous admission standards. A major impediment often cited is the lack of preparation in basic science courses such as mathematics, chemistry, and biology as well as low proficiency in verbal communication. The traditional baccalaureate degree in pharmacy involves what is called a "2-3" program in which two years are in a pre-pharmacy curriculum and the remaining three years are devoted to a pharmacy curriculum. In more recent years, several institutions have offered a "2-4 program" which leads to a Doctor of Pharmacy or "Pharm. D." degree. In addition, several institutions offer a M.S. degree in Pharmacy and fewer offer a Ph.D. degree.[14]

Admission to the B.S. in pharmacy program is generally based upon such criteria as grade point averages in pre-pharmacy college work. The GPA may be as low as a 2.5 or straight C in some institutions or to a C + or B − average in others. Since 1975, a substantial number of institutions have relied upon scores on the Pharmacy College Admission Test (PCAT) in assessing a candidate's application. Interviews are also employed as a means of determining motivation, interest, and commitment or special attributes possessed by the candidate.[15] Reading Comprehension and Chemistry scores on the PCAT may become central determinants of admission in some colleges of pharmacy. However, the use of this test is so new in several colleges that it is premature to make an assessment of the success of the test in differentiating the quality of applicants compared with the previously employed factors.

Where special admission programs have operated, they tended to involve a reduction in the expected grade point average by a few percentage points for otherwise desirable black students. In this study, there was no evidence that black and other minority students were accepted into the B.S. pharmacy program with less than a C average. As in the case of majority group students, a C average was offset by special qualities, such as motivation, perserverance, and interest in the profession that the student displayed during the assessment process.

The most commonly employed cognitive admissions requirement are college grade point average and, in twenty-nine institutions, scores on the Graduate Record Examination (GRE). Among the 56 institutions with graduate programs, in 1970-71, as in subsequent years, grade point averages expected for admission to graduate programs varied markedly from institution to institution. Thirty-seven institutions were on a four point (4.0) system. Twenty-one of these institutions required a 3.0 or B average but among the remaining colleges of pharmacy, the grade point average required ranged from a low of 2.5 to a high of 2.8 on a 4.0 system. Among those institutions on a 3.0 system or a 5.0 system or a 6.0 system, the GPAs expected were consistent with those institutions on a 4.0 system. By 1972-73, the minimum GPA expected had dropped to 2.25 on a 4.0

system but a GPA of 3.0 or above was still required of 35 of the 56 institutions with graduate programs. However, the GPA minimum ws raised to 2.5 on a 4.0 system in the mid-seventies and has continued to range from 2.5 to 3.0 since that time.[16]

In those institutions requiring the GRE, an average score of 1,000 or in the upper 50 percentile was usually expected until about 1975. Since about 1975, cut-off points on the GRE varied from a low of 530 to a high of 1250 in the 25 institutions requiring the GRE. The upper limits include a composite verbal and quantitative score. These requirements are most consistently expected of students who wish to pursue the doctoral degree rather than of students interested primarily in the M.S. in Pharmacy degree. About one-half of all graduate students in pharmacy enter with a previous degree in pharmacy. However, the proportion of those entering with a degree in some other field of study increased steadily throughout the seventies but seems to be leveling off around the fifty per cent mark in 1980.[17]

It is not possible to make distinctions between the qualifications brought by black students from those brought by white students in the application, assessment, and admissions processes. These is certainly no rational reason to assume that the races differed appreciably in qualifications. Wherever special admissions programs existed, they seemed to focus primarily on providing financial aid and special services to minority students rather than a reduction in academic requirements for graduate study.

Special recruitment efforts, availability of financial aid and scholarship monies, and the presence of black faculty continue to be the most powerful determinants of actual enrollment of black students.

Enrollment: B. S. in Pharmacy Programs

In 1971, the American Association of Colleges of Pharmacy (AACP) directed its Executive Committee to collect information on the racial composition of the student bodies in constituent colleges. Only one of the 54 colleges of pharmacy failed to report enrollment data by race. In 1971-72, the 618 black pharmacy students represented 3.75 per cent of all students enrolled in B. S. in pharmacy programs. However, 353 of these students were enrolled in the four historically black colleges of pharmacy. When these students are deducted from the 618, the 265 black students enrolled in the remaining 50 institutions represented a mere 1.0 per cent of enrollment in the traditionally white institutions. In other words, the average number of black students enrolled in pharmacy programs was 5.3 per institution at the white colleges. However, twelve of these institutions had no black students.

Among black colleges, Texas Southern University enrolled 88 men and 50 women (T = 138); Howard University enrolled 63 men and 33 women (T = 96); Florida A & M University enrolled 41 men and 18 women (T = 59); and Xavier University of Louisiana enrolled 31 men and 29 women or a total of 60 pharmacy students. Among historically white institutions, nine enrolled ten or more black students. These were: the University of California (11); Mercer University (14); University of Illinois (13); University of Maryland (11); Wayne State University (10); Brooklyn College of Pharmacy (26); University of North Carolina (11); Temple University (11); and Columbia University (15).[18]

The recruitment drive began to show enrollment increases in 1972-73 and in succeeding years. However, because of the overall effort by colleges of pharmacy to enroll significantly larger numbers of students, in general, as related to expansion under capitation grants, black students enrollment failed to keep pace with overall enrollment.

In fact, the primary inducement for the spiralling increases in colleges of pharmacy witnessed during the seventies appeared to be capitation grants. The quota requirements of the Health Manpower Act mandated participating colleges to increase enrollment in whatever class or year they defined as a "first class or year of study" by ten per cent. This definition referred to the third year of pharmacy in some institutions while in others it was defined as the fourth year of study. Some colleges of pharmacy ultimately exceeded their capitation quotas while many never did.[19]

In any event, in absolute numbers, black student enrollment increased annually throughout the decade. However, after 1972-73 when the proportion of black students in Colleges of Pharmacy stood at 6.6 per cent, the proportion declined dramatically. Thereafter, the proportion dropped by more than 50 per cent to 3.2 per cent of total enrollment in 1974-75 when 727 black students were enrolled. It rose slightly the following year to 3.8 per cent which represented an increase of 188 black students when 915 enrolled.

The upward climb continued in 1977-78 when the 984 black pharmacy students represented 4.2 per cent of total enrollment. It was unchanged in 1979 when 958 black students matriculated in pharmacy school. *Although significant increases in black student enrollment occurred in the historically white institutions, more than one-half (54.3 per cent) of all black pharmacy students are enrolled in the four historically black colleges.*[20]

The number of traditionally white institutions with a black student enrollment of ten or more reached 15 in 1979 but the proportion of blacks in 68 of these institutions is only about 2.3 per cent of total enrollment. The largest number of black students in the traditionally white colleges are enrolled at the University of Michigan (36 students), Arnold and Marie Schwartz (30), University of Houston (26), the University of Maryland (31), the University of California (26), and Purdue University (21).

Among the traditionally white institutions located in the "Adams States," the 1979-80 distribution of black students reported by AACP is as follows:

Figure 6. Distribution of Black Pharmacy Students in the "Adams States"

Institution	State	Total Number Of Black Students
Univ. of Arkansas Medical Center	Arkansas	4
Univ. of Florida	Florida	6
Mercer University	Georgia	12
University of Georgia	Georgia	12
Northeast Louisiana U.	Louisiana	15
Univ. of Maryland	Maryland	31
Univ. of Mississippi	Mississippi	13
Univ. of No. Carolina	North Carolina	15
Southeastern Okla. State U.	Oklahoma	4
Univ. Oklahoma	Oklahoma	4
Duquesne	Pennsylvania	2
Philadelphia College of P & S	Pennsylvania	5
Temple University	Pennsylvania	14
Univ. of Pittsburgh	Pennsylvania	3
Virginia Commonwealth U.	Virginia	9

Excluding black colleges, the University of Maryland ranks second in the nation in the total number of black students currently enrolled in B. S. in pharmacy programs. In most states, litigation has had a modest impact on the overall enrollment of black students in pharmacy. Clearly, a more effectively organized effort is warranted if these and other historically white institutions desire to succeed in recruiting and enrolling black students in pharmacy as they have been in the enrollment of white students. All of the institutions who responded to the inventory for this study indicated a goal of increasing the number of black students to a level commensurate with the proportion of blacks in their state populations. However, it is evident that a strong and aggressive recruitment program supported by black faculty in the college of pharmacy, adequate financial aid packages, institutional commitment, and a positive learning environment are all imperative and will continue to be so in the immediate future. The fact remains also that as recently as 1978, forty-eight of the 72 schools of pharmacy in the United States had fewer than ten black students and fourteen of them enrolled not a single black student.

The number of black faculty averages less than one per institution. Even that number is deceptive since most are teaching at the historically black colleges. According to studies conducted by Lars Solander, Director of the Office of Educational Research and Development of the AACP, full-time black faculty continued to be underrepresented on the faculties of the colleges of pharmacy. This shortage of black faculty remains despite increases in the number of black faculty noted in each of the Association's biennial surveys it has conducted since 1974. *Total black faculty in 1978 was 63 or three per cent of the total of 1,914 in the 72 schools of pharmacy in the United States.* In 1974, there were 41 black faculty out of a total of 1,558. In 1976, there was an increase of one black faculty to 42 out of a total full-time faculty of 1,621. (Forty-one of the 63 black faculty in 1978 are males and twenty-two are females).[22]

If an earlier distribution of black faculty holds true in 1980, it is apparent that certain areas of specialization are slightly more popular among black pharmacy faculty than are others. This distribution is depicted in Figure 7.

Enrollment: Pharm. D. and Graduate Programs

The Pharm. D. degree requires six years of training, including two years of pre-pharmacy and four years of course work in a school of pharmacy. It is not unusual for Pharm. D. degree student to have already received B.S. degree in Pharmacy and to continue for the fourth year of pharmacy training in order to be awarded the Doctor of Pharmacy degree. In 1969-70, sixteen schools of pharmacy offered the Pharm. D. degree. Of that number, five colleges offered it as the first professional degree. During that year, 871 students were studying for this degree as their first professional pharmacy degree and 154 Pharm. D. candidates already held the B.S. degree. Of that number 25 were in M.S. degree programs and 16 were seeking the Ph. D. Twenty-one of these students sought degrees in pharmaceutical chemistry.

Figure 7.   Distribution of Black Faculty by
Specialization in Pharmacy

Field	Number		Percent Of total in field	
	Men	Women	Men	Women
Pharmacy	11	1	2.5	.2
Clinical Pharmacy	4	4	1.5	1.5
Pharmaceutical Chemistry	7	0	2.5	.0
Pharmacognosy	1	1	.7	.7
Pharmocology	4	0	1.7	.0
Pharmacy Administration	3	2	2.9	1.9
Hospital Pharmacy	1	0	2.5	.0
Totals	31	8		

Source: March 18, 1974 Memorandum from Charles Bliven, Executive Secretary of AACP to Deans of Member Colleges

The least selected fields of specialization were pharmacy administration and pharmacognosy, each with two students.[24]

By 1973-74, the number of students pursuing a M.S. degree in pharmacy had increased to 1,035 and there were 957 enrollees in Ph. D. programs. The respective numbers of black students were 20 M.S. students and twelve Ph. D. students, or a net loss of thirteen black graduate students in one year. The M.S. black students were enrolled at the following institutions: University of California (1), the University of Illinois (1), Butler University (2), Purdue University (1), the University of Maryland (2), Wayne State University (1), the University of Mississippi (1), the University of Nebraska (1), Brooklyn College of Pharmacy (4), the University of Pittsburgh (2), the University of South Carolina (1) and the University of Houston (1).[25]

The Ph. D. students were enrolled at Arizona, Georgia, Illinois, Maryland, Missouri/KC, St. John's, Pittsburgh, Rhode Island, and Wisconsin. In toto, black students represented 1.9 per cent of the M.S. students and 1.2 per cent of the Ph. D. students. Among the "Adams States", only in Georgia, Maryland, Mississippi and Pennsylvania were black students enrolled in a graduate pharmacy program.

An analysis of minority student enrollment data for 1975 shows a pattern that was prevalent in most institutions both before and after that year. This analysis indicated that non-Americans account for 90.7 per cent of all students listed as minorities; black Americans represented 4.4 per cent; Spanish-surnamed equaled 2.4 per cent, and Asian Americans constituted 1.0 per cent of all persons listed as minority students enrolled in graduate programs. Even when these data are disaggregated by type of degree, non-Americans still constitute almost 89 per cent of doctoral students in the minority student population. However, the proportion of Asian Americans rose dramatically to 5.7 per cent of minority enrollment and Spanish-surnamed declined to 1.2 per cent.

Nine black students were enrolled in Masters degree programs. Of that number, one each was enrolled at the University of Florida and the University of Mississippi as the only two "Adams States" with black students in M.S. programs. Eight (8) black students were enrolled in doctoral programs. These students were distributed at the University of Southern California (1), the University of Florida (1), the University of Illinois (1), Purdue University (1), the University of Nebraska (1), the University of Pittsburgh (2), and the University of Wisconsin-Madison (1). Again, no institution enrolled a significant number of black graduate students.[26]

The number of black students in M.S. programs fell by one student to a total of eight and the number of doctoral students who were black rose to 12 in the Fall of 1976. The universities of Connecticut, Kansas, Michigan, SUNY/Buffalo, Ohio State, and Texas/Austin enrolled black doctoral students. These were in addition to Florida, Illinois, Purdue, Nebraska, and Pittsburgh. Florida A & M, Texas/Austin and Ohio State were added to the list of institutions that enrolled black students in M.S. programs.[27]

According to the latest available institutional data (Fall, 1978), seven black students are enrolled in doctoral programs and 20 are enrolled in M.S. degree programs in pharmacy. The doctoral students are scattered in seven different colleges of pharmacy; namely, Purdue, Kentucky, Michigan, Wayne State University, the University of Mississippi, the University of Kansas, and the University of Pittsburgh. Twelve of the twenty black students enrolled in M.S. programs are at historically black Florida A & M University. The remaining eight are distributed in eight different traditionally white institutions. Forty of the institutions that offer M.S. programs do not enroll a single black M.S. student. Thirty-three of the forty colleges of pharmacy that offer the Ph. D. degree have no black doctoral students.[28]

These data underscore in rather explicit terms the major problem of underrepresentation and token enrollment of black students in graduate degree programs in pharmacy. Once again, it is not possible to produce trained black pharmacists who can meet manpower needs in industry, educational institutions as university professors and administrators, and hospitals, or for the corner drug store unless they are successfully recruited and enrolled.

As previously argued, recruitment necessitates guarantees that adequate financial assistance will be available to black students, especially at the graduate level. By far, the most common method of financial support for graduate students is through the acquisition of teaching assistantships. This is followed in order by non-service stipends or scholarships, research assistantships, hospital interns, and instructorships. Scholarship money has been awarded to about a thousand graduate students over the past several years by the American Foundation for Pharmaceutical Education. These awards total about $175,000 per year and represent the largest single source of fellowships for pharmacy graduate students.[29] In addition, a range of loan programs of the type described in the chapter on dental education is available to students.

In general, the fellowships and scholarships, teaching, and research assistantships do not go to black students. Consequently, they must support their graduate education through repayable loans. But, successful enrollment is correlated with a good quality of financial assistance available to black students. Without it, irrespective of how sound the recruitment program appears to be on paper, institutions will continue to fail in efforts to enroll black students.

The pattern of enrollment of black students in graduate schools of pharmacy is particularly disturbing in 1980. It does not suggest that graduate schools in pharmacy actually made a firm commitment in the sixties and seventies to active and aggressive recruitment of black students into this profession. It is as if they expected the students to arrive at a decision by some mysterious osmotic process leading inevitably to their selection of pharmacy as a field of graduate study and in that particular graduate school. Therefore, no inducements were necessary. If knowledge about graduate opportunities in pharmacy were limited and if these institutions had no track record of having graduated a sufficient number of black students in the past, that type of institutional position is clearly untenable. Further, some administrators argue that recruitment and enrollment of black students into graduate pharmacy programs are hampered by salaries and wages that blacks holding an undergraduate degree in pharmacy can earn. Therefore, institutions cannot compete with the lucrative financial incentives offered by the job market. While this argument may have validity if the focus remains solely on the recruitment of students for purely academic positions, it does not have credibility when one begins to focus on the requirements for certain positions in industrial pharmacy, drug research in the pharmaceutical industry, hospital administration, and other occupational areas.

Inability to move beyond tokenism in the recruitment and enrollment of black students is an indictment of negative institutional behavior characterized by either the inability or unwillingness to translate what is tantamount to a "paper-constructed" program into concrete programmatic action. A serious commitment involves a clear articulation of goals and plans from the leadership structure, faculty participation and an understanding from those unsympathetic faculty that they will not engage in behavior which will ultimately undermine the program goals.

It necessitates an ability to *persuade* the ambivalent student that the opportunities possible in this field are significant, if indeed, they are. And, they are for some specializations in pharmacy. It may involve "selling the institution and its program", especially when little is known about the institution's commitment to the education of black and other minority students or in those instances in which previous experiences of black students were negative. It necessitates a sufficient financial outlay to support all aspects of the program. Most certainly, it means the enrollment of a critical mass of black students -- a missing element in all graduate schools of pharmacy with the exception of the one historically black college that offers the M.S. degree. Hence, the important question of the 1980s is: How committed are graduate schools of pharmacy to the goal of enrolling a critical mass of black students and to sustaining that enrollment so that racial parity will be commonplace?

Xavier University College of Pharmacy
A Case Study of Success

As previously noted, the four historically black colleges of pharmacy enroll over fifty per cent of all black pharmacy students in the United States. They graduate substantially more than one-half of all blacks receiving pharmacy degrees each year. One of the pioneers in pharmacy education for black Americans is Xavier University of Louisiana, located in the city of New Orleans.

Xavier University is a small, black, Catholic university, established by Mother Katherine Drexel as a high school in 1915. Mother Drexel also founded the Sisters of Blessed Sacrament whose mission has traditionally been service to American minorities. By 1917, Xavier had attained a normal school status and was committed to the training of black teachers at a time when State laws precluded them from receiving that training at the traditionally white public institutions. Xavier became a four-year liberal arts college in 1925. It established its college of pharmacy in 1927, and a Graduate School in 1933.[30]

The founding mission of this university was to offer higher educational opportunities to black Americans. Although white students were prohibited from attending the University by State law, the University's faculty and administrators have always been bi-racial. Consequently, its students, from its inception, were exposed to significant role models in an integrated academic environment and by its graduates who frequently returned to the institution for special occasions. The integrated learning environment often was an anomaly to the dual systems of education to which these students were exposed in public high schools. Xavier University, however, operated a "Preparatory" school which served as a feeder to the university. Both institutions were nationally recognized for high academic standards, a strong commitment to teaching and the training of students within the academic community, emphasis on self-discipline in the development of study habits, and the success rates of their graduates.

Prior to the *Brown* decision by the U.S. Supreme Court, the enrollment of Xavier University stood at 600 black students. Under the dual system of education, Xavier University was immensely successful in the training of black Americans for various occupational roles in a segregated society. It has continued that success rate. For example, a 1965 survey revealed that in the city of New Orleans, forty per cent of the teachers in the black segment of the public schools system were Xavier graduates. Approximately 75 per cent of the black principals and 100 per cent of the guidance counselors in these schools held at least one degree from Xavier University.

Following the *Brown* decision, Xavier University changed the racial composition of its student body. With the aid of federal funds, the University was able to recruit from an expanded pool of students. Between 1964 and 1979, the enrollment of Xavier University increased from 800 to 1,800 students. Most of them were full-time undergraduates.

The college of pharmacy was founded in 1927. From the beginning, its pharmacy curriculum was impressive and was bolstered by an equally outstanding pre-medical program and an excellent curriculum in the basic sciences of chemistry, mathematics, biology and physics. One measure of the strength of these combined programs is that over the years, more than 150 of Xavier's graduates have obtained degrees in medicine. It has an acceptance rate to medical and dental schools of 87 per cent of its graduates who apply. More than one-half of the black physicians now in private practice (22 of 38) and 39 per cent of the black pharmacists in the city of New

Orleans alone are Xavier graduates.

The College of Pharmacy is the only private, predominantly black college of pharmacy among the 72 accredited pharmacy institutions in the United States. From its inception, the mission of the college of pharmacy conformed to the mission of the University: "to increase the level, quality and scope of educational opportunities in pharmacy" for black and other minority students. Recruitment into pharmacy focused largely upon: (1) counseling and guidance programs with students in Xavier Preparatory High School, (2) dissemination of catalogs and bulletins to predominantly black high schools in New Orleans and elsewhere, and (3) the use of alumni in cities or States outside New Orleans and Louisiana, in high school career days, college fairs, and in special events.

During the sixties and into the seventies, recruitment programs expanded because of the input of federal funds and the attainment of private foundation financial support to increase and strengthen the University's offerings in health sciences. The University, under the new leadership of an articulate and imaginative lay President, sought much needed resources that would enable the University to prepare students in areas in which job opportunities existed. Pharmacy, like computer sciences and Business Administration, was one of the applied sciences targeted for growth and expansion. The funds obtained by Xavier University guaranteed the construction of a new College of Pharmacy building in 1970.

Students were recruited into a 2-3 curriculum which consisted of two years of pre-pharmacy training in the basic sciences and three years of training in the pharmacy curriculum. In 1970, the college of pharmacy shifted its emphasis to have clinical pharmacy as its central focus. This change was made in order to provide those students who wanted to immediatley enter the job market as well as those who planned to pursue a graduate degree in this area opportunities for work experience in a variety of settings. Hence, this curriculum now enables students to receive "on-the-job" training in local hospitals, community pharmacies, Drug Information Centers, and a variety of health clinics as a central component of their undergraduate program.

Despite high operating costs, the University has traditionally committed a substantial proportion of its resources to the college of pharmacy in order to sustain it as a much needed facility. This investment, however, reflects positive institutional behavior. Similarly, the emphasis placed on pharmacy as an important component of education in the health sciences and the encouragement given by counselors, and admissions and guidance officers, as well as by faculty in the recruitment process and during freshmen orientation activities to seriously consider pharmacy as one of its professional alternatives — all — indicate a quality of institutional behavior that is supportive, desirable , and effective.

Xavier's College of Pharmacy makes exceptionally fine use of its alumni chapters dispersed throughout the United States in the recruitment process). These chapters conduct programs on a career day at various high schools, invite students into the workplaces of their graduates in pharmacy, medicine, and dentistry in order to familiarize them with opportunities available in these fields, and distribute promotional literature to prospective students. Their active participation in the recruitment program has enabled the college of pharmacy to enroll students of all races and other minority groups from twenty states and several foreign nations. Hence, Special Grants, such as the monies received from the Office of Health Manpower Opportunities in 1972-75, were utilized to recruit students from an ever expanding potential pool.

Consequently, total enrollment in the college of pharmacy expanded from sixty-two students in 1968-69 to 225 students in 1978-79. (see Table 28).

Table 28. Enrollment by Race at the Xavier University College of Pharmacy 1968—1978

Year	Total Enrollment	Black Student Enrollment	White Student Enrollment
1968-69	62	62	0
1969-70	77	77	0
1970-71	88	88	0
1971-72	94	84	0
1972-73	116	95	11
1973-74	131	101	20
1974-75	152	106	46
1975-76	160	93	57
1976-77	213	104	72
1977-78	223	107	74
1978-79	225	129	66

Source: Warren McKenna, Dean, College of Pharmacy, Xavier University

The racial composition of the student body in pharmacy changed significantly during the seventies. As a result, the student body is multiethnic and multi-racial. Although white Americans comprise about 90 per cent of students listed in the category of white enrollment, that group also includes a few Asian Americans and Hispanic students. Almost thirty per cent of the enrollment in the school of pharmacy is non-black. The proportion had been as high as 45 per cent in 1976. In comparison to the historically white colleges of pharmacy, Xavier enrolls more black students than the four white institutions with the largest number of black students. The recruitment of a substantial number of white students is facilitated because Xavier has the only school of pharmacy in southern Louisiana. The only other pharmacy school in the state is 300 miles north.

Retention is facilitated by the implementation of a well-coordinated program that involves faculty, adequate financial assistance, academic and personal advising, counseling, tutorials, and a positive learning environment. Faculty members must extend themselves in the learning and teaching process and stimulate students to become active learners instead of passive recipients of information by professors who do not demonstrate the same enthusiasm for teaching as they do for laboratory research. Retention means having time to talk to students, to listen to their problems which evolve as they encounter the subject matter. Simultaneously, it demands that the rationale for exacting rigorous standards be communicated effectively, placed within the context of learning and developing a command of the tools of the subject matter instead of learning without meaning. This is precisely what the college of pharmacy, as other units within that university, has managed to accomplish over the years.

Financial assistance is an absolute imperative for the majority of students who attend the University. According to university officials, at least fifty per cent of Xavier's students cannot afford to pursue a college education without continuous financial aid. They come from families whose median family income is less than $7,500/year or at the poverty level in 1980 dollars. The gravity of their financial problems becomes more apparent with the realization that the average tuition at Xavier's College of Pharmacy, while substantially lower than that of major universities, is $1,800/year. However, total costs, including room and board, exceed $3,250 which is almost one-half of the annual incomes of fifty per cent of the families from which it draws its student body.

Xavier University receives about 2.5 million dollars per year in student aid which is distributed to more than 1100 students. The University's pharmacy college also benefits from special scholarships donated from private citizens. But, the college of pharmacy is purely an undergraduate operation. Consequently, it is not in a position to provide teaching assistantships, research assistantships, and graduate fellowships and scholarships that could be granted to black students at the traditionally white graduate schools. Unfortunately, black students are all too frequently unable to obtain teaching fellowships and research assistantships in these institutions.

In 1978, the Public Health Service of the U.S. Department of Health, Education and Welfare funded a "Comprehensive Pre-Health/Pharmacy Admission-Retention Program" at Xavier University. The twofold purposes of this program are (1) to facilitate access of minority and disadvantages white students into professional schools and (2) to reduce attrition from professional schools. The program is conducted by a staff of four key professionals who engage in a broad range of activities including recruitment through high school visitations, information dissemination, counseling, tutorials, and special workshops and who stimulate wider faculty involvement in special programs.

Although the program was initiated in 1978, visible payoffs were immediately observed the following year. For example, in 1979, fifty four black Pre-Health students who participated in the program, in contrast to forty-eight in 1978 were accepted to MODVOPP schools.[13] Minority student enrollment in the College of Pharmacy increased by 19 students over the 1978 figure. Attrition from the College of Pharmacy decreased by 38 per cent in 1979. If this type of success is sustained, even more significant changes will occur in stimulating access into the health professions and in reducing attrition in both pre-health programs as well as in the professional schools themselves.

The College of Pharmacy has graduated more than 1,000 black pharmacists since its establishment in 1927. In the period between 1969 and 1979, 226 black students and 147 white students were awarded the B.S. degree in Pharmacy from the University. During the same period, 53 "other-race students" received pharmacy degrees also. Although a major share of the historically white institutions are larger, have superior financial resources, and more faculty than does Xavier, none can match Xavier in the production of black pharmacists.

This case history exemplifies the kind and quality of success that an institution can achieve in the recruitment, enrollment, and graduation of pharmacy students in a pluralistic environment if the commitment and necessary resources are present. What happens at Xavier in the training of minority students in pharmacy, is, of course, repeated at Howard, Florida A & M, and Texas Southern Universities in the production of blacks in pharmacy. Full access of outsiders to pharmacy education will only be achieved when the traditionally white institutions operate sound programs for expanding equal educational opportunity.

Footnotes

1. During the 1960s, the American Association of Colleges of Pharmacy urged its member institutions to accelerate access of minority students in schools of Pharmacy.
2. Shirley M Malcom, *et. al., Programs in science for Minority Students: 1960-75.* Washington, D.C.: American Association for the Advancement of Science, 1976, p. 21-22.
3. *Ibid*
4. Cf. Jack E. Orr, "Report on Enrollment in Schools and Colleges of Pharmacy, First Semester, Term, or Quarter, 1972-73", *American Journal of Pharmaceutical Education,* Vol. 36, 1972, p. 138 ff; and the same reports for 1973-74, and 1974-75.
5. Abstracted from the Inventory submitted by the University of Georgia.
6. *Ibid.*
7. Abstracted from the Inventory submitted by the University of Florida.
8. Abstracted from the Inventory submitted by the University of Maryland.
9. *Ibid*
10. Abstracted from the Inventory submitted by the University of Wisconsin.
11. *Ibid.*
12. *Ibid.*
13. *Ibid.*
14. Orr, *Op. Cit.,* 1971-72.
15. *Ibid.*
16. *Ibid.,* 1972-73.
17. John Schlegel, *Graduate Enrollment Report, Fall, 1976.* Bethesda, Maryland: American Association of Colleges of Pharmacy, (no Date), p. 3.
18. Orr, *Op. Cit.,* p. 1201.
19. *Ibid.*
20. John Schlegel, *et. al., Report of Fall Undergraduate Enrollment in Schools and Colleges of Pharmacy.* Bethesda, Maryland, American Association of Colleges of Pharmacy, August 1975, p. 3, and Table IX.
21. American Association of Colleges of Pharmacy, *Enrollment Report on Professional Degree Programs in Pharmacy, Fall, 1977.* Bethesda, Maryland: American Association of Colleges of Pharmacy, (no Date), Table 7.
22. Lars Solander, "The Sex and Ethnic Composition of Faculty at Institutions of Pharmacy Education", (A Paper prepared in 1978 for the American Association of Colleges of Pharmacy). Unpublished.
23. *Op. Cit.,* 1971-72.
24. Martin I. Blake, and Jack Orr reports on "Graduate Enrollment Data" for 1969-70 and 1971-72, 1972-73 respectively. Both are published in the *American Journal of Pharmaceutical Education.* The respective volumes are 35 (February 1971), 108-127; 138-153 and 37 (February 1973).
25. Orr, Graduate Enrollment Data, 1973-74. *American Journal of Pharmaceutical Education.* Vol. 38.
26. John Schlegel and Christopher A. Rodowskas, *Enrollment Report of Professional Degree Programs in Pharmacy, Fall, 1975* Bethesda, Maryland: American Association of Colleges of Pharmacy, Vo. 40, (August, 1976), p. 1-5 and Tables IV and VII.
27. *Ibid.*
28. American Association of Colleges of Pharmacy, *Graduate Enrollment Report, Fall, 1978.*

Bethesda, Maryland: American Association of Colleges of Pharmacy, 1979, p. 1-3 and Table IV.

29. See Annual Enrollment Reports cited from 1970-78 on Undergraduate and Graduate Degree Programs.

30. The data in this section are based upon reports made available by officials of Xavier University of Louisiana College of Pharmacy in a communication, dated, February 8, 1980.

31. Refers to Medicine, Osteopathy, Dentistry, Veterinary Medicine, Optometry, Podiatry and Pharmacy.

CHAPTER 8

Blacks in Veterinary Medicine

The history of black students in veterinary medical education is a study of neglect, limited interest and institutional inaction. Of all the professional areas analyzed in this study, veterinary medicine is among the least successful in attracting black Americans. Were it not for the systematic recruitment efforts and demonstrated commitment to enrolling and producing black veterinary medical students evidenced at Tuskegee Institute, there would be almost no progress made by blacks in this field. In recent years, however, a significant proportion of the twenty-four schools and colleges of veterinary medicine in the United States have organized recruitment programs. Consequently, some of the responsibility for veterinary medical education of black students has shifted away from Tuskegee Institute. Nevertheless, a lion's share of that responsibility continues to rest with the College of Veterinary Medicine at historically black Tuskegee Institute in Alabama.

This chapter draws attention to the pioneer role of Tuskegee Institute in training black veterinarians. It examines recruitment practices and enrollment trends in the 1970s. It describes admissions criteria. Further, it identifies some of the persistent problems frequently associated with failure of the profession to attract more black Americans to it. Finally, it focuses attention on some of the concrete efforts to alleviate the problems of neglect and the underrepresentation of blacks in veterinary medicine.

Tuskegee Institute: Pioneer in Veterinary Medicine

The College of Veterinary Medicine at Tuskegee Institute was established in 1945. Since that year, Tuskegee has graduated approximately 85 per cent of all black veterinarians (holders of the Doctor of Veterinary Medicine [DVM] degree) in the United States. When this college was established, minorities were not generally accepted in veterinary medical programs throughout the U.S. as a consequence of prevailing

181

discriminatory, legal, and social practices and exclusionary educational policies. A few northern and western institutions admitted a token number of black students to their programs in agriculture, animal science, and veterinary medicine.

Therefore, the School of Veterinary Medicine at Tuskegee was founded to assure access of black and other minority group students to a profession that had not demonstrated interest in them. From its inception, the school was regional in scope. In addition to the State of Alabama, it recruited minority students from twelve Southern States which operated dual, segregated systems of education mandated by state law. Tuskegee soon admitted white students to its school of veterinary medicine. This act reflected the institution's policy of providing a multi-racial student body and pluralsitic learning environment for all of its students. Consequently, this school of veterinary medicine is the most racially and ethnically integrated school of veterinary medicine in the entire western hemisphere.

As a result of its sustained commitment to black Americans, minority students of this and other nations, and disadvantaged students, in general, it was estimated in 1974 that of the 400 black veterinarians then in the U.S., who represented 1.3 per cent of the total number of veterinarians, 93 per cent of them were Tuskegee graduates.[1]

Recruitment at Tuskegee:[2]

The recruitment program at Tuskegee can serve as a model for successful recruitment for all professional schools and colleges. Its motivating force is a firm, unswerving commitment to provide greater access and equality of opportunity in veterinary medicine. Over the years, recruitment at Tuskegee was predicated on four basic assumptions; that there is:

1) a dearth in career opportunity information regarding the profession readily available to members of minority groups;

2) limited knowledge about the profession of veterinary medicine and a lack of identification with the profession by members of minority groups. This situation of limited knowledge and low identification is a function of the apparent and real underrepresentation of minority groups in the profession. This paucity, in turn, reduces the chances that minority youth will have meaningful interactions with possible role models with whom they can identify or who may influence their aspirations and occupational choices;

3) lack of an organized recruitment program among other schools and colleges of veterinary medicine that could substantially in-

crease the number of black and other minority students in the field or which would offset active recruitment programs initiated and conducted by other professional and occupational groups, and

4) lack of information on financial aid programs available to minority students as well as inadequate understanding among some institutions about the extensiveness of the need of these students for financial assistance to support their educational pursuits.

Recruitment programs conducted prior to 1970 were largely supported from university funds. Because Tuskegee Institute is a state-assisted institution with academic programs generally characteristic of a comprehensive unversity, it has had to be particularly cautious in the allocation of its limited financial resources. Consequently, recruitment was not as extensive as desired as the decade of the seventies began. Those efforts that were undertaken generated approximately 60 to 65 applications from minority students per year.

In 1971, the College of Veterinary Medicine received a recruitment grant from the U.S. Department of Health, Education and Welfare. The three basic objectives of this grant were: (1) to establish a recruitment program specifically designed to search for, identify, and encourage qualified minority group members to seek admission and, ultimately, to be enrolled in schools of veterinary medicine; (2) to provide financial aid for a limited number of those students who enrolled in order to enable them to pursue a career in veterinary medicine; and (3) to assist these students in their efforts to achieve career objectives by means of a highly structured academic reinforcement program. The objectives were explicit in pointing to the intention of recruiting students who not only sought admission to Tuskegee but who may have desired to enroll in other schools of veterinary medicine.

Another grant was obtained in 1975 which allowed the institution to expand its recruitment efforts and to develop additional recruitment strategies. The staff included four part-time positions, one materials specialist, one full-time photographer and one coordinator-secretary. In addition, its 14 full-time faculty and its staff and students committed themselves to play an active role in the recruitment process.

Activities involved in the recruitment process included the following:

1) Preparation of brochures, slide shows and other materials for distribution to students, alumni, and recruiters. (These materials were utilized to heighten minority students understanding of the profession of veterinary medicine.);

2) Visits to 75 campuses of historically black colleges and to high schools that enrolled significant numbers of minority students;

3) Presentation of exhibits on veterinary medicine which showed minority students activities, at conventions of minority groups, State Guidance and Personal Conferences;

4) Sponsorship of two minority Recruitment Workships (each of 2-days duration) attended by representatives from all schools of veterinary medicine, the National Urban League (which sponsors a Street Academy program involving a substantial proportion of minority group students), recruitment organizations, colleges, and high schools. Workshop participants discussed methods and techniques for attracting minority group students to the veterinary medicine profession. Each institution shared its experiences in the development of a recruitment strategy and problems encountered in its individual efforts to enroll more minority students;

5) Continuing linkages established between Tuskegee and such organizations as the National Urban League committed to expanding educational opportunities in the profession for minority and disadvantaged students. The Urban League became a major facilitator of Tuskegee's recruitment drive by its dissemination of Tuskegee's film, "Because We Care", to members of its 100 Affiliates and educational directors;

6) Recruitment of Upward Bound Students in Summer Programs on the campus of Tuskegee Institute into the pre-veterinary medicine program of the college;

7) Sponsorship of a Health/Science Career Motivation Fair on the campus. High school students from surrounding counties who were interested in health career professions were invited to campus and to exhibit their own science work during the fair. Opportunities in the health professions were explored;

8) Linkages with regional and national programs of the State Guidance and Personnel Association, The American Guidance and Personnel Association were both established and strengthened through exhibits at their annual regional and national conferences. Meetings were held with high school counselors and advisors to acquaint them with the field;

9) Information dissemination activities including exhibits on veterinary medicine at national, annual and local conferences of the National Association for the Advancement of Colored People (NAACP), National Urban League, the Shriners, and the South Alabama State Fair;

10) News releases to the hometown newspapers of first and fourth year students. The purpose of these news releases is to inform local

residents of the students' enrollment in the school of veterinary medicine as well as to make note of the completion of all degree requirements. This is one method of establishing role models in the local community and of familiarizing local high school students with the profession of veterinary medicine;

11) The utilization of Tuskegee's film, "Because We Care" in recruitment. The importance of the film is multifaceted. For one thing, it portrays the overall educational programs and student life at Tuskegee. It also depicts black students and faculty at work in the school of veterinary medicine. In this way it has the impact of demonstrating to students that a full range of educational opportunities with minority faculty role models are available to them at the institution. And, finally, the film educates students specifically about veterinary medicine as a profession;

12) Establishment of linkages between the school of veterinary medicine and science departments of historically black colleges in the development of a "feeder program." By visiting faculty members in such departments as Biology, Chemistry, Physics, General Science, and Agriculture, as well as placement officials, their assistance is enlisted in the recruitment process by encouraging them to point to the option of veterinary medicine as a career choice;

13) The utilization of television in aiding recruitment. This activity involves several 60 second spots on stations in Birmingham, Montgomery, and Mobile, Alabama as well as in Columbus, Georgia. It also includes the showing of a 28 minute film about veterinary medicine on public television stations;

14) Exhibits and film showings at the annual meetings of the major black Greek Letter organizations;

15) Advertisements in publications that have a wide black readership. These include the major publications of the Johnson Publishing Company such as *Ebony* and *Jet* magazines. One series of advertisements in these magazines generated approximately 1500 inquiries about veterinary medicine;

16) The dissemination of thousands of leaflets and brochures about veterinary medicine to black and other minority students;

17) Involvement of its 90 alumni chapters in the recruitment program. The veterinary medical alumni association organized its recruitment committee to assist the overall recruitment activities of the school of veterinary medicine. These chapters participate in high school career day programs and develop their own special sources

to acquaint students with the profession. In so doing, they provide instant role models of the possibilities of success in a career in veterinary medicine;

18) A quarterly newsletter about activities in the School of Veterinary Medicine is mailed to all veterinary alumni, parents of students enrolled in the School, Tuskegee Institute General National Officers and Alumni Chapter Presidents as well as other schools of veterinary medicine;

19) Travel by its members to make presentations to several private groups sensitive to the need of providing financial aid to exceptionally needy minority students. Among the groups who pledged substantial increases in the level of their donations for scholarships/financial assistance were the Ancient Egyption Order Nobel of Mystic Shrine, Inc.; the Order of Eastern Star and the National Veterinary Medical Alumni Association.

20) Finally, in 1979, a new career motivational film released by Coca Cola, U.S.A. Since, a central focus of this film is the School of Veterinary Medicine at Tuskegee Institute, it is particularly important as a recruitment instrument.

The School also conducts a Summer Enrichment Program through which promising students are brought to the campus for an eight-week period. The program consists of an intensive diagnostic testing phase followed by training in "survival skills". Components of the latter phase include test-taking, reading comprehension, study habits, note-taking, time budgeting, and attitudinal aspects. Contact is sustained through the pre-veterinary medicine phase so that there will be continuity of contact with these students in all phases of their anticipated career in veterinary medicine. Tutorial and counseling services are provided by faculty and professionals with specializations in program areas.

As previously indicated, the School of Veterinary Medicine co-sponsored minority recruitment workshops with the National Urban League on the campus of Tuskegee Institute. These workshops served to facilitate an understanding of some of the major impediments in the recruitment of black and other minority students into the profession. They also demonstrated to the representatives of institutions in attendance what the possibilities of success were if they mounted aggressive recruitment programs. As a result of these conferences and the model provided by Tuskegee, the American Veterinary Medical Association (AVMA), through its council on Education, organized a Committee on Minority Affairs. This Committee is supposed to assist member institutions of the Association of American Veterinary Medical Colleges in their recruitment programs.

Under its HEW-sponsored recruitment program, the Council established affiliation with two educational assessment programs, the American College Testing (ACT) Assessment Program and the Students Search Service (SSS) of the College Entrance Examination Board (CEEB). These memberships helped to raise the number of minority students in the applicant pool by more than 100 students in a single year.

The four most tangible results of these recruitment activities are: (1) the number of applicants to the School of Veterinary Medicine tripled in the early years and has continued on the upswing since the program was initiated; (2) enrollment increased substantially; (3) stronger minority recruitment activities were stimulated at several schools and colleges of veterinary medicine throughout the nation; and (4) an identifiable recruitment model was established for other institutions to emulate.

Admissions Policy and Practices

Admission to the School of Veterinary Medicine is consistent with the philosophy of the institution's commitment to expand educational opportunities to minority and disadvantaged students. Tuskegee Institute was founded to specifically serve educational needs of black Americans. Over the years, this commitment embraced disadvantaged students in general, irrespective of race or ethnicity. Consequently, "The Committee on Admissions selects students who exhibit intellectual, personal, moral and social traits which are considered desirable in a Doctor of Veterinary Medicine. Selection is based on all of the data submitted by and on behalf of the student, not merely on achievement."

In toto, at least six factors enter into the admissions decisions; namely, (1) State of Residence, (2) the quality of the interview, (3) letters of recommendation, (4) academic trend, (5) the individual's expressed reason for selecting veterinary medicine as a profession and commitment to the profession, and (6) overall folder evaluation. Since Tuskegee's School of Veterinary Medicine is a regional school, it has contractual obligations to qualified applicants from participating States. Those students are given the highest priority. Next in priority are individuals from States without a college of veterinary medicine. The lowest priority is given to applicants from States that have a school or college of veterinary medicine.

Enrollment At Tuskegee's SVM

Since 1968, the School of Veterinary medicine has received approximately 5,101 applications. Of that number it accepted 481 students for admission. Of the acceptancees, 304 were black students, all of whom eventually enrolled at Tuskegee. Since 1970, the number of enrollees per year has grown from 37 to 60. Between 1976 and 1979, first year enrollment increased by approximately thirty-three per cent. (See Table 29).

Table 29. First Year Enrollment in the School of Veterinary Medicine by Race and Year, 1970-1978.

Year	Enrollment		Total
	Black Student	White Student	
1970-71	27	10	37
1971-72	30	12	42
1972-73	29	15	44
1973-74	29	16	45
1974-75	29	16	45
1975-76	30	15	45
1976-77	31	14	45
1977-78	34	16	50
1978-79	39	19	60*

*The total includes other, unclassified students.

Graduation Data

Since 1969, more than 350 black and white students have received Doctor of Veterinary Medicine degrees from Tuskegee Institute. Over two-thirds of all of its graduates are black and other minorities. (It is estimated that among the minority groups enrolled at Tuskegee, black students comprise approximately 95 per cent of the total).

A total of 63 females were recipients of Doctor of Veterinary Medicine degrees from Tuskegee between 1969 and 1979. Forty-one of these students were black and eighteen were white. Ten of the eighteen white females were awarded degrees in the two classes of '78 and '79. Fifteen of the black females received their degrees during the same period. However, black female graduates were more evenly distributed during the earlier years. It should be noted that eleven of the forty-six graduates in the 1979 class were black females. The increasing number of black females in graduating classes from schools of veterinary medicine is consistent with trends observed in human medicine, dentistry, and optometry.

Table 30. Number of DVM Degrees Awarded by Tuskegee Institute by Race and Year, 1968-69 through 1978-79.

Year	Total Degrees Awarded	Degrees to Black Students	Degrees to White Students
1968-69	30	26	4
1969-70	23	18	5
1970-71	24	19	5
1971-72	27	21	6
1972-73	16	10	6
1973-74	39	22	17
1974-75	33	21	12
1975-76	40	26	14
1976-77	38	22	16
1977-78	39	24	15
1978-79	46	32	14
Total	355	241	114

Financial Aid

Like students who enroll in Xavier University's College of Pharmacy, the majority of Tuskegee's students in veterinary medicine are from economically disadvantaged families. They require financial assistance to assure their continued enrollment in the institution. This need is substantially more critical in 1980 since the institution was forced to raise its basic tuition from $1,400 in 1975 to $2,150 in 1979-80. Since, approximately 85 per cent of all these students need some form of financial assistance, the institution provides as much help as possible. This aid is in the form of scholarships, and federal programs such as BEOGs (Basic Economic Opportunity Grants).

Impact Of The Bakke Decision

The *Bakke* decision has not had a measurable impact on recruitment and enrollment at Tuskegee Institute. Its program in the school of veterinary medicine is consistent with the spirit of the decision by the U.S. Supreme Court in the *Bakke* case, especially regarding opportunities for disadvantaged students regardless of race or ethnicity. It should be reiterated that the *Bakke* decision could serve as an impetus for the recruitment of black and other minority students. This is especially critical since the "race" variable may legally be taken into consideration, under certain circumstances, when making admission decisions. On the other hand, the *Bakke* decision may also provide a shield for those institutions that do not wish to increase or even include a minority student presence of any dimension. Tuskegee Institute is not one of the latter institutional types since it is the most thoroughly integrated schools of veterinary medicine in this hemisphere. It should be apparent to the reader that the case history of its School of Veterinary Medicine clearly exemplifies the type of progress that is achievable when an institution organizes a sound recruitment program, demonstrates a strong commitment to its objectives and provides leadership and support for it.

The enrollment data at Tuskegee provide ample support for the hypothesis that the presence of a significant or critical mass of black students in graduate and professional schools is positively correlated with the presence of role models on the faculty. There are fourteen black and other minorities on the faculty of the School of Veterinary Medicine. Students see them in action during recruitment, in films about the institution, in their daily work on the campus, in interaction with students of all races, and in leadership positions in the institution. That type of visibility is a major inducement to go into the profession since it heightens the level of students' aspiration for a career in a field in which a member of their own racial or ethnic group have achieved notable success.

The graduation data provide substantial support for the contention that the availability of an adequate program of academic and non-academic support services enhances retention and reduces attrition among students enrolled in a graduate or professional school. Although no data on utilization characteristics are available, impressionistic evidence leads to the conclusion that the availability of organized tutorials, study groups, faculty willing to offer assistance without demeaning the student's capabilities, and on overall positive learning environment are highly salient dimensions of a retention program. All of these factors contribute in some way to the retention of students through the completion of their program.

Desegregation Trends in Veterinary Medical Education

It is estimated that there are 35,000 holders of the Doctor of Veterinary Medicine in the United States in 1980. Of that number, about 600 are black, representing 1.7 per cent of the total number. Whereas, in 1974, it was estimated that about 93 per cent of 400 black American Veterinarians had received their D.V.M. degree from Tuskegee, it is now estimated that that proportion in 1980 stands at or about 85 per cent. This change reflects a small but increasing presence of black students in the remaining twenty-three schools of veterinary medicine in the United States located at:

1. Auburn University
2. The University of California/Davis
3. Colorado State University
4. Cornell University
5. The University of Florida
6. University of Georgia
7. University of Illinois
8. Iowa State University
9. Kansas State University
10. Louisiana State University
11. Michigan State University
12. The University of Minnesota
13. Mississippi State University
14. University of Missouri
15. Ohio State University
16. Oklahoma A & M University
17. Pennsylvania State University
18. Purdue University
19. The University of Tennessee
20. Texas A & M University
21. Tufts University
22. Virginia Polythechnic University
23. Washington State University

In addition to these institutions, a new college of veterinary medicine has been approved for the North Carolina higher education system. Five of these institutions have come into existence since 1972 and some have only recently graduated their first class of students. Seven of the ten "Adams States" are represented in the group of twenty-four institutions with schools or colleges of veterinary medicine. The only "Adams States" not included here are Arkansas, Maryland, and North Carolina but the latter state will soon have a fully-operative college of veterinary medicine.

Role of Associations in Veterinary Education for Black Students

Compared to other professional organizations, such as the Association of American Medical Colleges, Schools of Law or Dental Schools, the organizations affiliated with or which accredit schools or colleges of veterinary medicine have been particularly conservative in actively supporting minority group students. Neither the American Veterinary Medical Association (AVMA) nor the American Association of Veterinary Medical Colleges (AAVMC) has even endorsed a policy statement comparable to the type issued by the Association of American Medical Colleges or the Association of American Schools of Law. The policy statements issued by the latter organizations formally committed their members to increasing their individual and collective efforts to bring black Americans and other minority students into the mainstream of medical and legal education. The lack of pronounced and sustained organizational leadership, that is so essential in speeding up the process of access to equal opportunity in veterinary medical education, helps to account for the dearth of aggressive recruitment programs by schools and colleges of veterinary medicine.[3] Hence, some colleges of veterinary medicine reportedly have never enrolled and graduated a black student in their entire history.

This is not to minimize those actions taken by the AAVMC and the AVMA from which minority students have benefitted. In fact, some efforts were indeed initiated during the 1970s in minority recruitment but the results of these actions fell considerably short of expectations. In the main, it appears that the AVMA was instrumental in facilitating increased visibility of minority recruitment efforts. This effort was primarily in publications about programs in veterinary medical education. It provided a placement service for graduating veterinarians, but its main responsibility lies in the roles it performs as the accreditation body for the schools and colleges of veterinary medicine in the United States and Canada. Through its Council on Education, it considers minority recruitment and other issues relating to the access of minority students to the profession.

Representatives from the American Veterinary Medical Association attended both the 1974 and 1975 Minority Recruitment Conferences at Tuskegee Institute. An expression of commitment to minority concerns was articulated by its representatives. However, it is not clear how that commitment was translated into enduring, influential, significant, viable programmatic actions which lead to major changes in the rate of enrollment of black students to schools and colleges of veterinary medicine.

One of the resolutions the conference proposed in 1975 was that a minority affairs committee be established to deal with the broad problems of access, enrollment, and retention. Such a committee was formed under the ultimate responsibility of the American Association of Veterinary Medical Colleges. The same conference proposed that both the AVMA and the AAVMC waive the residency requirements for admission in all States. This proposal was prompted by the realization that enforcement of residency requirements was a major impediment to successful recruitment of black and other minority students. Similarly, participants agreed that these organizations should obtain "a commitment from veterinary colleges to admit a given number of minority students."[4] According to informed sources, only one of these proposals was ever implemented by the professional association (i.e., a Committee on Minority Affairs). However, the AVMA did continue to include minorities in some if its public relations materials, brochures and in the film "The Covenant."

In 1978, following the U.S. Supreme Court decision in the *Bakke* case, the Association of American Veterinary Medical Colleges co-sponsored a workshop conference, on Minority Representation in Veterinary Medicine, in Washington, D.C. The conference focused on such issues as the legal implications of the *Bakke* decision for minority students recruitment, the failure of the profession of veterinary medicine to attract black and other minority students, recruitment, admissions, and retention.

One of the most salient conclusions reached by the conference participants was the acknowledgment that "recruitment is a passive activity" at several schools and colleges of veterinary medicine.[5] Since the schools of veterinary medicine have traditionally tended "to act independently and in isolation," they have not developed, organized, systematic and aggressive recruitment strategies learned from the experiences or the successes of other institutions. The passive recruitment stance, it was reported, originates in the perception that the profession of veterinary medicine is free of discrimination.[6] Many persons, both members and non-members of the profession, do *not* accept the idea that it is discrimination-free.

The participants recognized the tremendous value in the presence of role models on the faculty of colleges of veterinary medicine. That importance is twofold. It provides evidence of the success possibilities for minorities in veterinary medicine. It also creates a more pluralistic and diverse faculty for the benefit of all students in the institution.

Attention was given to the types of actions that may discriminate against minority students, and, consequently, serve as a barrier to their enrollment. Among these were such things as: (1) excessive weights given to standardized tests, (2) high value placed on the grade point average (GPA), (3) intimidation and cold indifference that sometimes occur during the interview process, (4) misunderstandings and misinterpretations of behavior that may result from poor interviews, and (5) the rigidity of admissions requirements in some institutions.[7]

Consideration was given to the thesis that retention is, indeed, a multifaceted phenomenon. As such, it involves at least three aspects: (1) academic factors, (2) financial resources, and (3) the socio-cultural environment in which learning takes place.[8]

Among the major recommendations made at this conference and which were to have been sent to all the college Deans and to the President of the AAVMC were the following: (1) a National Coordinator of Minority Affairs should be appointed, (2) that each school or college of veterinary medicine should appoint at minimum a part-time Coordinator of Minority Affairs, (3) the meetings of the AAVMC Committee on Minority Affairs should be institutionalized and regularized, (4) stronger linkages should be established between all parties who have responsibilities for recruitment and admissions, (5) an examination of admissions procedures and the elimination of all those that lead to discrimination, and (6) decision makers in the admissions process should have a working understanding of the *Bakke* decision and its impact on institutional policies and practices. In addition, comparative data on admissions and recruitment, covering the years 1976, 1977 and 1978, were to be distributed with the Conference Report.[9]

No National Coordinator of Minority Affairs was ever appointed. Several colleges of veterinary medicine either appointed a local minority coordinator or expressed intention to draw more heavily upon university-wide minority affairs offices. Some institutions are taking a more holistic view of the applicants' portfolio when making assessments and rendering admissions decisions.

Recruitment of black students in historically white schools and colleges of veterinary medicine has indeed been hampered by the lack of knowledge among black youth concerning the profession itself. However, that is probably neither the sole nor the primary explanation for their underrepresentation in these institutions. That situation is undoubtedly more accurately attributed to the quality of recruitment programs at these institutions. Some institutions have no active recruitment program. Others are rather passive in their efforts. Few employ full-time recruitment staff. Often, recruitment programs that are initiated are not sustained for an appreciable period of time. Failure to continue follow-up efforts after visits to recruitment sites does not enhance those steps taken to increase minority enrollment in veterinary medical colleges.[10]

Ten of the twenty-four colleges of veterinary medicine completed the inventory employed in this study. Among these institutions, recruitment activities included (1) the development of informational materials (e.g., pamphlets, brochures) for use during recruitment drives; (2) mailings of informational materials to junior and senior high school students; (3) visits to junior and senior high schools to participate in career day activities; (4) participation in college fairs on campus and in high schools; (5) visits to predominantly black colleges; (6) the use of alumni in recruiting; (7) the participation of faculty members in visits to other campuses and in on-campus recruitment programs which sometimes include workshops and contact with students enrolled in other science programs; (8) banquets for pre-veterinary medicine students and tours of campus facilities in veterinary colleges, and (9) the use of Tuskegee's mailing list of prospective candidates for admission into veterinary medical colleges.

The only additional recruitment strategy employed by institutions which did not respond to our mailed inventory but which was listed elsewhere was summer employment with practicing veterinarians. This technique provides firsthand knowledge of some of the daily activities of veterinarians and may stimulate greater interest in the profession.[11] Irrespective of the strategy employed, the crucial variable appears to be the strength of the commitment of the institution, and its faculty and staff, as measured by the support of recruitment programs through adequate finance and with the employment of minority group role models. Without these, it is apparent that the recruitment of black students and other minorities is doomed to failure at the onset.

Admissions Policies and Criteria

Although admission requirements vary from institution to institution, there are also similarities between them. One similarity is the lack of "special admissions programs". Another is their reliance on both cognitive and non-cognitive or subjective criteria. The weight assigned to cognitive or traditional factors appears to be contingent upon the number of applications received by the institution in relation to available space in the first year class. Moreover, as the level of competition for limited space rises, one may normally expect stronger reliance on objective measures since quantifiable data permit cut-off points which may be more easily defended when selection and admissions decisions are called into question.

The college of veterinary medicine at the University of Illinois, for instance, announced in its applicant informatin bulletin for 1980-8l that there have been "approximately 400 qualified applicants" for the 91 available spaces in the college. Consequently, even if a person completes minimum academic requirements, admission to the college of veterinary medicine is not assured. This college reports a mean GPA of its applicants in recent years to be 4.5 on a 5.0 system. The mean number of credit hours achieved by applicants is 120. This implies that the average applicant is a college graduate. That is significant because it is possible, though increasingly unlikely, for a person to be admitted having only completed the two or three years of pre-professional training in a pre-veterinary medical program.

Selection criteria at colleges of veterinary medicine usually include: (1) college GPA, (2) scores on the Veterinary Aptitude test (VAT), and (3) State Residency Requirements. Other factors utilized are (4) letters of recommendation, (5) an interview at the college of veterinary medicine, and (6) motivation for a career in veterinary medicine. Some institutions look for the GPA in required science courses, performance on the Graduate Record Examination (GRE), the GRE-Advanced Biology score, animal experience, contact with veterinarians, extra-curricular activities, job experiences, leadership ability, effectiveness in communicating with others, maturity, and reliability.

The GPA expected varies from a C + to a B + average. VAT scores usually range from 75 to 80 per cent. However, it appears that the heaviest weighting of all traditional factors is assigned to the GPA. Several institutions reported a tendency to de-emphasize objective criteria in recent years, to look more toward evidence of personal development, and to place such factors as GPA, VAT or GRE test scores in a more balanced relationship to the non-cognitive factors employed in selection and admission decisions.

Some institutions have a definite point system that is utilized in determining crucial aspects of eligibility. Under one system, a maximum score of 70 per cent is given to academic criteria and 30 per cent or points are assigned to non-academic factors. In other instances, each selection category is assigned a maximum of 15 points. A composite score is determined by the total points accumulated in each category. In another institution, a maximum of 55 points may be awarded for objective criteria while the remaining 45 points are assigned to subjective factors included in the overall evaluation.

Bonus points may be given to an applicant for certain "ancillary factors". Such factors may include quality of course load, or taking an extraordinarily heavy course schedule for an extended period of time. On the other hand, bonus points may be awarded minority students as a method of eliminating test bias. Consequently, the number of bonus or special points allowed may be equivalent to the value assigned to aptitude test scores.

The majority of these institutions also reported to have had black faculty involved in the recruitment, selection and admissions processes or in one or all of these stages.

Each institution reported a strong desire to enroll many more minority students. Each maintained that lack of success arises from the competition for the better trained black students from other science professions. In effect, these institutions admit that they have not been successful in their overall recruitment program or in attracting blacks to veterinary medicine.

Albeit, there is some general evidence that heavy reliance on high GPAs and high VAT or GRE test scores has indeed militated against the selection of larger numbers of black and other minority students. It may very well be that in the future, as these colleges re-examine their selection and admissions criteria, they will readjust the points assigned to objective and subjective factors in such a manner as to reduce the hasty elimination of otherwise promising students.

Enrollment

Prior to 1975-76, the schools and colleges of veterinary medicine reported to the AAVMC enrollment data that were disaggregated by race and ethnicity. Thereafter, and for unclear reasons, institutions not only reported aggregated data on "minorities" but the institutions were scrambled by alphabetical designations which changed year after year. Consequently, it was not possible to discern institutional enrollment data from the files of AAVMC without a key. That was not available (See Table 31). This decision imperils efforts to engage in research on veterinary medicine that requires ethnic data.

According to Table 31, black student enrollment in the combined schools and colleges of veterinary medicine declined slightly between 1969-70 and 1972-73. However, the significant increase observed in 1973-74 has continued. In the early seventies, from 1970 to 1974, over 90 per cent of all black students in veterinary medicine were enrolled at Tuskegee Institute. Because of increasing attention given to the recruitment of black and other minority students, this proportion has declined to approximately 80 per cent in 1980.[12]

Following examples may illustrate the degrees of success in enrolling black students between 1969 and 1979. The total number of black students enrolled in this period was four at the University of Georgia; ten at Ohio State University; thirty-three at the University of Pennsylvania;* and two at Washington State University.

The School of Veterinary Medicine of Louisiana State University, which opened in 1973, has enrolled two of the nine black students who applied for admission since 1973. The college of Veterinary Medicine of Mississippi State University has enrolled three of the sixteen black students who applied since it opened in 1977. And, the College of Veterinary Medicine of the University of Tennessee has enrolled one of the two black applicants since the college opened in 1977. The problems manifested in these three new colleges reflect a fundamental concern commonly observed in most of the twenty-four colleges of Veterinary medicine: How to generate a sufficient number of applications from black students, or enlarge the applicant pool, so as to increase enrollment beyond the limited or token stage.

Total Enrollment

Total enrollment in veterinary medicine began to climb in the early seventies and continued its upswing throughout the decade. Similarly, an upswing in the enrollment of black and other minority students began in 1973. In that year, 111 black student veterinarians comprised 1.7 per cent of the 6,603 students enrolled in all U.S. colleges of veterinary medicine. In 1974, the 121 black veterinary medical students comprised 1.8 per cent of the 6,731 students enrolled in all institutions combined.

As a result of the decision in 1974 by AAVMC institutions to report aggregated data on minority students, there is no method of determining how the proportion of Black, Chicano, foreign black and American Indian students differ from each other unless HEGIS data provide it. Nevertheless, during the preceding year, the number of black students enolled increased by ten, that of American Indian students increased by two, but the number of Chicano students declined by seventeen, and the number of foreign black students declined by two students over the 1973 figures. Overall, there was a loss of seven minority students.

An examination of Table 31 shows a dramatic increase in the number of minority students enrolled in 1975-76, when aggregated data were first

---

*This member includes all minority students. Hence, the actual number of blacks is much smaller than 33.

Table 31. Total Enrollment in Doctor of Veterinary Medicine Programs by Race and Year 1969-78: All Institutions Combined*

Year	Total Enrollment*	Black Student Enrollment	Per Cent	Minority Student Enrollment	Per Cent
1969-70	—	92	—	106	
1970-71	—	91	—	106	
1971-72	—	90	—	107	
1972-73	—	90	—	119	
1973-74	6,408	111	1.7	143	2.2
1974-75	6,731	121	1.8	141	2.8
1975-76	7,030	—	—	272	3.8
1976-77	6,066	—	—	285	4.7
1977-78	7,641	—	—	273	3.9
1978-79	7,017	—	—	275	3.9

Source: Dr. Ellis Hall, Tuskegee Institute.
Total Enrollments also include foreign student enrollments.
*Beginning in 1975, the Association began to aggregate all data on minority students; thus, ethnic designations were discontinued.

reported. The absolute numerical increase was from 141 in 1974 to 272 in 1975 or an increase of 131 students. The net change in enrollment of black students at Tuskegee Institute was one student. Therefore, the remainder of 140 minority students was enrolled elsewhere. In all probability, total minority enrollment now includes Asian Americans and it is this policy change which accounts for the escalation of minority enrollment data. Those 272 minority students represented 3.8 per cent of the 1975 total enrollment of 7,267. The proportion of total enrollment represented by minority students was 4.7 per cent in 1976; 3.9 per cent in 1977; 3.9 per cent in 1978; and 4.1 per cent in 1979-80. In effect, increases in total enrollment of minority students have not kept pace with the rises in enrollment of white male and female students in U.S. colleges and schools of veterinary medicine. Although measured declines in total minority student enrollment appear to have been slight over the last four years of the seventies, a noticeable trend of decreasing minority representation is significant if no more than that they represent such a small percentage of total enrollment in the first instance.

The number of minorities in advanced training positions dropped precipitously from 43 in 1976 to 29 in 1978. This category includes persons seeking other graduate degrees, interns and residents. In terms of proportions in advanced training, that decline was from 10.7 per cent of 390 in 1976 to 8.3 percent of 359 in 1978.[13]

All the participants in this study indicated interest in enrolling larger numbers of black and other minority students. Only one has definite but minimum enrollment projections mandated by its Boards of Regents, and it is an "Adams State." Others do not express goals in numerical terms while the majority do not have any enrollment goals. Given past experiences that show a clear paucity in number of black students enrolled in veterinary medicine, it seems only logical for schools of veterinary medicine to establish attainable numerical targets or goals for increasing minority student enrollment. This may be the most effective stimulant for concrete action. Ultimately, one outcome may be a more pluralistic and diversified stu-ᵈent body as well as greater diversity in the profession itself.

Graduation Data

Institutional data on graduates from schools and colleges of veterinary medicine are all but impossible to obtain except from cooperating institutions. The HEGIS Reports are of limited value in this regard, for this time period, because in the early years these data were not always disaggregated in the manner required for this analysis. In subsequent years, the data are incomplete. Nevertheless, it can be reported with reasonable confidence that Tuskegee Institute still graduates more than 85 per cent of all black veterinary graduates each year. The University of Georgia graduated 75 per cent (3 out of 4) of its black students who enrolled in veterinary medicine during the 1970s. Washington State University graduated 100 per cent (2 out of 2) of the black students in enrolled during the same period. Ohio State University graduated five blacks with degrees in veterinary medicine in the decade. These few illustrations speak clearly and graphically to a compelling need to produce a sufficient number of black veterinarians necessary to assure even a modicum of representatin in the work force. Obviously, failure to enroll and to graduate black students accounts for blacks still comprising a mere 1.7 per cent of all veterinarians in the United States in 1980.

Problem Areas

The seven most commonly cited problems regarding underrepresentation of blacks in this profession are: (1) the absence of an adequate number of role models; (2) financial aid; (3) limited funding or absence of funding for aggressive recruitment; (4) disinterest of minority students in veterinary medicine; (5) lack of black applicants; (6) competition between the health science areas for the same pool of students; and (7) weak pre-professional training.

Top priority should be given to the recruitment of black faculty members in schools and colleges of veterinary medicine since their presence has consistently been the most powerful predictor of success in the recruitment and enrollment of black professional students. It is also evident that the larger the number of black faculty members and the more visible they are in recruitment processes the more successful the institutions are in attracting black students. Tuskegee Institute, as previously indicated, has more black faculty than any other college or school of veterinary medicine. According to data collected for this study, the following institutions do not have black faculty: Auburn, the University of Georgia, Oklahoma State University, Texas A & M University, and the three newest colleges of veterinary medicine: Tufts University, Virginia Polytechnic and State University, and North Carolina State University. The mean number of black faculty in the remaining schools and colleges of veterinary medicine is one per institution.

Lack of adequate financial aid is another problem which militates against successful recruitment and enrollment. Over half of the veterinary colleges are located in States with a high concentration of blacks in the state population. These are also States with high unemployment rates among the black population. High unemployment and comparatively low median family income origins make it imperative for these students to receive some form of financial aid even if they enroll in state-supported institutions in which tuition is normally low compared to private institutions.

Unfortunately, scholarships and fellowships are not always available. Of the institutions participating in this study, Ohio State University has had the best record of providing fellowships, scholarships and teaching assistanships to black students. The number of black student recipients in each of these categories has varied from one to three in each year since 1970. The University of Georgia provided two Regent's Scholarships of $5,000 each to black students in 1979. Only one or two institutions seemed to have been successful in confronting state legislatures in winning approval to waive resident requirements in order to enhance minority recruitment and, indirectly, assisting minority students to overcome the financial barrier to enrollment.

According to AAVMC, the median wage earned by DVMs is about $30,200 per year. This is substantially lower than the median income of physicians, for instance.[14] The veterinarian, too, has to think in terms of the initial costs of private practice such as the procurement of equipment, facilities and supplies and the payment of staff. Even when repayable loans are available, and those loans do not appear to be as readily accessible to veterinarian students as they are to students enrolled in human medicine and dentistry, there is extreme reluctance to subject oneself to long-term indebtedness, especially when the student comes from relatively poor economic backgrounds. This overall dearth in non-repayable loans, scholarships and teaching assistantships reduces the number of applicants from otherwise qualified and interested black students.

The "lack of interest" in veterinary medicine by black and other minority students to which much allusion has been made is a problem of image of the profession. It also reflects lack of knowledge about the profession and the disinterest on the part of white professionals to take the necessary action to alter negative perceptions. Unwillingness to appropriate sufficient funds for the recruitment efforts by interested white faculty at the traditionally white institutions serves only to exacerbate the problem and to perpetuate both the image as well as the underrepresentation of blacks and other minorities in veterinary medicine. Medicine, dentistry, and pharmacy will continue to be more successful, regardless of the prestige factor, so long as they provide more resources to attract these students. Veterinary medicine must become more competitive, more aggressive, and less conservative in its recruitment programs. In so doing, it can take more concerted action to help strengthen the pre-professional training and preparation received by black and other minority students. Undoubtedly, litigation for inclusion may prove to be another enabling tactic.

Although some of these institutions may have become more skeptical about their activities in view of the *Bakke* decision, others have probably hidden behind the decision as a shield for their own policies of inaction. Of course, the Supreme Court made set-aside quotas of the type formerly employed by the University of California/Davis unconstitutional. Importantly, it left the door wide open for university officials to mount recruitment and admission programs designed to increase diversity. It was clear in its support of "race," under certain circummstances, as one variable that could be a "plus" regarding admissions decisions. In any event, first and foremost, institutional commitment to the principle of equality of educational opportunity must be present in order for veterinary medicine to move toward noteworthy access to all students, irrespective of race.

Footnotes

1. Walter C. Bowie, D.V.M., "Opening Remarks," *Proceedings of Minority Recruitment Conference: Tuskegee Institute School of Veterinary Medicine* Tuskegee" Tuskegee Institute, 1974, pp. 10-12.
2. This discussion of the veterinary medical education program at Tuskegee Institute is based upon data made available by Dr. Ellis Hall of the College of Veterinary Medicine.
3. According to key informants on recruitment programs in colleges of veterinary medicine, the professional associations in this field never issued policy guidlines of the type articulated by counterpart associations in medicine and law. Individual representatives of the associations in veterinary medicine did express personal beliefs concerning the desirability of increased participation of minority groups in veterinary medicine on several occasions during the seventies.
4. *Proceedings of Minority Recruitment Conference:* Tuskegee Institute, 1975, p. 54.
5. Donald A. Abt., *Report of the Workshop Conference on Minority Representation in Veterinary Medicine* (Sponsored by The Association of American Veterinary Medical Colleges, The American Council on Education, and The Ford Foundation), November 1-2, 1978, p. 4.
6. *Ibid.*
7. *Ibid.*, p. 5.
8. *Ibid.*, p. 6
9. *Ibid.*, pp. 6-7.
10. *Ibid.*
11. *Proceedings of the Minority Recruitment Conference: Tuskegee Institute School of Veterinary Medicine, 1974.* Tuskegee: Tuskegee Institute, 1974. See also *Proceedings of the Minority Recruitment Conference: Tuskegee Institute School of Veterinary Medicine, 1975.* Tuskegee: Tuskegee Institute, 1975.
12. Based upon information made available by Dr. Ellis Hall of Tuskegee Institute.
13. Cf. "Trends in Minority Affairs at Schools of Veterinary Medicine", in Donald A. Abt, *Op. Cit.* (Appendix).
14. Based upon telephone conversations with Dr. William Decker, Washington Representative of the Association of American Veterinary Medical Colleges.

CHAPTER 9

Engineering and Architecture:
Pathways to Progress

In terms of absolute numbers of students enrolled, no professional area can legitimately claim as much success in changing the racial composition of its field as can the profession of engineering. None can claim as much direct involvement and participation of the corporate structure in concrete programs designed to assure the inclusion of blacks and other minorities in a profession. None can demonstrate so rapid an increase in the actual production of a greater number of minorities for the work force. Although profound and far-reaching changes occurred during the 1970s, much more is now required to assure racial parity in the field of engineering. Architecture, on the other hand, has not been by any measure as successful in mainstreaming outsiders. Consequently, its task is monumental compared to other professions.

This chapter delineates the role of the historically black colleges of engineering in the production of black engineers prior to the 1970s. It describes the primary role played by the corporate structure in accelerating the process of mainstreaming blacks in the field of engineering. It outlines recruitment practices and presents black student enrollment trends during this period. The relationship between enrollment, retention, and the production of undergraduate and graduate degrees and certain predictor variables is explained. Finally, this chapter focuses attention on the central problems of mainstreaming blacks in the field of architecture. All of these issues are explored in terms of their relationship to the theoretical notion that institutional behavior is a function of both internal and external pressures as well as societal conditions.

The Production of Black Engineers Before 1970:
Black Schools of Engineering.

In 1970, 2.8 per cent of the engineers in the United States came from four minority groups: Blacks, Chicano or Mexican-Americans, Puerto Ricans, and Native Americans. These four groups comprised slightly more than 14 per cent of the total U.S. population.[1] Blacks alone accounted for

204

11.1 per cent of the national total. Yet, they comprised only about one per cent of all the engineers in the nation. Clearly, using the proportion of blacks in the total population as the reference point, their representation in the engineering profession was negligible.

Prior to the 1970s, and especially before 1964, the major responsibility for the training and production of engineers in the black population was left to one or more of the six historically black colleges: (1) Howard University, (2) North Carolina A & T University, (3) Prairie View A & M (Texas), (4) Southern University (La.), (5) Tennessee State University, and (6) Tuskegee Institute. These six institutions played a unique role in training engineers from the minority population *before* court-ordered desegregation programs were implemented and *before* non-South public or private institutions committed themselves to large scale recruitment of black students in engineering. They achieved their success despite being persistently understaffed and underfunded. State legislatures did not have a history of treating the education of black students as a matter of high priority. Consequently, historically black institutions were in a constant struggle for funds to support strong educational programs.

The value of the historically black colleges and their schools of engineering in the production of engineers can be gleaned from the data which show that from 1969 to 1973, these six institutions awarded approximately one-half (47 percent) of all B.S. in engineering degrees earned by black students. Further, two-thirds of all bachelor's degrees conferred upon black students in engineering were from twenty-five of the 282 schools and colleges of engineering in the United States, including the six black institutions.[2] The proportion of blacks who received engineering degrees from these institutions was far greater in earlier years than is represented by the 1969-73 period. In 1969, for instance, the six historically black colleges of engineering awarded 189 bachelor degrees in engineering to black students. According to a survey of the twenty-five leading schools in the enrollment of black students in engineering, this figure was 93 per cent of all the degrees awarded to blacks. The distribution among the six black institutions was as follows: Prairie View A & M, 49; Howard University, 42; North Carolina A & T, 32; Tennessee State, 30; Southern University, 28; and Tuskegee Institute, 8. Purdue University, with a total of seven B.S. degrees conferred on black students in engineering, awarded more degrees to black students than did any other historically white institution that year.[3]

Although a more detailed profile of the role of black colleges in enrolling black engineering students will emerge later in this chapter, mention should now be made of their efforts to stimulate greater participation of black student in engineering programs. During the sixties, educators and civil rights leaders expressed special concern over the maldistribution of majors selected by black college students. Although black students were obviously underrepresented in literally all fields of study, they tended to concentrate in a few selected majors. Among these fields were Education, the Social Sciences, Home Economics and Physical Education. As efforts expanded to re-direct career choices among these students and as they realized career opportunities were growing, many black students began to shift to other fields such as engineering.

Partially because of the declining enrollments in the white schools of engineering during the 1960s and early seventies, several of these institutions were excited about the possibilities of saving their own engineering programs by enrolling black students in joint-degree programs with historically black colleges. Consequently, a number of joint-degree programs were established in the late sixties and early seventies. These programs are labeled "3-2 engineering programs". This designation means that the black student enrolled in a black college spends the first three years of a five year program taking the pre-engineering and liberal arts courses at the parent institution and the last two years of engineering training at a cooperating white institution. After the successful completion of all degree requirements, the student is awarded degrees from both institutions.

A partial list of dual degree programs located at black colleges is presented in Figure 8.

Cooperating traditionally white institutions included Georgia Tech, Vanderbuilt, Pennsylvania, University of Maryland, Tulane, Wright-Patterson, Louisiana Tech, and North Carolina.*

The implementation of dual-degree programs has enabled a significant number of black students to accomplish objectives of attending historically black colleges and, simultaneously, taking advantage of professional degree programs not available to them in the parent institution. These programs assist the parent institution to publicize expanded curricula offerings that are often vital for the recruitment of the better prepared students. Simultaneously, they helped earlier on to save engineering programs which were in trouble in the traditionally white institutions because of declining enrollments.

---

*Listing these institutions does not infer that the primary purpose for their participation was to "save" engineering programs that were in trouble.

Figure 8. Dual Degree Programs in Engineering at Selected Historically Black Colleges[4]

Atlanta University Center (Ga.) (four colleges)	Shaw University (N.C.)
	Tougaloo College(Miss.)
Bishop College (Texas)	Xavier University (La.)
Fisk University (Tenn.)	Wilberforce University (Ohio)
Hampton Institute (Va.)	Grambling State University (La.)
Le-Moyne-Owens College (Tenn.)	North Carolina Central
Lincoln University	University (N.C.)
Morgan State College (Md.)	Rust College (Miss.)
Norfolk State College (Va.)	Savannah State College (Ga.)

The Role of the Corporate Structure in Mainstreaming Black Americans in Engineering.

The progress observed during the seventies in the recruitment, enrollment, and production of black students in engineering can be attributed to both changes within colleges and universities and to forces external to the institution or within the larger American society. Among the *internal* or institutional factors, at least four stand out. They are: (1) an articulation of the need for expanding access of minority students to a higher education by faculty and administrators of traditionally white colleges and unversities, (2) the moral commitment to racial and social justice expressed by many constituents of these institutions, (3) pressures for change by black and white students already enrolled and who were inspired by the Civil Rights Movement, and (4) the need by some decision-makers to demonstrate that these institutions do not support policies of racial exclusion.

Four *external* factors can also be identified. They include: (1) demands made by civil rights leaders to expand opportunities in professional education and in graduate schools, (2) pressures explicit in new federal regulations which affected corporate hiring practices as well as access by universities to higher education funds, (3) affirmative action policies and the possibilities for heavy penalties for non-compliance with regulations and statutes, and (4) pressure exerted by the corporate structure on colleges and schools of engineering to increase the supply of engineers for possible hiring so that they could conform to governmental expectations.

The corporate structure was a prime mover in the rapid growth of black student matriculation in engineering colleges. Their concern was considerably broader than recruitment and enrollment of students from minority groups or with initial affirmative action issues. The fundamental problem was the *inadequate supply* of black and other minority group individuals who could either be hired for traditional engineering positions in the labor force or for managerial positions in the corporate structure. In the early seventies, three of every five employees in one representative firm, who occupied the highest positions in that corporation, possessed technical degrees. These top echelon positions were 99 per cent white and one per cent minority.[5] Inasmuch as competition for the few minorities who were eligible for top-level positions and for other jobs that required technical training was widespread and intense, an increasing number of corporations realized that a drastic change in the supply structure was imperative.

Initially, two popular methods for increasing supply were pervasive. One was assisting historically black colleges and cooperative traditionally white institutions to establish dual degree programs. Another involved financial assistance to the six black colleges of engineering to enable them to produce more engineers. In 1964, the Western Electric Fund made a grant designed to improve the overall "quality performance" of the historically black colleges with the ultimate goal of expanded production of black engineers. This program was coordinated by the Black Engineering College Development Committee of the American Society for Engineering Education (ASCE).[6] This Committee focused its attention on methods of expanding enrollment, faculty growth, strengthening the curricula, and ways of improving retention through graduation. Faculty exchanges were sponsored in order to assure adequate course offerings. Even at that time, other companies began to provide increasing financial assistance to black students who attented the engineering schools of the predominantly white colleges.[7]

In 1969, the Atlanta University Center, comprised of four undergraduate liberal arts colleges, received a grant of $265,000 from the Ohlin Charitable Trust Fund to establish a dual degree program in Engineering with the Georgia Institute of Technology (Georgia Tech). In 1970-71, Howard University was able to establish a chemical engineering facility with the assistance of a grant of $600,000 from the Esso Education Foundation. The Dupont Company made major grants in 1972 to each of the black colleges of engineering to help them strengthen their programs as well as to attract and produce a larger number of black students with engineering degrees.[8]

The major impetus for the organized and expanded involvement of the corporate structure in broadening engineering education opportunities among the minority group population came from General Electric. Troubled by the "supply imbalance" and the persistance of a condition of minority group underrepresentation in the population of practicing engineers, General Electric decided to take corrective steps.

In 1972, J. Stanford Smith, Senior Vice President of General Electric, addressed a meeting of 44 deans of colleges of engineering, at its Management Development Center in Crotonville, New York. He stated that it was necessary to increase the number of blacks in engineering by ten to fifteen times their present number in order to assure the inclusion of a significant proportion of blacks in the top ranks of industry before the century's end. The accomplishment of this goal, Smith asserted, required the combined cooperation of all sectors of the society — public and private — educational institutions at all levels, the entire business community, foundations, professional groups, and minority institutions themselves. He also urged some form of government involvement in what was termed by some a revolutionary enterprise. Otherwise, Mr. Smith admonished, professional education in engineering was entangled in a "formula for tragedy."[9]

Out of that "revolutionary action" proposed by General Electric came the Minority Engineering Effort. Shortly after Smith's urgent call for action, the Conference Board convened a group of executives as members of an informal Minority Engineering Working Committee. Key executives representing General Electric, IBM, Xerox, Western Electric, U.S. Steel, DuPont, and General Motors participated in this working group to study the issues raised by Smith and to ascertain what would be done to translate his ideas into concrete programs. Later, this informal working committee expanded to include the Executive Director of the National Urban League, the Engineering Council on Professional Development (ECPD), the American Personnel and Guidance Association, the Chairman of the Committee on Engineering of the National Academy of Engineering (NAE), and other representatives from the community of engineering educators. These persons became the Council of Minority Engineering of the Conference Board.[10]

ME[3]

Late in 1972, the Task Force of the Engineering Council for Professional Development (ECPD) founded ME or the Minority Education Engineering Effort. Although ME subsequently separated from ECPD, it was an essential component of the early movement to increase the representation of black and other minorities in the engineering profession. This group focused its attention on pre-college programs in order to acquaint students with opportunities in engineering and the requirements for admission into college programs. ME urged high school students to take courses in mathematics and the general sciences. It advised them to enroll in college engineering programs.[11]

The 1973 NAE* Conference

Another significant event in the minority engineering effort occurred in 1973 when the NAE sponsored a symposium in which 250 persons participated to discuss further plans for stimulating the effort to broaden the base of participation of minorities in engineering. More specifically, the organizer intended to determine the degree of access of blacks and other minorities to engineering, ascertain the extent and characteristics of existing access programs, promote the development of a coordinated national program supported by institutional and organizational commitment for action, and determine methods of creating new pathways for minority group success in the corporate world. The participants in this conference agreed to work toward a goal of achieving a ten-fold increase in the number of minority engineers within a decade.[12]

NACME

The Sloan Foundation was instrumental in funding the work of ME and the 1973 NAE Conference. This foundation provided an additional $12 to $15 million to advance the new efforts to produce more engineers from minority groups. It also funded the Planning Commission Task Force in December 1973. The Task Force urged the achievement of parity in terms of minority group participation in engineering education and professional development. It also produced an extremely important and influential study called *Minorities in Engineering: A Blueprint For Action.*

As a result of the 1973 Symposium, the President of the NAE established an advisory group called the National Advisory Council on Minorities in Engineering (NACME) and the Committee on Minorities in Engineering (CME) of the National Research Council.[13] The central function of the NACME was to advise the NAE president on all matters that pertained to the "specific objective of development career trajectories" for under-represented minorities toward parity. The concept parity referred to a number of minority group engineers which approximated their proportion

*National Academy of Engineering

in the national population. NACME also functioned to generate financial contributions to the engineering effort from its member companies which could then be utilized to support the work of the Committee on Minorities in Engineering. These funds are provided to the CME through its parent organization, the National Research Council (NRC).[14]

NACME, an informal organization comprised largely of senior executives of major companies, is also a major supporter of the National Fund for Minority Engineering Students (NFMES) which annually distributes substantial monies in scholarships. The 36 leading corporations and leaders in other business institutions use their influence to mobilize formal and informal support for the effort to create greater parity and to strengthen engineering programs.

However, the participation of member corporations in NACME is not uniformly strong. According to the Conference Board, participation in NACME activities by the 36 corporations can be more or less dichotomized in terms of company involvement. One group consists of those whose overall contributions are characterized as "modest". Companies in this category have representatives who infrequently attend NACME meetings and whose own engineering effort shows weak coordination. By contrast, the second group which contains the majority of the 36 NACME corporations, is especially active, has well-coordinated programs which involve a broad spectrum of supportive activities. For instance, they include the utilization of role models from the company as lecturer/teacher on college campuses. In turn, they facilitate recruitment and the overall engineering educational program of some colleges of engineering. Some corporations also assist in organizing community consortia.[15]

Seymour Lusterman points out that degrees of company involvement depend, at minimum, upon six external factors: (1) the size of the ethnic population in which the company is located (the assumption here is that the larger the size of the ethnic community the more active is the company), (2) the existence or absence of an engineering college in the community, (3) the degree of integration in the schools, (4) the degree of the company's own prominence in the community, (5) the number of companies located in the community and the relative importance of the NACME company or firm in relation to others, and (6) the historical pattern of relations between industry and institutions within the community. In any event, as Lusterman maintains, leadership from above, that is from the top echelon of the corporate structure, is essential for stimulating broad-scale and enduring commitment to the minorities in engineering effort in any firm. Where that commitment is absent, suspect, tentative or in doubt, subordinates do not tend to display much enthusiasm for the program. Strong commitment produces equally sound results.[16] All of the six external factors are interrelated and affect the course of action taken by firms in any given community. Similarly, they may also be affected by regulatory guidelines issued by the federal government and affirmative action mandates.

CME

The Committee on Minorities in Engineering was offically established in 1974 as a coordinating body for efforts to increase access of minority students in engineering. Its four basic functions are : (1) to define needs, identify resources, and make recommendation regarding specific activities that may facilitate its ultimate objective; (2) to construct models of workable approaches for stimulating greater minority group access to engineering; (3) to serve as a resource center for those interested in learning more about successful approaches to increasing minority group participation; and (4) to serve as a disseminator of information, data, and research studies concerning the minorities in engineering effort.[17]

The Committee consists of two specific classes of members. On the one hand, there are volunteers who are characterized as more deliberate and passive in their approach to the CME's mission. The second group, on the other hand, consists of paid professional staff members who engage in specific actions supportive of the programs proposed and directed by the volunteers. The staff also interfaces with NACME and provides assistance for its endeavors. Volunteer activities embrace a range of programs including conferences, symposia, and promotion of research on critical issues germane to the overall mission of increasing minority group representation in the engineering profession. The volunteers make a special effort to influence public policy at the federal, state, and local levels through the development of position papers and other documents.[18]

Supportive Organizations

The following list of organizations, corporations, and institutions is intended to be illustrative rather than exhaustive of all pioneers in the effort to broaden participation of blacks and other minorities in Engineering. Omission of other groups here in no way diminishes the value of their roles in this process. Unquestionably, this effort was assisted by such groups as: National Executive Committee on Guidance (NECG), the Junior Engineering Technical Society (JETS), the Engineering Joint Council (EJC), the National Scholarship Service and Fund for Negro Students (NSSFNS), the U.S.

Chamber of Commerce, the National Association of Manufacturers, the National Alliance of Businessmen, the American Association for the Advancement of Science (AAAS), the National Science Foundation (NSF), the National Institutes of Health (NIH), and the Department of Health, Education and Welfare.

These groups performed a number of important functions which immediately and in the long run led to significant increases in the enrollment and graduation rates of black students in the field of engineering. Among these were the following:

1. coordinating efforts by engineering educators and representatives from business and industry and technical organizations in the engineering profession as well as organizations in minority group communities in working toward the goal of expanded engineering opportunities;

2. identifying and motivating minority group students to select engineering as a professional field;

3. developing motivational materials that are used in the recruitment process, (including films, slides, pamphlets, brochures, leaflets, circulars, and exhibits), and which may assist counselors and advisors to pre-college students;

4. implementing guidance programs to stimulate engineering careers;

5. forming high school chapters of student clubs comprised of students interested in the engineering profession, (There are approximately 1500 of these chapters in high schools across the country);

6. conducting the National Engineering Aptitude Search (NEAS) to indicate "possible success" in college engineering programs. (Approximately 8,000 students in grades 9 through 12 take this test each year);

7. providing campus level support through counselling, tutoring and advocacy for minority students;

8. providing role models from the corporate structure for campus recruitment activities, classroom lectures, and temporary faculty in selected pre-engineering and professional degree programs;

9. visiting college campus fairs, high school career days and general support for recruitment and retention in schools and colleges of engineering;

10. providing summer jobs and internship programs to further stimulate interest in the engineering profession;

11. engaging in a broad range of pre-college programs including promotion of interest in general science, math-

ematics, and high school physics, as important preparation for engineering programs;

12. identifying social, economic and institutional obstacles to
the successful recruitment and enrollment of black and
other minority students in schools and colleges of engineering;
13. assisting high school faculty to improve their own
capabilities for providing quality instruction in science and
mathematics curricula;
14. working with parents to promote their own interest in
engineering as a possible career option for their children;
15. helping minority faculty in historically black colleges to re-
tool themselves for up-dated curricula in engineering and
in the supportive science and mathematics fields;
16. facilitating a more supportive learning environment for
black and other minority students;
17. conducting and /or supporting pertinent research relative
to access and expanding equality of educational opportunity in engineering programs; and
18. helping to eliminate the financial barriers to enrollment
and retention in schools and colleges of engineering.

Recruitment of Black Students in Engineering

Special interest in expanded recruitment of black students, as previously stated, can be traced to efforts in the 1960s to strengthen the engineering
curricula of the historically black colleges of engineering. However, these
efforts were limited, tentative, and severely underfunded in relationship to
the magnitude of the problem. Immediately, they proved inadequate to
meet the increasing demand for more blacks in engineering and, later, for
the goals of parity within a decade. It soon became apparent that this task
exceeded the capabilities of the black colleges. Hence, widespread involvement of white colleges was absolutely imperative even to move one step
beyond token levels of minority student access to the engineering profession.

Institutional Strategies

Among the early institutional leaders in predominantly white universities in the effort to increase the enrollment of black and other minority students in engineering were the New Jersey Institute of Technology, the University of Bridgeport, Carnegie Mellon, the Illinois Institute of Technology and the University of Illinois/Urbana. Their efforts were supported by the business and industry communities as well as from institutional funds.[19] Recruitment programs in these institutions were initiated between 1967 and 1969.

The program at the University of Illinois/Urbana began in 1969. From its inception, its primary focus was on stimulating wider participation in engineering among those minorities who were grossly underrepresented in the profession. Specifically, the target groups included Black, Chicano, Puerto Ricans, and Native Americans. Special minority recruiters were hired. A pre-college outreach program became an essential program component for stimulating interest and attracting junior and senior high school students to engineering. The University invited these students to spend two summer weeks on the campus. During this visit, high school students were not only exposed to college life but they had an opportunity to study mathematics and to engage in analysis of engineering problems. This summer program became the prototype for the Minority Introduction to Engineering program (MITE) which the Engineering Council for Professional Development subsequently established. On campus, a support service system was established for students who needed various types of assistance toward the completion of degree requirements. Tutorials were at the core of this program.

From the late 1960s into the mid-and-late-seventies, special minority recruitment programs for undergraduate engineering degree programs continued to expand to the point that almost half of the 282 schools of engineering were actively committed to the recruitment of black and other minority students. In other instances, the degree of involvement ranged from the total absence of a special recruitment program; a passive stance characterized by reliance upon the institution to admit students who might drift into engineering; to relatively moderate, uncoordinated and disorganized efforts.

Those institutions that were heavily involved in special recruitment employed a number of different strategies and frequently a combination such as those listed on pages 213-214.

In addition to these and other types of recruitment strategies, a number of state and/or regional organizations provided early identification programs reaching into the junior high school levels, and scholarship monies. Principal among such groups were The Committee to Increase Minority Professionals in Engineering, Architecture and Technology (CIMPEAT) of

Atlanta; Philadelphia Regional Introduction for Minorities in Engineering (PRIME); California Consortium for Minorities in Engineering (CCME); Committee for Institutional Cooperation-Midwest Program for Minorities in Engineering, Inroads, Inc.; Engineering Consortium for Minorities; Texas Alliance for Minorities in Engineering (TAME); and the Southwest Consortium for Minorities in Engineering. Generally speaking, all regions of the country were covered by such programs as those listed above.

Heavily involved institutions also drew upon currently enrolled minority students to assist in recruitment. They utilized lists of potential students generated by the ME program, the Student Search List of the College Entrance Examination Board, the American College Test (ACT) list, the National Achievement Scholarship List, and similar sources of potential students.

The salience of a high quality of recruitment is registered in enrollment patterns. The data clearly show that wherever strong, well-coordinated and organized recruitment programs exist, the enrollment of black students is equally significant and considerably beyond the token access stage. On the other hand, if recruitment is non-existent or passive, enrollment is unimpressive or non-existent.

Even among the largest schools of engineering in the nation, such as the University of Pennsylvania, aggressive action is necessary. This is so even when the institution enjoys high prestige in the engineering profession. High school students rarely know anything about an institution's reputation in the same sense as academicians do. To be passive means that, despite a good reputation, an institution may still enroll only a handful of minority students.

Despite the obvious progress made in the enrollment of black students in engineering during the seventies, the fact remains that a number of institutions only have good "paper" programs to accelerate equality of opportunity. The most fundamental problem is their failure to commit adequate resources, human or financial, for the implementation of these programs.

Admissions Criteria and Policies

Undergraduate Admissions

Admission to undergraduate colleges of engineering is not generally done by the engineering colleges. Students are admitted to undergraduate colleges and declare a major field either in the freshmen year or at some subsequent stage. The tendency of many students to explore possible major fields of study before declaring a specific major compounds the issue of comparisons between first-year entry and entry in subsequent years. Nevertheless, institutions most frequently rely upon such quantitive assessment factors as SAT Scores, the high school GPA, standing in high school graduating class, and rating of the high school from which the student was graduated in determining eligibility for admission.

Qualitative factors utilized include a wide range of criteria which impact upon the discretionary powers of assessment, selection and admissions officers in determining who is worthy or desirable. These non-cognitive factors are fundamentally the same as those identified in earlier chapters. However, at the undergraduate level, they may include prowess and special skills such as in athletics, music, or debating. Motivation, determination, perserverance, and any number of extra-curricula activities may be considered. Hence, special admissions in the traditional sense may not be as significant a factor in determining admission as would be the case in some graduate programs. Where it may be salient is in whether or not the student has completed "pre-engineering" requirements, especially in mathematics and science credits and with a sufficiently high grade point average.

Graduate Admissions

At the graduate level, special recruitment and special admissions programs for black and other minority students are rare. Admissions decisions are most frequently based upon GRE scores, especially in the sciences; GPAs in undergraduate colleges; letters of recommendation and sometimes personal interviews. Almost all of them require a B.S. in Engineering, Physics, or a related field. Cut-off points on GRE scores are institution-specific and no particular pattern is discerned. The minimum GPA required ranges from 2.5 to 3.0 on a 4.0 system. However, many institutions seem more concerned about the quality of academic performance during the last two years of an undergraduate program in engineering. They want to ascertain patterns of growth in academic performance and many are willing to take a chance on "late bloomers." Hence, some degree of flexibility in admissions criteria is apparent. In fact, increasing flexibility may become more widespread as larger numbers of graduate schools in engineering begin to be affected by declining enrollments created by financial inducements for B.S. in Engineering graduates from high technology industries, the business community, or the corporate world.

Enrollment of Black Students in Engineering

Unquestionably, the efforts initiated by General Electric, the corporate structure in general, and the external influences on both the business community and institutions led to major progress toward expanding engineering training opportunities for black and other minority students during the seventies. While the goal of a ten-fold increase within the decade was not attained, significant progress was made toward achieving a greater balance between demand and supply. Once again, increases in enrollment patterns reflect the quality of recruitment, scholarships, financial aid in general, and, most significantly, the presence or absence of black faculty. State totals by year show that being under litigation to dismantle dual systems of education is increasingly correlated with both first year enrollments and with total enrollment of black students. In this analysis, our attention is first focused on trends in enrollment in B.S. in engineering degree programs. The B.S. in engineering is here regarded as the first professional degree in engineering.

Undergraduate Enrollment Trends

In 1970-71, there were 1,289 black students enrolled in first year B.S. engineering programs in the United States. They represented 1.7 per cent of the total enrollment in first year classes of 71,661. In the following academic year, engineering colleges experienced a shocking decline in first year enrollment when they lost approximately 13,000 students. This decline in first year enrollments continued through the 1973-74 academic year and was followed by a dramatic upturn in 1974 that has since continued.

Although engineering educators cannot fully explain this alarming downward trend with complete confidence, they advanced certain possible explanations. This particular trend was attributed most frequently to the Vietnam War. Some students voiced their opposition to science and engineering due to the presumed role this field performed in the continuation of the military effort. Other, eligible students were drafted during this period. Still other students believed that a degree in engineering was not marketable. For whatever reasons, the decline in first year enrollments continued for three years. This drop was partially off-set by an increase in total enrollment in 1972-73 which probably resulted from large numbers of engineering junior colleges transfers into four year colleges and universities (See Table 32 and figure 2).

Table 32. First Year Enrollment in B.S. in Engineering Programs by Race, Year, 1968–79, and Percent: All Institutions Combined.

Year	First Year Enrollment	Black Student Enrollment	Per Cent Black Students in First Year Enrollment
1968-69	77,484	—	—
1969-70	—	977	—
1970-71	71,661	1,289	1.7
1971-72	58,566	1,289	2.2
1972-73	52,100	1,477	2.8
1973-74	51,920	2,130	4.1
1974-75	63,440	2,848	4.5
1975-76	75,343	3,840	5.1
1976-77	82,250	4,372	5.3
1977-78	88,780	4,728	5.3
1978-79	95,805	5,493	5.7
1979-80	103,724	6,339	6.1

Source: Engineering Manpower Commission.
Blank spaces indicate data unavailable for that year.

As first year enrollments dropped off during the early seventies for all institutions combined, the actual number of black first year students began a steady increase that had not abated at the end of the decade. The proportion of black students in the first year classes climbed in every year of the decade and by 1979-80 stood at 6.1 per cent of all first year engineering students. Translated into actual enrollment figures, this means that in 1979-80, there were 6,339 black first year engineering students out of a total first year enrollment of 103,724 in all engineering colleges.

These data also suggest that one of the central explanations for the comparative ease with which black and other minorities entered colleges of engineering during the seventies was that engineering colleges had a great deal of space for anyone interested in the field. These colleges were not filled to their capacities like medical and dental schools, were not pressed for space, and some had difficulty in enrolling a sufficient number of students to maintain credibility within their own institutions. This lack of competition for a resource that was not scarce meant that many institutions welcomed this new source of students, not entirely out of "a new social consciousness" but as a mechanism for halting downward spiral in enrollments.[20] This fact, coupled with the attractive salaries offered engineering graduates and assurances of jobs, at a time when unemployment rates continued to rise in the nation made engineering more attractive than many career alternatives.

Moreover, the applicant pool among blacks to colleges and universities began to expand significantly during the early part of the decade and black students entered post-secondary education in unprecedented numbers. In fact, the traditionally white institutions soon outdistanced the historically black colleges in enrolling black students. This should not be surprising since white institutions outnumber black institutions by almost fifty to one. Even if all of them went very little beyond limited access, their total enrollment of black students would be significantly greater than that found among historically black institutions. The crucial variable in the access-outcome equation is the difference between these two institutional types in their rate of productivity of black student graduates. In this regard the historically black institutions do a far superior job in terms of absolute numbers produced.

Total Enrollments

Total enrollments refer to the combined enrollments of all classes. Table 33 depicts the trends in these enrollments. According to these data, the declines evident in first year enrollments were similarly reflected in total enrollment for the period between 1969 and 1973-74. With the exception of the 1972-73 academic year, total enrollments dropped alarmingly during this period. There was a loss of approximately 16,000 students between 1969-70 and 1970-71; a further loss of about 8,000 students occurred during the following year. Inexplicably, there was a gain of about 15,000 students in 1973-74. This upswing in total enrollments continued for the remainder of the decade so that by the 1979-80 year the 340,448 students enrolled in all colleges of engineering combined represented an increase of more than 150,000 over the 1970-71 figure.

In absolute numbers, total enrollment for black students also showed a steady climb during the seventies. In 1969-70, there were only 2,757 black students enrolled in the nation's colleges of engineering. As we have previously stated, a disproportionate number of these students were enrolled at the six historically black colleges. Because of coordinated and well-organized recruitment programs, the availability of financial aid to assist

Table 33. Full-Time B.S. in Engineering Enrollment Programs by Race, Per cent of Total and by Year, 1969–79: All Institutions Combined.

Year	Total Enrollment	Total Black Students	Per Cent Black of Total
1969-70	210,825	2,757	1.3
1970-71	194,727	4,136	2.0
1971-72	186,705	4,356	2.2
1972-73	201,099	5,508	3.0
1973-74	186,700	5,508	3.0
1974-75	201,100	6,827	3.4
1975-76	231,379	8,389	3.6
1976-77	257,835	9,828	3.8
1977-78	289,248	11,388	3.8
1978-79	311,237	12,954	4.2
1979-80	340,488	14,786	4.3

Source: Engineering Manpower Commission; reports from 1969-1979/80.

impoverished students, particularly, and the willingness of colleges of engineering to enroll them, black students entered this field in heretofore unprecedented numbers. As a result, in 1979-80, the number of black students in engineering represents an approximately six-fold increase over the 1969-70 figure (See Table 33 and figures 3 and 4).

The 2,757 black students enrolled in 1969-70 represented 1.3 per cent of total engineering enrollments. The 14,786 enrolled in 1979-80 means that there has been an enrollment increase of approximately 600 per cent over the proportions noted a full decade earlier. However, it also means that just as black students entered engineering in exceptionally larger numbers than ever before in their history, so did white students (and, of course, students from other minority groups). Essentially, the entire field of engineering expanded.

Capacities continued to increase but the demand for engineering graduates in the late 1970s outdistanced supply once again. Consequently, engineering is a critical need area with substantial career opportunities for all those who qualify. A changing technology has now dictated certain manpower needs that must be met. The result is that students trained in engineering, especially in computer, petroleum, and mechanical engineering, are probably the most highly sought bachelors degree recipients in the United States in 1980.

Who enrolls black students in engineering? Aside from the six historically black colleges, our data show that the following twenty institutions* were the leaders in the enrollment of black undergraduate students in engineering during the 1970s:

Georgia Tech	Michigan State University
Purdue University	Illinois State University
University of Houston	University of Michigan
University of Illinois/Chicago	North Carolina State University
New Jersey Institute of Technology	University of Illinois/Urbana
Wayne State University	University of Missouri/Rolla
Pratt Institute	University of New Mexico
General Motors Institute	Detroit Institute of Technology
Ohio State University	Drexel
University of Tennessee/Knoxville	Texas A & I

In terms of proportion of black students in the engineering college, the latest available data by State, as opposed to institutional data, show that the following jurisdictions have the largest proportions of black students in their schools or colleges of engineering: the District of Columbia (40.6%), Illinois (11.4%), Michigan (11.1%), and Mississippi (6.8%).

In the "Adams States," the latest available data provide the ranking in terms of proportion of black students enrolled in all schools and colleges of engineering in the state shown in Table 34.

Using the definition of total access as a number of black students enrolled that approximates or surpasses proportion of blacks in the state's population, it is evident that black students are far from attaining parity in enrollment. Clearly, at the present rate of matriculation, a massive effort will have to be mounted in order to achieve even the second level of access.

Graduate Enrollment

In 1969-79, there were 55 black students enrolled in M.S. in engineering programs in all of the United States. The number more than doubled to 112 in 1971-72 and continued its upward climb in every year of the seven-

---

*Only seven of these institutions were included in the top fifty engineering programs listed in the September 8, 1980 issue of *The Chronicle of Higher Education*.

Table 34. Current Rank of "Adams States" by Proportion of Blacks Enrolled in Engineering Schools by State and Proportion of Blacks in State Population, 1979: All Institutions Combined (B.S. Programs).

State	Per Cent Black in all Engineering Schools (B.S. Programs)	Rank	Proportion of Blacks in State Population (Per cent)
Mississippi	6.8	1	35.9
Maryland	5.2	2	20.1
Pennsylvania	4.7	3	8.8
Virginia	4.7	4	18.7
Georgia	4.6	5	26.1
Florida	4.1	6	14.2
Oklahoma	2.9	7	7.1
Arkansas	2.4	8.	16.9
North Carolina	2.4	9	21.9
Louisiana	2.1	10	29.8

Sources: Engineering Manpower Commission and U.S. Bureau of the Census.

ties. The 165 black students in M.S. degree programs in 1971-72 comprised only 0.74 per cent of all M.S. engineering students. By 1976-77, the 321 black students enrolled in M.S. programs represented 1.2 per cent of the 25,516 students in all institutions seeking that degree. According to the 1979-80 fall enrollment data, there were 357 black students seeking the M.S. degree out of a total of 27,171 or 1.3% black students in the total M.S. enrollment. Despite increases in absolute numbers, black students are grossly underrepresented at this level of training.

Among the early leaders in the enrollment of black students at the M.S. level are such institutions as the following: Stanford University, University of California/Berkeley, the University of Southern California, George Washington University, Howard University, Georgia Institute of Technology, Massachusetts Institute of Technology, the University of Detroit, Michigan State University, the New Jersey Institute of Technology, the New York Polytechnic Institute, and Cornell University.

Later in the decade, the University of Pennsylvania, the University of Illinois, and Pratt Institute were added to this list.

In the "Adams States," Georgia Tech led all institutions in the training of blacks at the M.S. level. Only Pennsylvania, among all remaining colleges of engineering in the "Adams States," has enrolled a significant number of blacks at the M.S. level.

The number of black students studying for a Ph.D. degree in engineering in 1979-80 is five times greater than the number enrolled in this program in 1969-70. However, the overall trend in the enrollment of black doctoral students did not show a steady or continuous incline. For instance, the number of black doctoral students doubled from 22 to 44 between 1969 and 1971 and increased by eight students in 1972. However, as the national totals of all students enrolled in doctoral degree programs dropped between 1972 and 1973, so there was a decline in the absolute numbers of black students enrolled in engineering doctorate programs. The erratic nature of the pattern is once again observed in the more than doubling of the number of black students studying for the doctoral degree in 1974, followed by a decline which resulted in a loss in excess of sixty per cent by 1976-77 when only thirty black students enrolled in doctorate programs. The 109 black students enrolled in doctorate programs in 1979-80 represent 0.81 per cent of the total number of 13,461. Clearly, black students are substantially below parity at the Ph.D. level.

Among the institutional leaders in the enrollment of black Ph.D. students are such institutions as: Stanford, M.I.T., the University of California at Berkeley, George Washington and Howard University. Only George Washington and Howard are not listed in the top 50 engineering colleges in 1980.*

With regard to the "Adams States," not a single institution enrolled a significant number of black students in Ph.D. programs throughout the decade. The notable exception was the University of Pennsylvania and that occurred during the second half of the period.

At the graduate level, black students tend to enroll in the following engineering specializations: Chemical, Civil, Computer, Electrical, Mechanical, and Nuclear Engineering. They do not seem to be attracted at all to such specializations as: Ceramic, Architectural, Construction, Drafting and Design, Environmental, Metallurgical, and Mining Engineering.

These enrollment data depict an overall trend of continuous progress in both first year and total enrollment of black students in schools and col-

---

*This statement is based upon a ranking determined by "achievements of graduates" of the 282 colleges of engineering. (See *Chronicle of Higher Education,* September 8, 1980, p.2).

Table 35. Master's Degree Enrollment in Engineering by Race and Year, 1969 - 79 All Institutions Combined (Full-Time)

Year	Total in M.S. Enrollment	Total Black Students in M.S. Enrollment	Per cent Black of Total
1969-70	—	55	—
1970-71	—	112	—
1971-72	22,405	265	.74
1972-73	22,588	221	.98
1973-74	—	220	—
1974-75	—	200	—
1975-76	—	251	—
1976-77	25,516	321	1.26
1977-78	26,107	326	1.30
1978-79	25,360	322	1.27
1979-80	27,171	357	1.31

Blank spaces indicate data unavailable for that year.
Source: The Engineering Manpower Commission.

leges of engineering throughout the 1970s. Similarly, students of all races have entered engineering in numbers never previously conceived. Yet, there remains a major shortage of engineers for the labor force because of increased demands in the high technology sector. As a result, graduate enrollments are significantly below the level desired, primarily because of the market situation. There is little or no need for most B.S. degree recipients to think in terms of enrolling in M.S. or Ph. D. programs when the entry level salaries of this group is often far greater than what a newly minted Ph.D. can expect to receive as an assistant professor in most universities and colleges. Many companies in the corporate structure also offer the possibility of graduate training as a fringe benefit in order to attract much needed skills. This fact holds true for all races and both sexes.

Consequently, it takes an exceptionally motivated student to defer social and economic gratification to study for an advanced degree on a stipend that is from one-seventh to one-fourth of what the starting salary of a B.S. degree holder commands in 1980. Nevertheless, one must take note of the increase by approximately one thousand students in the total number

Table 36. Doctorate Engineering Degree Enrollment by Race and Year, 1969 - 79

Year	Total Doctorate Enrollment	Total Number of Black Doctorate Students	Per cent Black of Total
1969-70	—	22	—
1970-71	—	44	—
1971-72	14,100	49	.35
1972-73	13,460	52	.39
1973-74	11,904	38	.32
1974-75	—	71	—
1975-76	—	67	—
1976-77	10,963	30	.27
1977-78	12,369	101	.82
1978-79	12,321	100	.81
1979-80	13,461	109	.81

Blank spaces indicate data unavailable for that year.
Source: The Engineering Manpower Commission.

enrolled in doctorate programs in 1979-80 over the number enrolled in 1978-79. However, the increase in the number of black doctorate students was about nine per cent but their representation among the doctorate population is insignificant.

Retention

There is a general agreement that the major problem of the 1970s was the retention of those black students who were enrolled so that the supply side of the equation could be more balanced.[21] The gravity of the situation may by underscored by a single example. The numbers of black graduates with a B.S. in engineering increased only by 77 students between 1974 and 1977 despite the pronounced increases in first year and total enrollments in years that should have produced more graduates. We have no precise account of retention rates for every one of the 282 engineering institutions in the nation. However, the Engineering Manpower Commission of the

Engineering Joint Council has demonstrated in its research that the graduation rate of minority students is about 54.8 per cent in contrast to 77.8 per cent for white students.[22] This is a highly significant disparity of more than 23 per cent. For black students, the rate may be higher or lower, depending upon the institutions selected for analysis. At the current rate of production of black engineers, who comprise about two per cent of all engineers, massive increases in the total number of black students both enrolled and graduated, will be imperative over the next twenty years.

According to a study conducted by the Task Force of the Committee on Minorities in Engineering, the Assembly of Engineering, and the National Research Council and reported in 1977, the causes of the high attrition rate among blacks and other minorities are indeed multi-dimensional and many-faceted. The five most serious factors associated with low retention or high attrition are: (1) inadequate pre-college preparation, (2) inadequate motivation, (3) inadequate financial assistance, (4) low confidence and self-doubts about one's own ability to perform adequately in engineering programs, and (5) personal and family problems.[23]

In addressing these problems, *it must be reiterated that all black and all minority students do not fall in the category of "under-prepared, under-financed and problem-inflicted" students. A significant proportion of these students are more than adequately prepared, often exceptionally-well prepared for college programs, regardless of the field of study.* A smaller proportion, as previously stated, come from well-to-do homes and do not ordinarily require substantial financial assistance other than parental resources. *It is a tragic mistake to assume that black or any other minority students represent a homogeneous type and it is worse to treat them as if all required special, and all too frequently, demeaning and paternalistic attention.*

Nevertheless, we cannot deny certain demonstratable realities. The school systems of the United States, especially urban public schools, have failed black and other minority pupils. As a group, these students do not have the same quality of preparation in science, mathematical or computational skills, or in the fundamentals of the English language. Self-esteem, that is often high in the home environment, is forced to new depths by the environmental and learning conditions within the school. Expectations of some teachers are low regarding black students and many of them seem anxious to fulfill the prophecy of low performance communicated by teachers. Besides, many black students are literally turned off by racism within the school system.

The home environment, especially parental motivation, may not be conducive to many of these students to develop sound skills. That condition may be a function of many factors, including parental disinterest; lack of parental guidance and influence on the child's behavior; the debilitating consequences of poverty or near poverty, which demand that both parents spend excessive amounts of time away from home; and the absence of learning tools within the home itself.[24] These situations coalesce to create a total environment in which under-preparedness might be an expected outcome.

This situation is also related to the issue of motivation for a career as an engineer. Without adequate pre-college programs and excellent academic counseling in junior and senior high schools, it is unlikely that a sufficient number of black students will learn the essentials required for succes as an engineering student. Students frequently do not receive the information even when institutions mail voluminous amount of leaflets and flyers to high schools. Counselors may not always believe that a black or poor student has the capacity to be an engineer. Therefore, persuaded by his or her own biases, the pupil is encouraged by the counselor to opt for a less strenuous curriculum leading to a less prestigious and less demanding job upon graduation from high school.

Finance

A particularly serious impediment to retention through graduation is the financial plight of a disproportionate number of black students. This problem goes considerably beyond having sufficient funds for tuition and books. It extends to worry and the psychological difficulties engendered by excessive anxiety over *how* to obtain the money required to remain in school.[25]

As a general rule, students meet their financial needs in one or a combination of at least four methods: (1) parental contributions; (2) federal and state programs, including loans, scholarships, and grants; (3) students'/spouse's own earnings; and (4) funds from foundations, private business and industry, and individual philanthropists.[26] However, according to a study conducted in 1978, minority students do not generally have the resources required to meet the average tuition costs and other education expenses. The ECDP estimated in 1978 that average tuition costs in public engineering schools (in-state only) was $624.00 while that of private institutions was $3,099.[27] Total costs amounted to $2,927 in public institutions and an average of $5,620 in private colleges of engineering. Given the high proportion of black students who come from families whose median family income is under $10,000 per year, it is evident that most require some form of financial aid. The problem is especially critical for blacks attending private institutions since their costs are considerably higher.

Recognition of the magnitude of the financial aid problem is not a recent phenomenon. At the inception of the minorities in engineering effort, it initiated programs to assist black and other minority students to meet their financial requirements for engineering education programs. Earlier on, substantial contributions were made by members of the corporate structure, foundations, and private individuals which provided much needed scholarships for black and other minority students.

When the Planning Commission for Expanded Opportunities in Engineering issued its highly influential report, *Minorities in Engineering: A Blueprint for Action* in 1974, it called for the strengthening of financial aid contributions to specified minority groups. It also recommended the establishment of a new fund-raising organization which would be the conduit for all funds raised to support the minority in engineering effort. Out of that recommendation, based upon its belief that the quality of financial aid was the major obstacle to accelerating minority students access to engineering, came the National Fund for Minority Engineering Students (NFMES).[29] This organization was established in 1975, just as financial aid resources overall began to decline, especially for students in other professions. Its essential functions are to "raise and distribute scholarship funds."[30] Early in 1976, the organization had raised over $2 million, one-half of which was distributed to 1,000 minority students.[31]

The 54 NACME members, including 34 corporate chairmen, renewed their pledge for support for the engineering effort in 1976. Of 24 NACME members, in a survey conducted under the auspices of the Conference Board and reported in 1978, this commitment was already evident. More than $5 million in total corporate contributions were made in 1977 alone. (In addition, these firms had given generously for the purchasing of school supplies, laboratory and classroom equipment and books.)

They also contributed 88 per cent of the total amount given by NACME companies to the NFMES. Pledges rose to 9.3 million by June 1979 from some 89 corporations. Private foundations pledged an additional $2 million. It should be noted, however, that three companies gave over a million dollars each and that this represented 42 per cent of all contributions from the corporate structure. The top companies accounted for 67 per cent of the total contributions. Without a doubt, these contributions generated considerably more money in scholarship dollars for black and other minority students.[32] Precisely how much more is not known; however, since over 14,000 black students are currently enrolled and since most minority students rquire financial support, much greater help is presently needed.

The NFMES initiated a system of "incentives grants" for a school or college of engineering which "committed itself to challenge goals." These goals are enrollment targets mutually agreed upon by the Fund and the institution. Awards are granted as these targets are met or approximated and scholarship monies are made available to support the target group. Seventy engineering colleges, including Massachusetts Institute of Technology (MIT) and the Illinois Institute of Technology (IIT), participate in the program. A thousand students are presently receiving NFMES funds. The awards granted to individual students range from $250 to $2,000. The criterion for an award is now "the most qualified students who also need financial aid." When the Fund was first established, the criterion was listed as "a needy student."[33] The average award is about $1,000 per year.

The Fund also sponsors a Summer Engineering Employment Project. This financial aid strategy further enables many black and other minority students to obtain invaluable work experience in an engineering setting such as in industry and in business. Many of these companies offer job opportunities to students who participated in summer programs following the completion of the B.S. degree in engineering.

Self Confidence

The low level of self-confidence that is also offered as an explanation for the drop-out rate among black and other minority students is not only related to underpreparation but to other more personal and social factors. While it is quite true that many students do not receive adequate preparation in science and mathematics, many all too early in their lives internalize negative attitudes about these subjects. They come to regard such subjects as too difficult or that the subject matter that can only be mastered by the bright and gifted students. Similarly, in many instances, those students who excel in science and mathematics are placed on a pedestal by their teachers and set apart for special treatment. Those who do not demonstrate the same degree of skill or apparent competence may inadvertently be regarded as somewhat "less brainy" or sometimes unable to master the most rudimentary science problem. If the student constantly confronts such attitudes, without proper home reinforcement of his/her own capabilities, he/she is much more likely to internalize these doubts and to develop a lowered sense of self-confidence than would otherwise be the case.

Estrangement

Personal and family problems combined with efforts to adjust to a different social milieu may generate complications in the learning process. When lifesytles encountered in the college environment, especially for students living in residence halls, are widely varied from those of the home situations, new adjustments are necessary. Making this transition is sometimes difficult for students. Their reaction may be withdrawal and further isolation from their peers. That isolation is a manifestation of institutional alienation which, in turn, may become so severe that the response deemed by the students to be most appropriate is simply to withdraw.[34]

Attrition can be controlled. It can be reduced if the early signals of difficulty are immediately recognized and properly handled. But this requires imaginative programs, staff, financial aid, role models, counseling, academic support, a positive and supportive learning environment, reduced racism, and elimination of smothering paternalism disguised as liberalism. It demands recognition of the reality that each black student is a distinct and unique personality—separate and apart from all others. In effect, broad-based retention programs which avail academic and affective support services to all students are vital.

Graduation or Productivity Rates

Although about three times as many black undergraduate students received the B.S. in Engineering degree in 1978-79 as was the case in 1969-70, black students receive only about two per cent of all such degrees awarded in the United States. The number of B.S. degrees earned by black students dramatize clearly the need for stronger retention programs. The disparity between the numbers of degrees conferred to white students and those awarded to black students shows *that the gap between these two groups is widening rather than narrowing.* Consider that the total number of engineering degrees earned in 1978-79 is approximately 10,000 or more than the 1969-70 figure. Observe the fact that the 901 B.S. degrees earned by black students represent an increase of only 522. This means that with all the programs developed and implemented for assuring greater access, the nation's engineering colleges are producing only 522 more first professional degrees in engineering among black Americans than they were a full decade earlier. *During that period, a total of 430,123 persons received B.S. degrees in Engineering. Slightly under 7,000 (6,967) were black Americans who earned a paltry 1.6 per cent of all those degrees conferred between*

1969-70 and 1978-79. Clearly, blacks are not catching up as fast as they must despite the notable progress they experienced during the decade (Table 37 and figure 5).

Table 37. Bachelor's Engineering Degrees by Race and Year, 1970-79: All Institutions Combined

Year	Total B.S. Degrees Awarded	Total B.S. Awarded Black Students	Per Cent Black of Total
1969-70	42,966*	378	.8
1970-71	43,167*	407	.9
1971-72	44,190	579	1.3
1972-73	43,429	657	1.5
1973-74	41,407	796	1.8
1974-75	38,210	734	2.0
1975-76	37,970	777	2.0
1976-77	40,095	844	2.1
1977-78	46,091	894	2.1
1978-79	52,598	901	2.0
Totals	430,123	6,967	1.6

*Figures for black students are understated because they do not include data from non-reporting institutions.
Source: The Engineering Manpower Commission.

Graduate Degrees

During the seventies, graduate schools of engineering in the United States conferred a total of 163,372 M.S. degrees. Of that number, 1,240 went to black students. In other words, *black Americans earned less than one per cent (0.7) of all M.S. degrees conferred in Engineering during the seventies.* (Table 38 and figure 5).

Progress made cannot be denied, but, progress is relative and must be placed in the context of the total number of degrees conferred. While the total number of such degrees conferred per year averaged about 15,000, the total for black students averaged around 120 per year. The highest number

Table 38. Master's Engineering Degrees Conferred by Race and Year, 1970-79: All Institutions Combined

Year	M.S. Degree Total	Number Awarded Black Students	Per cent Black of Total
1969-70*	15,548	50	.3
1970-71*	16,383	47	.8
1971-72	17,356	78	.4
1972-73	17,152	104	.6
1973-74	15,885	158	.9
1974-75	15,773	141	.8
1975-76	16,506	154	.9
1976-77	16,551	147	.8
1977-78	16,182	202	1.2
1978-79	16,036	159	.9
Total	163,372	1,240	.7

*Figures for black students are understated because they do not include data from non-reporting institutions.
Source: The Engineering Manpower Commission.

of M.S. degrees earned by black students was in 1977-78 when the 202 degrees conferred on them represented 1.2 per cent of M.S. degrees. As shown in Table 38 however, there was a sharp decline in the numbers of M.S. degrees conferred in both totals as well as on black students.

The leading institutions* in terms of places from which black students earned the M.S. degree in Engineering were:

1. Howard University
2. Stanford University
3. M.I.T.
4. George Washington University
5. Ohio State University
6. University of Ill./Urbana
7. University of Calif./Berkeley
8. University of Michigan
9. University of Southern Calif.
10. New York Polytechnic
11. Tuskegee Institute
12. Columbia University
13. Cornell University
14. University of Florida
15. Air Force Institute/Ohio
16. Oklahoma State University
17. Drexel University

*Eleven of the 15 institutions are included in the Chronicle's list of 50 top schools of engineering in 1980.

Among this group of institutions, the range in the number of M.S. degrees earned was from 215 at Howard University to 21 from Drexel. The relative position of these institutions might change slightly if all of them had reported in every year of the decade. However, the pattern of graduation observed in relation to the number of enrolled students suggest confidence with the present ranking of institutions.

Among the "Adam States", the following institutions are included in this top group: (1) the University of Florida ranked fourteenth and (2) the Oklahoma State University ranked sixteenth. No other institution in an "Adams State" is included in this top twenty-five institutions.

The number of black students who earned the Ph.D. degree in Engineering rose steadily from one degree in 1969-70 to nineteen degrees conferred on black doctorate students in 1978-79. Again, this total is less than one per cent (0.5%) of the total number of degrees earned in 1979 compared to .03 per cent of the degrees conferred to the lone black engineering doctorate a decade earlier. Prior to 1979, a peak of seventeen black students who earned the doctorate was attained in 1974-75. This was followed by a sharp decline to only 10 degrees the next year. However, the present trend shows an upswing the duration of which we venture not to predict. It is apparent, however, that this paucity in graduate degrees, at both the M.S. and Ph. D. levels, has serious and significant implications for affirmative action. Depending upon the number of positions that actually require a graduate degree in Engineering, there may very well indeed be an escalating problem of inadequate supply. Blacks are not entering Ph. D. programs and blacks with Ph. D.s are not being produced. (Table 39).

Even after considering the fact that all institutions did not report degrees awarded to the Engineering Manpower Commission for every year, a ranking of institutions is still possible based upon reported data. Accordingly, the leading institutions* in terms of Ph. D. degrees in Engineering earned by black Americans during the seventies were:

1. Stanford University
2. Cornell University
3. University of Calif./Berkeley
4. M.I.T.
5. University of Ill./Urbana
6. UCLA
7. University of Southern Calif.
8. Ohio State University
9. George Washington University
10. Howard University
11. University of Michigan
12. Harvard University
13. Carnegie Mellon University
14. John Hopkins University

15. University of Oklahoma

In this group, the total number of Ph. D. degrees earned by black

*Ten are included in the top 50 colleges of engineering.

Table 39. Engineering Doctorates Conferred by Race and Year, 1970-79: All Institutions Combined

Year	Total Number of Doctorates Awarded	Total Awarded Black Students	Per cent Black of Total
1969-70*	3,620	1	.0
1970-71*	3,640	8	.2
1971-72	3,774	13	.3
1972-73	3,687	13	.3
1973-74	3,362	12	.3
1974-75	3,138	17	.5
1975-76	2,977	10	.3
1976-77	2,814	16	.5
1977-78	2,573	15	.5
1978-79	2,815	19	.6
Total	32,400	122	.3

*Means that these figures for Black students are underestimated because they do not include data from non-reporting institutions.
Source: The Engineering Manpower Commission.

Americans during the seventies, ranged from 41 at Stanford to 11 at John Hopkins and the University of Oklahoma.

The "Adams States" are presented in this group by only the two institutions ranked fourteenth and fifteenth—John Hopkins and the University of Oklahoma.

Problems

One of the most important tasks before the engineering education community is to provide more role models for black students in colleges of engineering. It appears that fewer than twenty institutions have black faculty members among all 282 schools and colleges of engineering in the United States.[35] Six of them are the historically black colleges. Especially in highly technical and scientific fields to which so much prestige is attached, it is imperative for young people to have first-hand knowledge that members of their racial or ethnic group have become successful. The absence of black faculty members at the graduate level, particularly, may be a major explanatory factor. Black graduate students do not have the option of working and studying with black mentors as do white students. It is in fact a tribute to the recruitment programs and the minority recruiters that they have done as well as they have in producing what is tantamount to a phenomenal increase in the overall number of black students in engineering during the decade. The problem here is not only the diminishing momentum observed between the completion of the baccalaureate degree and ability to encourage enrollment in graduate degree program. It is a serious indictment of the academic community and State legislators for failing to make academic salaries more competitive with industry and the corporate community.

Another major problem is the disjunction between entry level enrollment figures and the actual production of engineers among black Americans. The rate of production lags far too much behind enrollment. That is a function of attrition and failure to mount and sustain a sound retention program. Undoubtedly, the two primary aspects of this problem are the quality of preparation in science and mathematics and the availability of a high quality financial aid package. To resolve the difficulties implicit in these two variables requires an essential linkage between all pre-college programs and changes in institutional conditions. Obviously, expanded financial aid is needed now more than ever before.

Most of the institutions surveyed claim that the *Bakke* case will have absolutely no impact on their minority engineering effort.

Architectural Education

A more detailed study of architectural education and the access of black and other minority students than provided here is imperative. The Association of Collegiate Schools of Architecture did not provide useful data required to establish ten year trends. According to one source, "only 20 to 30 per cent of the 87 schools" approved by the National Architectural Accrediting Board (NAAB) report enrollment and graduation data by minority group on a regular basis.[36] This truncated discussion is based largely upon information on the period between 1975 and 1978.

Apparently, during the sixties, professional architects made some efforts toward consciousness raising regarding the status of women and minorities in the profession. Like so many professions, environmental design and general architecture had been and still remain, although to a lesser degree, a white male preserve. Conferences were held on the subject of expanding educational opportunities for these groups but there is very little information concerning formal programs which went beyond the talking stage.[37]

It is estimated that less than one per cent of the architects in the United States are black.[38] Many of them were trained at one of the seven departments and schools or colleges of architecture located at historically black colleges: (1) Florida A & M (FAMU) University, (2) Hampton Institute, (3) Howard University, (4) Prairie View A & M, (5) Southern University, (6) Tennessee State University, and (7) Tuskegee Institute. As of the Summer of 1979, the program at Prairie View was not accredited by the NAAB and the program at Tennessee State University was in Architecture and Engineering Technology.[39] Further, FAMU's School of Architecture is regarded by some as a white architectural school in a black institution, established primarily by white faculty.

Many of the historically black colleges are currently experiencing difficulties in the recruitment of students for their programs. Consequently, they have embarked upon aggressive recruitment efforts in order to attract more students and to protect their programs. Part of the problem stems from the change in clientele especially in institutions, black and white, that are in transition from an undergraduate to a graduate school.

Although a few changes may have occurred since *Architectural Schools in North America* was issued, some insight may be gleaned from enrollment data on "ethnic and minority students" reported by the historically black colleges. At that time, the number of students so classified were (1) 23 at FAMU, (2) 15 at Hampton Institute, (3) 21 at Howard University, (4) no data on Prairie View A & M University, (5) 142 at Southern University, (6) 96 at Tennessee State University, and (7) 108 at Tuskegee Institute. Since we normally assume that about 90 per cent of students classified in the category of "racial and ethnic minority" are black, it is safe to assume that Southern University in Baton Rouge enrolled more black students in Bachelor of Architecture programs than any institution in the nation. In fact, of the 52 institutions included in this study, 39 enrolled fewer students in this category than did Hampton Institute.[40]

Efforts to enroll more black students in schools of architecture throughout the nation seemed to have peaked during the mid-seventies. It is claimed that only about one-third of the 87 NAAB accredited schools now make a special effort to attract black students to this profession and most of them are private institutions. In general, public institutions have larger number of applicants and, judging from past history, have not been as inclined to expend the necessary energy to assure a more heterogeneous mix among the student body of colleges of architecture.[41]

Nevertheless, according to 1977-78 estimates, of the 14,000 students enrolled in the first professional degree programs (Bachelors of Architecture, B.A. in Architecture) across the country, 1300 or 8 per cent are members of *minority groups*. Some 138 minorities or six per cent of the 2,233 students reported are enrolled in 4-2, M.A. programs. The largest *minority students enrollment,* once again, appears to be at those institutions that have minority faculty members.

In 1977-78, there was an average of 2.5 minority faculty in the 52 colleges of architecture covered by this special report. There was an eight per cent increase in the number of minority faculty hired over the 1976-77 year. It is extremely difficult to establish one's own firm and, further, there are so many risks involved when one does not have a good supply of clients. Consequently, most graduates either attempt to find jobs with established firms *or* they obtain employment in the governmental sector *or* in college and university teaching. In 1978, there were approximately 1500 full-time and 1,000 part-time faculty in schools and colleges of architecture. Of that number, 177 were members of minority groups.[42]

This sketchy overview raises a number of questions concerning access and production of black students in architecture which cannot be addressed here. We need far more disaggregated data and more specific institutional information concerning precise programs that these colleges operate now or have conducted in the past in order to fully assess their roles in the mainstreaming of black Americans in architecture.

Footnotes

1. *Minorities in Engineering: A Blueprint For Action* (A Report of the Planning Commission for Expanding Opportunities in Engineering). New York: Alfred P. Sloan Foundation, 1974, p. 1.
2. *Ibid,* Pp. 96-97.
3. *Ibid,* p. 99.
4. *Ibid,* p. 188.
5. Seymour Lusterman, *Minorities in Engineering: The Corporate Role.* New York: The Conference Board. 1979, p. 2.
6. *Ibid.*
7. *Ibid.*
8. *Ibid.*
9. *Ibid.*
10. *Ibid.,* p. 3.
11. Ibid.; See also: *The Committee on Minorities in Engineering: Scope and Activities.* Washington, D.C.: National Research Council (no date given), p.3.
12. *Ibid.*

13. *Ibid.*
14. Committee on Minorities in Engineering, *Supra.*
15. Lusterman, *Op. Cit.,* p. 3.
16. *Ibid.*
17. *Op. Cit.*
18. *Ibid.,* p. 3-7.
19. Interview with Levoy Spooner of the Committee on Minorities in Engineering, June, 1979.
20. *Minorities in Engineering: Blueprint for Action.* p. 9.
21. Cf. *Rentention of Minority Students in Engineering,* Washington, D.C.: National Academy of Sciences, 1977.
22. *Ibid.,* p. 2.
23. *Ibid.*
24. James E. Blackwell, "Social Factors Affecting Educational Opportunity for Minority Group Students," Chapter I in *Beyond Desegregation: Urgent Issues in the Education of Minorities.* New York: The College Board, 1978, 1-12.
25. *Op. Cit.*
26. *Financial Aid Needs of Undergraduate Minority Engineering Students in the 1980s.* New York: The National Fund for Minority Engineering Students, December 1978.
27. *Ibid.*
28. *Ibid.*
29. *Minorities in Engineering: A Blueprint for Action,* pp. 134-35.
30. Ruth G. Schaegger, "Corporate Leadership in A National Program", *The Conference Broard Record.* Vol. XIII (No. 9), September 1976, p. 9.
31. *Ibid.*
32. Lusterman, *Op. Cit.,* p. 42.
33. *Retention of Minority Students in Engineering.*
34. *Ibid.*
35. From an interview with Dr. Lucian Walker, Dean of the Howard University School of Engineering, June 1979.
36. Personal interview with Dr. Hugo Blasdel of the National Architectural Accrediting Board, September, 1979.
37. *Ibid.*
38. *Ibid.*
39. *Ibid.*
40. *Architectural Schools in North America.*
41. *Op. Cit.*
42. *Ibid.*

CHAPTER 10

The Legal Profession and Black Americans

The primary focus of this chapter is on legal education for black Americans during the 1970s. It examines the impact that policy changes in the legal professional associations enunciated during the sixties had on the inclusion of black and other minority students in the seventies. It describes enrollment trends and some of the more salient factors associated with these changes. Special attention is given to issues of conflict and confrontation evidenced in special admissions and the *DeFunis* case; the role of the Council on Legal Education Opportunities Program in mainstreaming blacks, and persisting problems which impede progress in increasing the representation of black Americans in the legal profession.

Historical Overview

Any assessment of changes in the status of black Americans in the legal profession or of efforts to increase their representation in schools of law must be placed in the context of overall enrollment changes during the period under review. In that regard, it should be stressed that law schools have experienced phenomenal growth and an unparalleled demand for space since the end of WW II. This exceptional demand for space in schools of law is at the very core of many of the dominant issues regarding the paticipation of blacks and other minority students in legal education. It is the cutting edge of issues pertinent to admission to the practice of law. For this growth in demand stimulated the movement to develop quantitative measures for a more restrictive determination of eligibility.

The instrument of *exclusion* sought was attained with the development of the Law School Admission Test (LSAT) in 1947. (Others may perceive the LSAT as a determinant of *inclusion;* however, that is a point of major contention especially since so many persons regard the LSAT, like all objective admission test, as a primary method devised for the benefit of professional schools to exclude those persons they do not want.) Albeit, the LSAT was not immediately embraced by American schools of law. In fact, according to Franklin R. Evans, only 18 law schools utilized this test in its assessment processes in 1948, the first year it was administered.[1] However, with the realization of its immense utility, its popularity grew widely and spread to cover approximately 90 per cent of all entering law schools students by 1967. In 1980, the American Bar Association, which sets the standards for accreditation for law schools and accredits them, requires them to *demand* that students seeking admission to law schools present their LSAT scores.[2] Even with this requirement, the power of the test for either inclusion or exclusion was never quite revealed until the unprecedented numerical growth began about 1953.

As will be detailed in a subsequent section, total enrollment in schools of law, approved by the American Bar Association, has about tripled since 1950 when it stood at 43,685 compared to a total enrollment of 122,860 in 1979-80. Since 1962 or even 1965, total enrollment has more than doubled in ABA approved law schools.[3] Although the enrollment of black students in schools of law was highly significant during this period and also more than doubled, it did not keep pace with the upward spiral witnessed among white applicants who were successfully admitted to law schools.

Notwithstanding, there can be little doubt about the veracity of the viewpoint that as pressure for space in law schools mounted, the utilization of quantitave determinants as a defensible solution to the problem and as justification for decisions to *exclude* accelerated. This position has never achieved unequivocable and universal support. To the contrary, opponents to this exclusionary quality of cognitive measures attacked them with the same vehemence employed by their defenders, who proclaimed their unimpeachability as the gatekeepers of the meritocracy. The latter group ultimately argued their sacrosanct character as the primary safeguard against the lowering of admissions standards and as the protectors of quality of the product graduated by schools of law.

Proponents of the so-called "meritocracy" often advocated cognitive measures as a front-line defense against affirmative action, special admissions, special programs, and special scholarships for black and other minority students. Some opponents to this perception of a "strict meritocracy" based solely upon objective or mathematically constructed criteria countered with the assertion that this sudden attachment to cognitive devices was a mask for either a resentment of the efforts to include blacks and other minorities or for subliminal racism.

Despite the decision by the U.S. Supreme Court in the *Bakke* case, the arguments remain, organizational and individual polarization persists, and suspicions of motives of various groups pervade social relationships and intergroup interactions. Minority students suffer because of the racism in American society that permits individuals to asume that simply because the person is a lawyer from a minority group, the credentials he or she has legitimatley earned, frequently with honors and distinction, are necessarily suspect. Hence, these persons are called upon to *prove* their mettle in ways never demanded of other groups in this country.

Another aspect of the historical context is the central role performed by schools of law located at historically black colleges and universities in the training of lawyers from minority groups. These institutions arose during the pre-Brown period; that is, prior to 1954. They were established largely by State legislative mandates, often hurriedly, as a means of keeping the existing public schools of law "for whites only." Some came into existence at the public historically black institutions upon the insistence of black educators in the State who wished to elevate their colleges to university status by the establishment of professional colleges or schools such as law.

Over time, schools of law were developed at Texas Southern University (Houston), Southern University (Baton-Rouge, Louisiana), Southern Carolina State College (Orangeburg), North Carolina Central University (Durham), Lincoln University (Jefferson City, Missouri), Florida A & M University (Tallahassee), and Howard University* (Washington, D.C.). Of these institutions, only four remain as viable schools of law in 1980: (1) Howard University School of Law which is the oldest and perhaps most prestigious of this group, (2) The Thurgood Marshall School of Law at Texas Southern University, (3) Southern University School of Law, and (4) the School of Law of North Carolina Central University. Although established primarily to serve the needs of black Americans during the period of de jure segregation, each one of these law schools is considerably more racially integrated than is any other institution, public and private, in the state or jurisdiction in which it exists. Generally, each one produces more white graduates than is the number of black law students graduated from the traditionally white law school in the same jurisdiction.

Their graduates and faculty members are among the more distinguished members of the legal profession in 1980. They include judges, prosecutors, public officials, law school professors, college Presidents, members of the President's Cabinet (Hon. Patricia Harris), and a member of the United States Supreme Court (Associate Justice Thurgood Marshall). Without a doubt, these historically black schools of law, like their counter-

---

*Note: The School of Law at Howard University is one of the oldest in the U.S. It was established in 1869.

parts in medicine, have performed a major service to the nation as a whole. Without them, the proportion of blacks in the legal profession, minute as it is, would be considerably less than the two per cent representation of blacks in law in 1980.

The Impact of Professional Associations on Mainstreaming Blacks in Law

The major thrust of the sixties regarding equality of access was on how to include more black Americans and other minority group members *within* the system. That was indisputably an integrationist perspective which demanded that groups who had been denied democratic privileges of citizenship now be granted all such rights and privileges. These groups demanded to be "in" and being "in" meant access to the entire institutional fabric of the social system as well as equality of rewards for services rendered.

Including previously excluded groups necessitated major policy changes and unequivocable leadership among the various professional associations which would direct the course of action taken by their appurtenant or constitutent sub-groups. By the early 1960s, the American Bar Association had dropped its membership barriers to black Americans. The persistence of these barriers had much earlier on led to the founding by black lawyers of the National Bar Association. The ABA committed itself in policies promulgated at this time to ending discrimination based upon race or ethnic origin. What these policies meant was that blacks and other minority groups were no longer to be denied membership, by official decree, in law schools or any ABA-sanctioned activity on the basis of their race or national origin.[4]

The American Association of Law Schools (AALS) issued an "Anti-Discrimination Article" that was included in the Articles of Association in 1963. This article demanded member institutions to provide or maintain "equality of opportunity in legal education" and to terminate all aspects of discrimination based upon race or color. This position was affixed in the "Approved Association Policy" that was annexed to that provision. This policy left no doubt that its central concern was with ending discrimination in *admission* to law schools.[5] In effect, this policy was a reaffirmation of some of the recommendations made earlier on by the AALS Committee on Racial Discrimination to end restrictive barriers to equity in the admissions process. That Committee also warned in 1963 of current and impending problems blacks faced in being admitted to practice law and to the future problem of persuading blacks even to apply to law schools.[6]

Under the leadership of the AALS President in 1963, Walter Geilhorn, a number of far-reaching changes were in the offing. In his presidential address in Los Angeles in December, 1963, President Geilhorn advanced a number of what could then be described as "bold plans" for more effective inclusion of black Americans in legal education and for the removal of impediments to the practice of law. Three of his eight proposals had a direct bearing on the legal education of black Americans. They were projects which the AALS could support immediately. They were: (1) a study of the major social, economic, financial, education, and other structural impediments to the enrollment of black students in schools of law; (2) a study of the inbalanced delivery of legal professional services to the needy and disadvantaged population; and (3) a study of the strengths, weaknesses and problems associated with part-time legal education. A grant was obtained from the Ford Foundation to support the study envisioned in the first project. A Special Committee on Provision of Needed Legal Services was created to implement the second project and a special project on part-time legal education was organized with the appointment of a full-time project director.[7]

Both the ABA and the AALS are among the sponsors of the Council on Legal Education Opportunity Program (CLEO) which has assisted an average of 200 minority students per year since 1968 to enroll in schools of law. In addition, both groups established special committees to address primary concerns of minority groups. The ABA's legal consultant issues an annual report which up-dates enrollment data on blacks and other minority students by class for all approved schools of law. *A Review of Legal Education in the United States*, published annually, also describes specific kinds of institutional data. These data may include enrollment and graduation statistics, tuition per year, admission requirements, and other information of special interest to a potential applicant. *The Pre-Law School Handbook* is considerably more detailed in this regard since it often expands on the specific institutional characteristics.

The AALS established a Standing Committee on Minority Groups. This committee has over the years addressed a number of issued related to the AALS policy enunciated in the "Principals of Non-Discrimination" and its "Equality of Opportunity" provision that was amended on December 30, 1970. Of special concern are issues such as conformance among the various law schools to the principles of unbiased admissions procedures and policies, unbiased allocation of financial assistance, the recruitment and hiring of minority group members as law professors and assurance from the Committee on accreditation that standards of fairness, and unbiased actions are indeed upheld by all ABA-approved institutions. These concerns suggest a high degree of suspicion that conformity to these principles has not always been uniform from law school to law school.[8]

Minority group members serve on a number of standing and special committees of the AALS and, under the auspices of the Committee on minority groups, they have played a special role in enhancing the overall policy of equality of opportunity and full participation of minority lawyers in the affairs of the Association. One of the Committee's major efforts, in addition to stimulating greater access of black and other minority students to law schools, is the *Registry of Minority Faculty* which has been published for approximately six years by Professor Derek Bell, Jr., formerly of the Harvard Law School. This annual publication serves a dual purpose. First, it enables the AALS to have a quick reference guide to minority school faculty by institution. Second, it enables the Committee to monitor the hiring and promotion practices of institutions with special attention to the status of racial and ethnic minorities in law schools.

In addition to these activities, the professional associations played a major role in the establishment of the Law School Data Assembly Service (LSDAS). Through this service, applicants may have one set of their admissions profile, transcripts, test scores and so forth centralized for circulation to interested institutions that participate in this service. This program, in turn, enables applicants to save enormous sums associated with paying for multiple transcripts when they apply to several law schools during the course of a single year.

One special case deserves extended discussion at this juncture. That is the case of *DeFunis* v. *Odergaard* previously described in Chapter I. This case was important because of the issues it raised regarding both constitutional law and its focus on special efforts undertaken in the 1960s to accelerate the process of achieving equity in the admission of minority students to law schools. When the case reached the Washington State Supreme Court, about thirty *amicus curiae* were filed.[10] Reaching a consensus on the position to be taken by the AALS involved consideration of several issues of enormous importance for higher education; for instance, on such issues as preferential admissions, discretionary authority of law schools admissions committees, the appropriateness of judicial intervention on "single-factor (e.g. race) issues," the question of how standardized tests, such as the LSAT that had been in use since 1948, should be characterized, and how to handle the question of the validity of the LSAT. Ultimately, dissensus on those matters was resolved to the satisfaction of the members of the Executive Committee of the AALS and its name was attached as one of the four organizations represented in the brief sponsored by the Council on Legal Education Opportunity.[11]

It should be pointed out, however, that the AALS never opposed a hearing on the *DeFunis* case, neither in the Washington Courts nor before the United States Supreme Court. Whereas the Executive Committee of the AALS had joined four other groups in an amicus brief in the lower court, it presented a separate brief before the U.S. Supreme Court. When the Supreme Court handed down its decision in this case on April 23, 1974 in which it held that the case was moot since DeFunis was already enrolled in law school and was about to be graduated within two months, the AALS felt that its position was vindicated.

The AALS also filed one of the 62 *amicus curiae* briefs in the *Bakke* case. In effect, it reiterated its position on the continuing need to rectify past practices of segregation and discrimination which resulted in the present underrepresentation of blacks and other minorities in the legal profession. However, the AALS appears to be quite deliberative in its approach to the difficult issues and some individual members continue to be conservative and to promote principles of gradualism.

The ABA assumed a new posture on affirmative action at its 1980 meeting in Honolulu. Its House of Delegates passed Amendment 212 which requires law schools, as a condition of accreditation, to substantiate evidence that they "provide" full opportunities" to minority students who wish to study law. This Amendment thus becomes incorporated into the ABA's standards for the Approval of Law Schools. This new amendment, passed upon the recommendation of the ABA's Section on Legal Education, represents a particularly powerful instrument for enforcing an implicit educational policy of long standing.

The National Bar Association, the historically black association of lawyers, was formed to carry out essentially the same functions of the originally all-white American Bar Association. It holds annual meetings which serve as a forum for addressing current issues that black lawyers confront in the profession. It sponsors seminars through which practicing lawyers can be up-dated on new elements of the law which affect their practice and its members engage in a number of social action programs. The NBA has been quite supportive of efforts to increase the representation of black Americans in the legal profession initiated in the early sixties. It should also be stressed that among its members are those black lawyers who argued most of the cases spearheaded by the National Association for the Advancement of Colored People and the Legal Defense and Educational Fund which led to far-reaching decisions by the U.S. Supreme Court. Its members have, therefore, served as role models of special significance to minority students. By itself, this function has had an immeasurable impact on successful endeavors to attract more black Americans to the legal profession.

Council on Legal Education Opportunity Program (CLEO)

CLEO was organized in 1967 for the expressed purpose of strengthening efforts to enroll more minority students in the legal profession and to increase the probability that those accepted for admission would have an excellent chance for successful completion of a law school career. The establishment of CLEO came as a result of a study by the Committee on Minority Groups of the AALS, sponsored by the Ford Foundation, which demonstrated a dismal profile of the representation of minority group students in the law school student population. The study also provided demonstrably significant data which showed that the low production rate of minority group lawyers was an outcome of certain forms of institutional behavior as manifested, for instance, in admissions practices.[12] When CLEO was founded, less than 1 per cent of all lawyers in the nation were black Americans. Further, due to maldistributions, some states had one black lawyer for every 30,000 black residents.[13]

CLEO operates on the premise that the traditional methods of assessing the potential of students for study in schools of law and for their possible success as lawyers do not always work for all groups of minority and disadvantaged students. It further maintains that the successful completion of special pre-admission and pre-enrollment programs, may be a more effective and appropriate assessment technique for making selection decisions than strict reliance on LSAT scores and GPAs. The third aspect of the CLEO program is its emphasis on the value of financial assistance to minority students who seek admission to law schools so that they may be encouraged to actually enroll once they are admitted.

When CLEO was established in 1967, it was sponsored by the American Bar Association, the American Association of Law Schools, the National Bar Association, and the Law School Admission Council. Later, in 1972, a new sponsor was added to this list, the La Raza National Lawyers Association. The membership of the Council is comprised of delegates from these bodies including student representatives.[14]

The major program efforts of CLEO are the identification, selection and enrollment of minorities and disadvantaged students in Summer Institutes; the conduct of these institutes, and the awarding of annual stipends to students who participate in the program. Seven regional summer institutes are held each year. These sessions are designed to familiarize students with the demands of law school, to introduce them to the study of law through mini-courses and to help them correct certain identifiable deficiencies in their own educational backgrounds.[15] Each of the seven institute sites represents one of approximately seven geographic areas into which the country is divided for administrative purposes by CLEO. The institute sites vary each year within a region but their host institution is always an ABA-approved school of law. Courses are taught in the main by law school professors who are committed to the CLEO goal of strengthening the capabilities of these students to become successful matriculants in the law school of their choice.

Student participants in CLEO programs and activities are primarily, but not exclusively, from the underrepresented minority groups: Blacks, Chicanos/Mexican Americans, Puerto Ricans, and American Indians/Native Americans. The average number of participants per year since 1967 has been about 200 students. Between 1968 and 1978, some 2,626 students had participated in the CLEO program. Of that number, 2,550 had completed the summer institute program. By 1978, 2,415 of this group had entered law school. In the first four years, 69 per cent of CLEO students graduated from law school. The present retention rate is approximately 95 per cent. They have enrolled in over 140 ABA-approved law schools. About fifty per cent of CLEO students are black.

The general status of these students, regarding the admissions process, varies and may be even mixed within any particular cohort. That is to say, among the 200 institute participants may be students who have already been accepted to law schools but who want the institute experience. There may also be students who have a conditional acceptance with the final decision awaiting the outcome of their performance in the CLEO program. Also included are students who may be accepted into a law program but who are in the process of applying.[16]

Since 1967, CLEO estimates that approximately 50 per cent of its students have been accepted outright by schools of law and that the remaining fifty per cent are divided between those with conditional acceptances and those who were applying but subsequently accepted. In the Post-Bakke era, it appears that more and more law schools are looking to the CLEO program and its summer institutes to assist them in reaching final decisions regarding certain students. More specifically, the CLEO Performance Evaluation or the CLEO endorsement serves as that *additional* factor needed to achieve a positive selection or admission decision.[17] What is apparent here is that many schools of law are increasingly reluctant to take the chances they once took regarding the use of certain predictive measures. Others are cautious, because of the ever present danger of legal entanglements, that their selection procedures can be rationalized. Still, other law schools may very well be using the *Bakke* decision as a justification for retrenchment or inaction regarding the recruitment and selection of minority students for law school.

The funding of the CLEO program has always been inadequate to fulfill the goal of providing *adequate* financial assistance to its students. Funds to support program operation and administration, including salaries for professional staff and program activities such as pre-law recruitment and the summer institutes, were initially supported by annual grants from the Legal Services Division of the Office of Economic Opportunity (OEO). In 1971, this grant was supplemented by another one from the Special Student Services Division of the Office of Education in the Department of Health, Education and Welfare. The annual federal appropriation ceiling for CLEO's entire operation has been one million dollars since 1967. In fact, the monies from the two sources meant that a total of $950,000 was received from the federal government for its operations. As a result, CLEO, which provides annual stipends of $1,000 to each of its students, is at precisely the same funding level in 1980 as it was in 1967 and the federal government has made no allowances for depreciation of the dollar. However, the federal appropriation generated three million dollars in institutional support.[18]

CLEO officials have been forced to make annual treks to Congress to justify its existence and to fight for its one million dollars, despite all of the evidence of the program's success in achieving its objectives. There were years in which its continuation was in doubt largely because of congressional indifference and in one year it probably would have been sacrificed had it not been for timely intervention from then Senator Walter Mondale. The program is funded through 1981 pursuant to provisions of Title IX, Part D of the Higher Education Act of 1965 as amended. But, once again, CLEO stipends are limited to $1,000 per year per student. This is in sharp contrast to the stipends of $4,500 per student per year permitted to other programs funded under the same Act, such as the GPOP and Public Service Scholarships. However, there is a present possibility that CLEO scholarships might be increased as Congress reconsiders the 1965 Higher Education Act once again in 1980.* CLEO has requested a ceiling of $5 million which would enable it to raise the level of its stipends consistent with the present dollar value and to continue to strengthen its program as well as to

---

*In October, 1980, Congress approved an increase in the CLEO ceiling to be set at $5 million for FY 81 and 82, $7 1/2 million for FY 83 and 84 and $10 million for FY 85. This change will enable CLEO to *expand* its operations.

broaden the scope of its activities to include a more representative number of minority students.[19]

None of the sponsoring organizations have made substantial financial contributions to CLEO over the years even though they have provided other advice and support of immeasurable value. On the other hand, several major foundations and corporations have given direct financial assistance for the maintenance and support of CLEO's program. Among these are such groups as the following:

Ford Foundation
Rockefeller Brothers Fund
American Bar Endowment
General Electric Foundation
Celenese Foundation

Standard Oil of Indiana
Alcoa Corporation
Standard Oil
P.P.F. Industries
Philip Morris Corporation

IBM
RCA Corporation
Xerox Corporation
General Motors
International Mining
& Chemical
Ford Motors Company Fund
Singer Corporation
Western Electric
Jones & Laughlin Corporation
CBS Corporation

Despite the constant struggle over funding, it is evident that CLEO has functioned as a major organizational structure for increasing the number of black and other minorities in the legal profession.

Recruitment of Black Students

In the mid-sixties, following the articulation of positive policy changes and enunciation of Articles of Anti-Discrimination by the ABA and the AALS, most law schools began to seek out students from previously excluded populations. For many of these institutions the process of actively recruiting students was an entirely new experience. Earlier on, they had relied on their reputation or on their position as the flagship public institution as inducements for students to seek admission to their respective institutions. As a result, the law school community, like the sub-communities of the entire institution, was a homogeneous unit comprised of a population dominated by white males in the student body, white male faculty, administrations, and white male boards of trustee. The 1960s demanded a change in this situation—the creation of more racially and ethnically heterogeneous communities in which men and women of all groups could have equal access.

The various steps taken to increase the representation of black students in schools of law may be better understood by focusing on specific examples of institutional activities reported for this study.

The University of Wisconsin law School at Madison established its Legal Education Opportunities Program (LEOP) in 1967 to assist minority groups and disadvantaged groups to obtain legal education. One of its major activities is the recruitment of minority students, both resident and non-resident. As a result of its recruitment program, the annual enrollment of black students in this law school rose from six per year in 1968-69 to an average of eighteen per year from 1972 to 1980. Recruitment here involves information dissemination, visits to various colleges and universities, use of role models from faculty and minority student bodies, provisions for financial assistance, campus visits and many other activities described in other chapters. In terms of proportion of enrollment, black students comprised 2.4 per cent of enrollment in 1968. However, in 1979-80, they represent 6.2 per cent of enrollment in the school of law at Wisconsin. This change is the direct consequence of the recruitment component in the LEOP program. Black student enrollment reached an all-time high of thirty-five students in 1973-74 which then represented 11.7 per cent of enrollment.[20]

An affirmative action program was inaugurated at the Ohio State University College of Law in 1968. It embraced active recruitment as one of its central components. The College of Law carried on much of its recruitment activities during the seventies at historically black colleges and at other institutions which had a significant concentration of minority students. In early years, recruitment responsibilities were shared by the Dean of Admissions, faculty members of the College of Law, and representatives of the Black Americans Law Student Association (BALSA). In 1977, the College of Law added another staff person to share the recruitment effort, assist in enrollment, and to strengthen the retention program. The use of the interview became more pronounced in the recruitment process, primarily to supplement information provided in the applicant's file. The College also recruits students enrolled in the midwest CLEO Institute and it now extends its efforts to the pre-college level.

The success of its recruitment is also measured by the change in enrollment of black students between 1968 and 1978-79. In 1968, the three (3) black students enrolled in the College of Law represented 1.8 per cent of enrollment of 166 students. In 1978-79, the twenty-one enrolled black law school students represented 8.2 per cent of the total enrollment of 254 students. The largest number of black students ever matriculated was twenty-seven in 1974-75. This cohort comprised 11.4 per cent of enrollment at that time.[21]

In addition to these types of activities, others include such strategies as open house sessions (University of Oregon), participation in recruitment conferences and follow-up activities (Case Western Reserve University), counseling prospective applicants about law careers (the University of Washington), the use of the 13 College Program and the College Board's Locator Service (American University), the utilization of a law school pre-start program which was a prototype for CLEO (Emory University), and continuing use of CLEO programs (the University of Richmond).

In general, recruitment programs were immensely successful during the seventies in seeking out, identifying, and encouraging significantly larger numbers of black and other minority students to enroll in colleges of law. However, success was neither constant nor uniform from institution to institution. Nor did it lead to racial parity in access to legal education. The presence of black students remains at the token and limited access levels in the overwhelming majority of American schools of law. Further, as we shall see in the section on enrollment, a downward trend has begun which will have to be re-directed in order for black Americans to attain parity in admissions and productivity even by the end of the century.

Admissions

The major controversy in law schools during the 1970s regarding black and other minority students arose over the issue of admissions standards. Heated debates centered on special admissions programs, preferential admissions procedures, which presumably favored minority students, the utilization of the LSAT as a major determinant of admissions, the relative weights assigned to both *quantitative* assessment measures, admissions goals and set-aside programs, and other special programs for minority group students. At some point during the decade, each of these aspects of the admissions issue came under relentless attack by their opponents which transcended mere intellectual discourse. There were outright demands for the immediate dismantling of all programs organized to strengthen minority group access.

In order to redress past inequities in admissions policies which had clearly prevented the selection and enrollment of minority students for law schools, it is estimated that approximately sixty per cent of all ABA-approved law schools contructed special admissions programs of one kind or another during the seventies. The specific characteristics of these programs varied from law school to law school. In most instances, the development of these programs was motivated by identifiable educational values, and for explicit political, economic and social reasons. In terms of their operation, they ranged from set-aside arrangements, or clearly defined goals for the number of black and other minority students who should be admitted to each entering class, to more flexible arrangements which included adjustments in standard admissions requirements.

Among the prominent colleges of law with special admissions programs during the period were such institutions as the University of Washington, Temple University, Boalt Hall of the University of California/Berkeley, Ohio State University, the University of New Mexico, New York University, the University of Denver, the University of Colorado, Florida State University, Indiana University, the University of Iowa, the University of Wisconsin, Rutgers, and the University of Texas/Austin.

Although it is not possible to describe each program here, a brief discussion of three of these programs will illuminate processes involved in special admissions.

The College of Law of the University of Washington initiated a special admissions program in 1973. Its goal was to include "a reasonable representation of minority students" in each entering class and to assure increasing diversity in the legal profession. Under this program black and other minority students were evaluated in a separate tract from all other applicants. LSAT scores and undergraduate GPA's were modified in order to achieve the goal of admitting "academically qualified minority applicants despite records of less strength" on these measures than other applicants. Other, non-cognitive factors that were strong in the minority applicant's portfolio enabled them to advance above the mathematical cut-off levels of rejection. The process was effective in changing the racial and ethnic composition of the law school. However, since the *DeFunis* litigation the number of black students in entering classes had fallen from a high of fourteen in 1970-71 to only five black students in 1978-79.[22]

The admissions programs at Ohio State University gave special consideration to minority identification and disadvantages experienced by virtue of that status, and experience with test-taking and its impact on the LSAT score. Compensations derived from these considerations enabled the college to admit more disadvantaged students than otherwise would have been the case.[23]

The University of Wisconsin used recommendations from both its LEOP Committee and from its Admissions Committee. The Admissions Committee considered the applicant's file independently of the recommendations made by the LEOP Committee but then made its final decision on the basis of recommendations submitted by both committees. A number of both academic and non-academic factors entered into the admissions process for all students. However, minority students benefitted, but not unfairly, from the employment of "nontraditional" criteria such as diversity of experience and background factors.[24]

The admissions process at the University of Texas Law School is especially noteworthy. Its admissions committee, like many during the seventies, is composed of law school faculty members, members of the student body, and regular admissions or administrative personnel. The college employs what it calls an "administrative mode" and a "committee mode" in determining eligibility for admission. The administrative mode is based upon the "presumption of qualification". In this mode, the two objective factors, that are weighed, assigned a numerical value, and combined for an index number, are the LSAT score and the undergraduate GPA. An administratively pre-determined range of acceptable index scores is established. If a student's index score falls within that range, qualification is assumed.[25]

On the other hand, the committee mode is utilized to consider "evidence of qualification". In this situation, either the Admissions Committee or a sub-committee analyzes all the data accumulated in the applicant's file and evaluates this information to determine suitability for its program and to make an assessment of the potential of that student to serve the legal needs of society.[26]

Admissions decisions are made after careful scrutiny of all relevant factors, including the number of available seats, fairness to all applicants, and the importance of diversity in the educational setting. All factors, traditional and non-traditional, are subjected to intense evaluation. For example, the LSAT score is evaluated in relationship to scores received on other objective tests such as the SAT, the ACT or the GRE. In this way, it is possible to ascertain relevant information about test-taking ability and the degree to which test scores are consistent or inconsistent with performance on LSATs and in undergraduate courses. Hence, a suggestion regarding the overall potential for success in law school may be made in a rational manner. The academic record is also evaluated in terms of the curriculum chosen by the applicant and the relative strength or reputation of the institution attended. The committee looks for growth, progress, change, maturity, and other factors that are indicative of quality in the academic performance.[27]

Subjective measures also include a sense of creativity, leadership in extracurricula activities, post-graduate experiences, and evidence of energy and determination. The applicant's own personal history is subjected to a systematic and rigorous assessment in order to give the candidate benefits of any doubt.[28]

As a result of this process, the total number of black students specially admitted to the University of Texas School of Law has been significant in every year since 1973.

For instance, in 1976, special admissions offered to black students numbered thirty-five while sixty-five were made to Mexican Americans. In that year, three blacks and thirteen Mexican Americans were "regularly admitted". Eighteen of the black students admitted under special admissions and two admitted under "regular" procedures actually matriculated. (The numbers for Mexican Americans were 45 and 16, respectively). In 1977, 47 special admissions were awarded to black students and 57 went to Mexican Americans. The numbers of regular admissions increased for both groups. For black Americans the number was seven and for Mexican Americans it rose to sixteen. Of the 47 black students admitted under special admissions arrangments, 23 actually matriculated while 35 of the 57 Mexican Americans matriculated in the school of law. Three of the seven regularly admitted blacks and ten of the 16 Mexican Americans regularly admitted matriculated in 1977.[29]

According to objective data provided from the School of Law at the University of Texas, approximately 318 of the 6,000 students admitted during the past decade were members of minority groups. *More than 98 per cent* of the minority students have succeeded as law students.[30] Although this is significant evidence of the exceptional quality of a sound special admissions program, total access was not attained since blacks comprise 12.5 per cent of the population in Texas.

The experience of these four institutions with special admissions programs is instructive regarding a number of issues pertinent to the entire admissions process. For instance, it is apparent that the LSAT, while a highly useful and important predictive device for performance in the first year of law school, should not be relied upon in and of itself as a reliable predictor of overall success during the entire law school career. Certainly, no valid claim can be made that LSAT scores have significant predictive power regarding success as a practicing lawyer.

According to a study made of performance on LSAT tests, black and Chicano/Mexican American applicants, showed lower LSAT and lower undergraduate GPAs than did white students as a group. However, both black and Mexican American students were accepted at higher rates than white students even though the "rate of acceptance" for white applicants was higher, overall.[31] This difference reflects institutional consideration in the main, of applicant's racial or ethnic background when selecting students during the seventies. However, as Evans asserts, one cannot regard the accepted minority students as anything less than fully qualified for law school. When employing the major objective predictors of success in the first year of law school, the mean GPA and the LSAT scores for minority students are higher than the mean for *all* (emphasis added) who were enrolled in 1962.[32] The meaning of this observation should not be dismissed. What is suggested here is that, as competition for limited space accelerated, law schools raised the cut-off points for undergraduate GPAs and LSAT scores higher and higher. Consequently, some students who were highly admissible in 1962, are often below eligibility levels by 1980 cognitive standards. And, that is where the major problem with these measures lies.

Albeit, Evans maintains that blacks would experience a reduction rate of 60 per cent and Mexican Americans a reduction rate of 40 per cent, if they were accepted at the same rates as majority students "at the same level of LSAT" and undergraduate grade point average.[33] Law schools report that if the same levels of expectations were universally applied, all things equal, and without consideration to subjective measures of evaluation, the percentage of black students in the first year classes would fall to less than one-half of what it was in 1977 or from about 5.3 per cent to between 1 and 2 per cent. And, this would have a national impact and widespread ramifications not only for matriculation but for among other things, the structure of financial aid programs.[34] That is one reason why the adjustments made by law schools in the wake of the *Bakke* decision are of special salience in view of the possible "chilling effect" that misinterpretations of the Supreme Court's decision could have on the enrollment of minority students.

Responses to the Bakke Decision

There is evidence that several institutions have found creative methods of assuring that the full range of their assessment, selection and admission procedures conforms to the legal mandates of the Bakke decision. For example, in November, 1978, the law school faculty of Rutgers University voted to modify its admissions program by expanding to 30 places the number of seats allocated for its special admission program. This was done to assure greater inclusion of *disadvantaged* white students in the program. As a result, all seats in the entering class are open to all groups irrespective of racial or ethnic designations.[35]

The University of Washington eliminated its two-tiered or separate evaluation system but does take race into consideration as one of many factors employed in the evaluation of an applicant. Ohio State University utilizes several of the factors reported by the School of Law at the University of Texas/Austin. So does Stanford University employ a combination of quantitative and non-quantitative factors while stressing the primacy of the former over the latter in making these decisions.[36] The impact of the collective weight of special and regular admissions program can only be assessed in the cumulative growth or changes in the number of students enrolled during the decade. Our focus is solely on the number of black students. Although many of the trends demonstrated for black students were observed for other minority students, space does not permit an equal treatment here of these groups in this study.

Enrollment of Black Students in Law Schools During the 70s

First Year Enrollment

First year enrollments of black students in law schools almost doubled in the period between 1969 and 1979-80. (Table 40).* However, observations of the absolute numbers of black students portray a condition of steady progress, followed by a downward trend and another upswing. As the seventies began, 1,115 black first year students were enrolled in schools of law. They represented 3.8 per cent of all first year enrollees. The peak in proportion of black students of all students in first year classes was not achieved until 1978-79 when it reached 5.9 per cent. However, the highest absolute number was in 1976 but this was followed by a decline of some eighty-five black students in the next first year class. In 1978-79, there was a gain of seventy-nine black students but this was followed by another loss of nineteen black students in the 1979-80 first year enrollment.

Undoubtedly, it is this inability to *sustain* a steady, upward trend which led to allegations of a retreat by law schools from the commitment to move forward toward parity. This assertion underscores the immense disillusionment among many black Americans over what they perceive to be failures in legal education. It is apparent that, despite the progress made and the high quality recruitment programs established, law schools in the United States are currently enrolling only 887 more black students in first year classes than they did in 1969-70. Hence, the progress achieved is substantially less than what it may appear to have been at first glance.

---

*See also Figures 1-4 in Chapter IV.

Table 40. First Year Enrollment in J. D. Programs in Approved Law Schools by Year and Race, 1968-79: All Institutions Combined

Year	First Year Enrollment	Black Student First Year Enrollment	Per Cent Black of Total
1968-69	23,652	—	—
1969-70	29,128	1,115	3.8
1970-71	34,713	—	—
1971-72	36,171	1,715	4.7
1972-73	35,131	1,907	5.4
1973-74	37,018	1,943	5.2
1974-75	38,074	1,910	5.0
1975-76	39,038	2,043	5.2
1976-77	39,996	2,128	5.3
1977-78	39,670	1,943	4.9
1978-79	34,118	2,021	5.9
1979-80	34,632	2,002	5.7

Source: *A Review of Legal Education in the United States - Fall 1979.* Chicago: American Bar Association Section on Legal Education and Admissions to the Bar, 1980.

Total Enrollment

The total number of black students enrolled in the 164 ABA-approved law schools more than doubled in the period between 1969-70 and 1978-79. Specifically, the total number of black students enrolled increased from 2,128 in 1969-70 to 5,257 in 1978-79 (See Table 41). There was a steady rise in tht total number of black students with increases in every year through 1976-77. In the following year, there was a loss of 202 black students but this was followed by a slight increase of some 45 students. However, as the decade closed, there was another loss of 93 black students in total enrollment. These losses are accounted by both declines in first year enrollment and by attrition beyond the first year of law school.

In 1976-77, the proportion of black students in total enrollment peaked at 4.6 per cent. Since that year, that proportion has continued to decline to a level of 4.2 per cent ot total enrollment in 1979-80. This continuing decline may or may not be significant in terms of what it portends for the

1980s. However, it does suggest that there cannot be a withdrawal of demonstrable commitment to produce a sufficient number of black attorneys. Otherwise, the nation will never advance beyond the level of calculated tokenism regarding opportunities for minorities and, therefore, accentuate the disillusionment of the present moment.

Table 41. Total Enrollment in J. D. Programs in Approved Law Schools by Race and Year, 1968-79: All Institutions Combined

Year	Total Enrollment	Total Black Student Enrollment	Per Cent Black of Total
1968-69	62,779	—	—
1969-70	68,386	2,128	3.1
1970-71	82,041	—	—
1971-72	93,118	3,744	4.0
1972-73	101,664	4,423	4.3
1973-74	106,102	4,817	4.5
1974-75	110,713	4,995	4.5
1975-76	116,991	5,127	4.4
1976-77	117,451	5,503	4.6
1977-78	118,557	5,305	4.4
1978-79	121,606	5,350	4.4
1979-80	122,860	5,257	4.2

Source: *A Review of Legal Education In the United States - Fall 1979.* Chicago: American Bar Association Section on Legal Education and Admissions to the Bar, 1980.

Because of all the accusations that black and other minority students entered professional schools, such as law schools, at the expense of white students, it is imperative to stress the rapid acceleration of the total number of students enrolled in law schools during the decade. As Table 41 shows,* the total enrollments for *all* students in law schools rose in every year of the study period. Specifically, it jumped from 62,779 in 1968-69 to almost 123,000 students in 1979-80. Not until 1979-80 did minority students represent 8.1 per cent of total enrollment. That proportion is neither equivalent to nor tantamount to equality of opportunity and equality of ac-

*See also Figure 3 in Chapter IV.

cess in law schools. Consequently, the issue of "reverse discrimination" looms larger and larger as a red herring constructed to forestall efforts to assure equality of opportunity to underrepresented minority groups.

Enrollment in Historically Black Law Schools

As noted earlier on, the historically black schools of law tend to enroll proportionately more white students than is the case regarding black students in the traditionally white institutions in the same jurisdiction or State. In 1970, for example, of the 61 black students enrolled in law schools in Louisiana, 45 of them were enrolled at Southern University while only four were enrolled at the Louisiana State University School of Law. In the same year, 92 of the 99 black law students in North Carolina were enrolled at North Carolina Central University while only 7 were enrolled at the University of North Carolina/Chapel Hill. About 150 of the 177 black students in Texas law schools were enrolled at the Texas Southern University School of Law.[37]

In 1979-80, 147 of the 210 students enrolled at Southern University were black while only 23 of the 876 law students at Louisiana State University were black and other minorities. Loyola and Tulane Universities enrolled 49 and 29 *minority* students, respectively. Over 90 per cent of these students were blacks. Although these numbers represent improvements in the representation of black students, their overall distribution remains fundamentally unchanged. In North Carolina, North Carolina Central University School of Law is sixty-four per cent black but 36 per cent white. The University of North Carolina Law School at Duke University is 95 per cent white and five per cent black. Each institution enrolled larger numbers of students during the decade; however it is self-evident that the historically black institutions have done a better job of increasing access to all students than have the historically white institutions. Further, integration remains at essentially a token level of access in most traditionally white institutions.

Special Programs

Once black students matriculate, *some but not all* of them require either academic or non-academic support systems or both as ways of fostering retention through graduation. Academic support is provided in part by faculty members who take an interest in sharpening the intellectual tools of students. Many institutions offer formal and informal tutorials. Learning centers assist students to sharpen their communications skills. Special orientation programs are conducted for newly matriculated students to help them attain a better understanding and appreciation of what is involved in the study of law, how to complete registration procedures, to become acquainted with other students, and to meet their first year professors.

Some institutions offer special courses, such as the one on Legal Methods at Ohio State University, to students who have completed the first year of study but who have an apparent need to sharpen writing and oral communications skills. Sometimes these students are permitted to carry a reduced academic load. Special study sessions and tutorial and voluntary review sessions are conducted by BALSA and/or the minority student organizations in the law school. BALSA also offers test-taking exercises.

Non-academic support activities encompass a range of services including scholarships and financial aid and psychological support. Many law students receive varying amounts of direct and indirect financial aid. These monies may be provided through the CLEO program, grants from the local chapter of Bar Associations, special institutional grants, faculty contributions, State authorized tuition waivers, federally subsidized work-study programs, and repayable loans.* Many institutions have been especially instrumental in finding new sources of financial assistance as original sources dried up or were reduced. However, much more has to be done, given the economic disabilities of a substantial proportion of black students.

BALSA, Black Student Unions, black faculty, and other minority group organizations perform an immensely valuable service in the affective support they provide some students. For this group, it is comforting to know that there is a critical mass of students of one's own racial or ethnic group or there are black law professors who care about their welfare and who are willing to spend time in informal discussions with them as individual students. There is special strength gained from the informal social contacts provided in social interaction with black students at social functions, the cafeteria, lunch rooms, or in intramural sports activities. All of these activities form a cohesive network of non-academic support services which have a positive impact on the overall well-being of black students in a predominantly white institution.

Problems

Nevertheless, major problems remain. Although the topic of applications to law schools was not explored here, there is no question about the

*In addition, since 1972, the NAACP-LDF has awarded over 2,500 law scholarships.

need to expand the application pool among black students. The pool question is inextricably related to both the expanded career options that open other opportunities for black college graduates and the specific need to enroll and graduate larger numbers of black college students. Even if this were done, other structural inequities which impede access would have to be eliminated and the economic problems of financing three years of legal education would have to be confronted head-on.

Comparatively few black students can afford to support themselves in law schools through their own funds. Many potentially successful students are unwilling to subject themselves to a major debt through repayable loans in order to attend law school when other options are available. Others pay a heavy price in lowered performance when they are compelled to take on substantial employment in order to provide for themselves and their families while enrolled. Grades often suffer and many professors are not inclined to take the psychological problems or pressures induced by home situations into consideration when reviewing the student's performance. This situation may lead to abandonment of the immediate goal of a law degree. The financial aid problem could be alleviated in a large measure through more governmental intervention - federal or state - with broader coverages under the Graduate and Professional Opportunities Program (GPOP), and a graduate system similar to the undergraduate BEOGs and the SEOGs.

Although a far greater number of exceptionally well-qualified black students could be recruited for law school programs than currently, the fact remains that all too many black students are still victimized by weak preparation in elementary, secondary, and collegiate education. Too many suffer from deficiencies in oral communication and writing skills. There are problems of personal confidence, of sophistication, and of lack of ease in dealing with other persons. Without a doubt, the deliberate intimidation of students by prejudiced, insecure, and power-hungry professors does not allay the fears that some students already have of active participation in classroom discussions. Further, the lingering problem of racism, subtle as it may be from time to time, must still be confronted. Racism is not always subtle! In fact, it sometimes is bold and direct. It is evidenced in the attitudes that some professors have toward black and other minority students and faculty, in their treatment of minorities, in their deliberate attempts to subject minority students to public embarrassment or ridicule, in their harsher grading of minorities, in their unwillingness to make the same kinds of exemptions or special dispensations for black students as they freely grant to white students, and in their beliefs that all blacks students are necessarily less competent than even the average white student.

That schools of Law have made some in-roads in dealing with the problem of the underrepresentation of black faculty in them cannot be dismissed as liberal rhetoric. However, that progress is but one step in the right direction. The need for increasingly larger numbers of black faculty and administrators is self-evident, particularly since there are only approximately 352 black and other minority faculty in all schools of law in the United States.[38] But, the issue of the black faculty does not end with their hiring. It extends to promotions and the granting of tenure. Comparatively few law schools have tenured minority faculty. Observation of the special difficulty for members of minority groups to receive tenure is a reality that does not attract black lawyers to the academic life. It is incumbent upon law school deans, other administrators, and professors to socialize black students into the fundamental requirements for positions as law school faculty early in their careers and to do all they can to motivate those who show a special interest toward teaching as a career. However, it may be necessary to reassess the reward structure so that rewards associated with academic life may compensate for whatever sacrifices that may be made by abandoning private practice, wherever that is required.

Retention, of course, persists as a major issue and it is currently being addressed, however unevenly, in methods discussed under the heading of special programs. The point is, nevertheless, that larger numbers of black lawyers ought to be produced each year. It appears that, from an examination of the numbers of black law students in third year classes, *law schools are producing about 1,300 black law graduates per year.* But that number is reduced each year by the disproportionate number of black students who fail to pass Bar examinations. That failure further hampers the productivity rate and delays substantial increases of black lawyers in the labor force. *Nevertheless, it is estimated that in 1981, there are some 10,000 black lawyers* * *in the total number of 574,810 lawyers in the United States.* Clearly, it is necessary to move far beyond this educational paralysis that has so far guaranteed a negligible representation of black Americans in the field of law.

*This is slightly less than 2 per cent of the total.

Footnotes

1. Franklin R. Evans, *Law School Admissions Research* Princeton, N.J.: Law School Admission Council, 1977, p. 569-71.
2. *Ibid.*
3. In the interest of space, this assessment is based solely upon events within ABA-approved institutions and not on the non-ABA-approved schools of law dispersed throughout the United States.
4. Michael Cordoza, *The Association Process, 1963-1973* Washington, D.C.: American Association of Law Schools, 197, p. 18.
5. *Ibid.*
6. *Ibid.*
7. *Ibid.,* p. 47.
8. Cf. Walter Leonard, "Report on the Committee on Minority Groups" *Proceedings the 1972 Annual Meeting, Section I,* Washington, D.C.: The American Association of law schools. Pp. 65-69. Appendix to this Report carries the reference Amendment to the Principals of Non-Discrimination. This Amendment connotes a more inclusive policy statement than the policy statement than the policy enunciated in 1963.
9. Special Services of the ABA and the AALS also include task forces and committees established to offer concrete resolutions to the problems of underrepresentation and retention among minority students.
10. Michael Cordoza, *Op. Cit.,* p. 61.
11. *Op. Cit.,* p. 62.
12. *Op. Cit.,* p. 66 Also from President Walter Leonard of Fisk University; former Chair of the AALS Committee on Minority Groups.
13. Wade J. Henderson, *Statement on Behalf of the Council on Legal Education Opportunity Before the Subcommittee on Post Secondary Education.* U.S. House of Representatives, June 13, 1979.
14. *Ibid.*
15. *Ibid.*
16. *Ibid.* Cf. Also, "Report of Activities of the Council on Legal Education Opportunities for the Period Ending, 1972" in *AALS Proceedings of 1972 Annual Meetings,* Pp. 123-124.
17. Henderson, *Ibid.*
18. *Ibid.*
19. *Ibid.*
20. From data provided by the University of Wisconsin/Madison.
22. From data provided by the University of Washington and from Allen P. Sindler, *Bakke, DeFunis and Minority Admissions.* Pp. 31-38, 41-42, 118, Passim.
23. From data provided by the Ohio State University.
24. From data provided by the University of Wisconsin/Madison.
25. From data provided by the University of Texas/Austin.
26. *Ibid.*
27. *Ibid.*
28. *Ibid.*
29. *Ibid.*
30. *Ibid.*
31. Franklin Evans, *Op. Cit.,* p. 566.
32. *Ibid.,* p. 567-568.

33. *Ibid.*
34. *Ibid.*
35. *New York Times,* November, 1978.
36. From Date Provided by Stanford University.
37. James E. Blackwell, *The Participation of Black Students in Graduate and Professional Schools.* Atlanta: The Southern Education Foundation, 1977.
38. Section on Minority Groups, Association of American Law Schools, *1979-80 Directory of Minority Law Faculty* (Derek Bell, ed.), Washington, D.C.: Association of American Law Schools, 1980.

Chapter 11

The Social Work Profession:
Parity Attained and Lost

The only profession in which black Americans have ever attained parity of access and equity in graduation rates is that of social work. This chapter explores the approaches taken by the social work profession to provide equality of educational opportunity. It examines enrollment and graduation trends, and faculty distribution. It also offers explanations to account for the current decline in the enrollment of black students in graduate schools of social work.

Historical and Societal Context of Mainstreaming

Unlike other major professions, social work began to take steps to correct the racial inequities in its profession in the same year that the *Brown* decision was pronounced by the U.S. Supreme Court. In 1954, the Commission on Accreditation for Schools of Social Work "adopted a mandatory standard" which required assurance of non-discriminatory practices be assured in all schools of social work. According to this policy change, all schools of social work were to conduct their programs without any form of discrimination based upon race, ethnic origin, creed, or color. This policy had universal applicability in that it covered selection and admission of students, and conduct in the classroom and in field practice assignments as well as the organization of a school's program.[1] This fundamental principle of non-discrimination in social work has been national policy since 1954.

With the promulgation of the new policy, and given the nature of the profession itself, an influential leadership role assumed by this profession in the mainstreaming of blacks should move it forward. As the civil rights movement picked up momentum. social work practitioners, educators, and policy makers vocalized the urgency of producing more black social workers. In principle, social work had always embraced the idea of inclusion and openness for those persons who wished to be trained in the

266

technology of human services delivery and in the methodology of helping others to realize their greatest potential. But principles and practices are sometimes inconsistent and strangely contradictory.

The profession, however, suffered from a credibility gap which centered upon its image with black and other minority groups. On the one hand, its practitioners were committed to the principle of providing assistance to the disadvantaged and to persons in need of various types of human services. It aimed to acquaint others with the strategies designed to help them organize their lives for more effective and orderly living. The population, in general, had an image of the social workers, as Andrew Billingsley once labeled, "the public assistance worker"[2]—an image that was sometimes favorable and sometimes unfavorable, depending on the contacts and types of experience that potential students had had with social workers. Social work as a profession did not enjoy the kind of prestige and status ordinarily accorded science and technology fields by the American public. In the fifties and early sixties, some of the misinformed expressed a view that social work was, like elementary school teaching, "woman's work," "something people did when they could not do anything else." This image problem did not reflect a view unique in the relationship between the profession and the black population. It was shared by others who knew little about the profession and the human services objectives it encompassed.

However, the recruitment of black Americans into the profession necessitated a confrontation with the image issue by social work educators and practitioners. Success of the community of social work educators, administrators, and practitioners in addressing this problem is evidenced by the profound changes in the racial composition of social work schools during the 1960s. Prior to the early sixties, the primary hope that black Americans had for admission to a graduate school of social work rested upon actions taken by the two historically black institutions that had Schools of Social Work. These two institutions, Atlanta University, headed for several years by Whitney Young, and Howard University, led the way in the training of black graduate students in this field. Had it not been for their efforts, there would have been considerably fewer social workers from the black population than ordinarily expected. Early failures to expand social work opportunities underscored the inconsistencies between the practice of middle-class whites administering social services to the poor and minorities and the importance of having more representatives from the social work clientele involved in decisions that affected their own well-being.

The problem of underrepresentation of black Americans in white colleges of social work was critical although in 1960, 13 per cent of all first year graduate students in social work were black.[3] In fact, few white institutions admimitted black students and in most of those who did permit black students to enroll the representation in the total school population was at best a token one. Therefore, the Council on Social Work Education began to urge its member institutions to become more aggressive in the recruitment of black and other minority students. In some instances, because of the insistence of black faculty, individual Schools of Social Work took their own initiative to develop constructive "manpower production programs" designed to produce a greater supply of black Americans with graduate degrees in Social Work.

The Berkeley Model

Although a number of institutions took their own initiatives, we shall describe the program at the University of California at Berkeley as a proto-type of efforts initiated in the 1960s for the integration of blacks in social work which, in turn, strengthened the mainstreaming process and led to parity in both access and in production during the seventies. This descrip-tion draws heavily upon the historical overview provided by Andrew Bill-ingsley, a former member of the faculty of the School of Social Work at the University of California/Berkeley, who is now President of Morgan State University in Maryland. His seminal article, which forms the basis of this analysis, is entitled "Black Students in Graduate Schools of Social Work."[4]

Prior to 1963, there were many years in which black students were not included in the yearly social work admissions. In 1963, only two black students were enrolled in the schools of Social Welfare at the University of California/Berkeley. This token enrollment was inconsistent with the ideals and goals of the Graducate Council at the University which expressed a commitment to demonstrably expand educational opportunities in its graduate programs for minority group students. This commitment was ex-plicated in a memorandum issued by the Administrative Committee of Graduate Council in the Spring of 1964 to Deans and department heads. This memorandum called for special efforts to seek, identify and encourage the enrollment of "disadvantaged minorities" in graduate programs on that campus. As Billingsley states, most schools and departments ignored the recommendation by taking no concrete action that would change the status quo. However, the Dean of the School of Social Welfare appointed a com-mittee of faculty members in the Fall of 1964 to take what would become pioneer efforts to include more black and minority students in graduate schools of social work.[5]

It should also be noted here that during this period a number of terms presumed to characterize the conditions, status, and learning abilities of black and other minorities were either coined or popularized. Among these were such as the following: "culturally deprived," "educationally deprived", "culturally neglected," "culturally disadvantaged," and "educationally disadvantaged." Although deeply embedded in the professional jargon, the popularity of these terms diminished over time as their negative connotations were increasingly rebuked. Many rejected the notion of "cultural or educational", if not "individual", *deficit* implicit in them. Unfortunately, academicians and administrators were so vociferous in their labeling of blacks and other minorities as "educationally disadvantaged" that the American public tended to mistakingly and sometimes conveniently perceive *all* black Americans as educationally inferior.

The development of a minority recruitment program in the School of Social Welfare at the University of California/Berkeley was an innovation in education. From its inception, it enjoyed three of the essentials for success. First, it had unqualified support from the top echelon of administrative leadership within the University. Second, it had the commitment of the immediate leadership of the unit which housed the program. In this case, it meant the Dean of the School of Social Welfare, And third, it established a faculty committee to guide and direct the recruitment process. This "racially balanced" faculty committee was committed to the principle of widening the door of opportunity for minority students.[6]

The Berkeley faculty, at one point, was trapped in conceptual difficulties which dramatized the terminological dilemmas and ambiguities to which intellectuals are prone: Who is a minority? Who is "educationally disadvantaged"? Should all blacks be admitted because of their past history of segregation and discrimination? Are middle-class blacks to be given the same treatment as "the authentic ghetto types," since they may not be as "educationally disadvantaged" as lower class blacks in the ghetto? Should the focus be on the "authentic ghetto type" who is the epitome of the "culturally deprived" and the personification of the "educationally disadvantaged"? Or, should the term minority be percieved in much broader terms so as to include certain segments from the population as a whole which might have shared similar experiences to black Americans because of ethnic status or socio-economic status?

With these questions resolved to the Faculty Committee's satisfaction, the Committee established important linkages with the institutional structures whose actions affected outcomes of their own deliberations and policies. These were such groups as the Executive Committee of the School of Social Welfare, the Admissions Policy Committee of the School of Social Welfare, and the minority faculty of the school who ultimately played a crucial role in the program's overall development[7].

In organizing this program, the Minority Student Committee dealt with a number of concerns described in previous chapters. These include (1) the image and knowledge of the profession in the target population, (2) the lack of finance sufficient to support graduate or professional education, (3) academic qualifications for graduate or professional school admissions, (4) motivation for a career in the profession, and (5) special admissions and special support services to strengthen the possibilities of graduation for those in need of academic and special assistance. The latter involved finding ways to support financially the special programs of counseling, guidance, and other special services that might be required.

Upon the recommendation of the Minority Students Committee, the School of Social Welfare established a minority recruitment program. This action exacted not only a change in policy but fostered a major shift in faculty attitudes. This meant a shift away from the notion of "selecting only from those who seek us out" to actively recruiting students not generally included in the School's population. This list of recruitment activities is consonant with those strategies followed by the School of Veterinary Medicine at Tuskegee Institute, especially in the utilization of faculty, alumni and minority students in the distribution of promotional materials, the high school visitation program, and special activities at the historically black colleges.

The Committee's recommendation for establishing counseling and advising services as an integral part of the program was also approved by the faculty. However, the minority faculty, who served as important and invaluable role models in the initial recruitment efforts, played a vital role in this component of the program.

Upon the recommendation of the Minority Student Committee, the School of Social Welfare made the critical decision to change its admissions criteria. For instance, a minimum GPA of "B" in college work was no longer required for admission. In fact, according to Billingsley, a few students with a GPA of "C", but who possessed other qualities desired in Social Welfare students at Berkeley, were also accepted. In effect, there was a major shift from the utilization of purely cognitive determinants to non-cognitive factors in making admissions decisions.[8]

At the suggestion of the Minority Student Committee, the School of Social Welfare hired more minority faculty members. The number of black faculty in the school rose from two in 1963 to seven in 1968-69, and the first faculty of Mexican American origin was hired in 1968-69. In addition, black professionals served as visiting lecturers or as part-time faculty during this period. Since the total number of faculty within the school at this time was about 60 to 64, the minority faculty represented a significant proportion or a critical mass within the school. Hence, an essential program component was assured.[9]

However, it is important to note that, despite the admissions of students with less than a "B" average from college, "virtually no academic assistance" was provided to minority students during the period from 1964-68. There was no organized tutorial program, for example. These students were successful in completing the program primarily through their own efforts. Notwithstanding, it should be stressed that there was a critical mass of black students at Berkeley and the black faculty comprised a significant segment of the total Social Work faculty. Further, there were other psychological support mechanisms within the school environment on which students could draw as needed.

Financial support for the program came initially from the California State Department of Social Welfare which provided stipends to those students who agreed to work for the State following the completion of their degree requirements. Some students were supported through the Work-Study program of the Office of Economic Opportunity of the U.S. Department of Health, Education and Welfare. Others received fellowships under funds made available from NIMH, the Children's Bureau and other federal agencies, and some were financially supported through funds provided by private foundations such as Carnegie, the San Francisco Foundation for Aged and the Stern Family Fund.[10]

The outcomes of these efforts were particularly gratifying in terms of increasing access to black students. Over a five year period, 1964-69, total enrollment in the School of Social Welfare increased from 272 in 1964-65 to 375 in 1968-69. The total enrollment of black students climbed from seven (7) in 1964-65 to 52 in 68-69 or from 2.5 per cent to 13.9 per cent of total enrollment. Minority enrollment jumped from 5.0 per cent to 23.7 per cent within that five year period.[11] Clearly, parity was attained in the School of Social Work and in a relatively short period of time. This was accomplished on the strength of the program and the commitment to it as manifested in the leadership, faculty, the presence of role models, financial support, and the adjustments in traditional admissions requirements that did not compromise the integrity of the institution despite initial repudiation by some faculty members. Minority graduates of the School of Social Welfare at the University of California/Berkeley during this period now hold some of the more important positions in the professions.

During the seventies, profound changes occurred at Berkeley. For one thing, black student enrollment in the School of Social Welfare dropped by almost 50 per cent between 1970 and 1978. The once sizable number of black faculty declined to two persons in 1978-79. These changes raise the question of whether or not the initial commitment and programatic strengths were sustained and if the black faculty were victims of a revolving door syndrome: hired but lost through denial of tenure.

Recruitment, Admissions Criteria and Special Admissions in Schools of Social Work

According to a study reported by Arnulf Pins for the Council on Social Work Education in 1963, the 326 black students enrolled in all schools of social work in the United States that year represented 13 per cent of total enrollment. Further, there was an upward turn in enrollment in graduate schools of social work during the remainder of the sixties. But, the enrollment of black students, while continuing to grow, did not keep pace with overall enrollment. Harold Greenberg and Carl Scott reported that by 1968, the black students comprised only 10 per cent of total enrollment in social work schools.[12]

Given that decline, demands grew for mainstreaming outsiders, and for integration and inclusion of a diverse student body; as a consequence, most social work schools initiated programs to assure increases in minority student enrollment. Many of these institutions, like the University of California at Berkeley, quickly grasped the apparent compatability of their professional goals and the commitment of black and other minority students to social change. They realized the basic issues involved in meeting fundamental survival needs as well as enhancing upward mobility aspirations of a large segment of the population, who comprised such a substantial proportion of their clientele. It was imperative to include larger numbers of these groups in order to fulfill a wide array of social and educational goals. Numbers of students from these groups demonstrated that the traditional social work curriculum, for instance, was not always appropriate in addressing the problems of the real world. Therefore, the educational benefits resulting from the infusion of students from the social work clientele could be enormous and far-reaching. Consequently, many institutions initiated special minority recruitment programs. They could no longer rationalize the apparent contradictions between basic social work philosophy and existing educational policies. How could a helping profession with a focus on urban problems and the conditions of the poor justify the exclusion of students from among its clientele? Could not substantial institutional and professional benefits be reaped by their inclusion?

According to a 1969 survey, forty eight or 75 per cent of the 64 schools of social work then in existence had recruitment programs in operation or were planning them for implementation in the 1969-70 academic year. Over eighty per cent of these institutions had already designated a specific person to be in charge of recruitment programs. In about one-fifth of the cases that person was either the Dean or a member of the Dean's immediate staff. Slightly less than half of the institutions established special committees that were chaired either by a member of the administration or by a special recruitment officer secured for this purpose.[13] Essentially, this pattern continued throughout the seventies, that is, recruitment was conducted under the auspices of the Dean, a special recruitment officer, or a minority recruitment committee with faculty participation.

However, during the seventies, greater efforts were made toward the development and implementation of coordinated recruitment programs. These ranged from individual institutional efforts, such as the program implemented in the School of Social Work at Western Michigan Univesity, to a consortium model established at Texas.

At Western Michigan University, the first recruitment committee had an "ad hoc" status as a "committee on minority concerns" when established in 1972. A year later, the Committee was changed to a "standing" status with recruitment designated as one of its central functions. Full-time faculty members carried on recruitment responsibilities in addition to their regular teaching load. The limited funding made available to the Committee had to be utilized primarily for recruitment trips. An optimistic goal of one-third minority student enrollment was initially set but never attained. However, enrollment did increase but declined again to about 6 per cent in 1978. Once again, the School reaffirmed its commitment to the goals of the program. Since 1978, a faculty member, with released time and aided by a graduate assistant, assumed primary recruitment responsibilities.[14]

The University of Texas/Arlington participates in the Texas Consortium of Graduate Schools of Social Work which was organized in 1968-69. This consortium was established for the expressed purpose of expanding opportunities for minority students in graduate schools of social work. Since its inception, it has been supported by the National Institute of Mental Health. Under the shared funding arrangement, participating institutions receive funds for recruitment programs and special stipends. For instance, in 1972-73, the first year in which the University of Texas/Arlington participated, the institution received approximately $40,000 to support student stipends and a full-time recruiter. Since that time, the University has not only had a full-time recruiter but it provides from three to eight stipends per year.[15]

Over half of the schools of social work use flexible admissions criteria. The implementation of these criteria is not viewed as a compromise of standards; rather it is a mechanism for creating equity in the admissions process and for a more careful and systematic evaluation of all strengths that a potential student brings to the institution. The Graduate Record Examination (GRE) may or may not be required; it may be waived for students whose college GPA is 3.0 or above. It may be required but not weighted disproportionately among all the measures employed for evaluation. Few institutions actually employ inflexible, absolute, quantitative admissions standards. Frequently, schools of social work scrutinize very carefully and give greater weight to the quality of academic performance during the last two years of undergraduate work and the performance exhibited in courses in related fields (e.g., sociology and psychology).

Schools of social work stress evidence of ability to perform at the graduate level and potential for success as a social worker. The former may mean an overall college GPA of C; or a C + in the last two years of undergraduate study, or a cumulated B average or above, or a C + average with good GRE scores; or a C average combined with excellent potential as revealed through non-cognitive measures of evaluation. The latter may include motivation for a career in a human services profession; leadership ability; evidence of having overcome undue hardships and the maintenance of motivation in the face of great adversity; relevant work experience which demonstrates a commitment for helping others; sound judgment; and the maturity to understand the complexities of interpersonal relationships and interactions with diverse groups of people. These attributes become a major focus of the applicant's personal statement, the interview, and references as each of these is evaluated.

In order to minimize expenses to students, many schools of social work draw upon their alumni for asistance in local or regional interviewing. Frequently, these persons are in a much better position to evaluate, for instance, the nature of the work experience which the applicant reports because of their knowledge and proximity to the locale in which the experience occurred. It is, therefore, likely that any number of very promising students who may not ordinarily have been accepted were admitted because of this participation of alumni.

Sources of Financial Assistance

During the late 1960s and up until the middle of the 1970s, funds to support educational programs were reasonably adequate. Schools of Social Work obtained significant funding through direct grants, research projects which carried monies for students stipends, assistantships and fellowships, and substantial foundation support. Various branches and agencies of the federal government were particularly supportive. Among these were NIMH, the Children's Bureau, the Social and Rehabilitation Service, the National Institutes of Health, the National Institutes of Education, the Offic of Economic Opportunity, and the various manpower branches of the U.S. Department of Labor. State health and welfare departments, as well as alumni groups, also provided major financial support for efforts to expand opportunities for black and other minority students in social work.

Many of these efforts, as we have already reported, came under a brutal attack by individuals and some formal organizations who charged that they discriminated against white students, or were tantamount to "reverse discrimination" and that their constitutionality was highly suspect. The same allegations were levied against the Minority Fellowship Program of the American Sociological Association funded initially by the National Institute of Mental Health (NIMH). In fact, a few members of the American Sociological Association either resigned or threatened to resign in protest over the establishment of its minority program.

Coincident to these attacks, institutions with professional programs witnessed a significant decline in financial aid funds, scholarship monies and general financial support. This curtailment, if not total dissipation, of financial support had a deleterious impact on recruitment and enrollment efforts. In fact, in some instances, as in social work, many of the gains achieved in the late 1960s and early 1970s were either lost or difficult to maintain. While this retrenchment of funds may not have been the sole reason for the decline in black student enrollment apparent in the late 1970s, it was indeed, a principal contributor.

Enrollments of Black Students in Schools of Social Work

At no time in this century has the census-counted black population of the United States reached 12 per cent of the total U.S. population. Throughout the 1970s, the reported black population ranged from 11.1 per cent to about 11.6 per cent.* Using our definition of total access or enrollment parity, black students in first year social work classes not only reached parity but exceeded it in all but the three final years of the decade. (See Table 42).

*Many persons, lay and professional, have argued that every U.S. census count misses a significant number of blacks. If that is so, the proportion of blacks in the total population may exceed 12 per cent.

. First Year Enrollments

First year enrollments in schools of social work increased at a steady rate for all students between 1969 and the 1972-73 academic year. The peak for first year enrollment among black students was reached in 1972-73 when the 1,226 first year black students represented 15.7 per cent of all students enrolled at that level. In both absolute numbers and in proportionate enrollment in first year classes, the increases observed for black students in the three preceding years were noteworthy. As the decade opened, black students comprised more than fourteen per cent of all first year MSW students.

First year MSW enrollments declined for all students for a three year period between 1974 and 1977 and enrollments still appear to be sporadic. By contrast, the downward trend in absolute numbers as well as in proportion of all students in first year classes witnessed among black students in 1973-74 has continued unabated. As a result, there were fewer black students in first year MSW classes in 1979 than there were in 1969. The percentage of black students in first year MSW classes dropped significantly from 14.2 per cent in 1969-70 to 9.3 per cent in 1979-80 (See Table 42).

Hence, the optimism of full integration in schools of social work reflected in percentage distributions through 1975-76 has been replaced by trepidations and alarm that some institutions have already reneged on their original commitments and that the dissipation of funds for stipends, scholarships, and general financial aid is having a negative impact on black students presence in graduate schools of social work. Another explanation for the declining enrollment of black students in first year MSW classes lies in expanding career choices for black students, or options that some students, who would ordinarily have selected social work, have taken in other, more prestigious and lucrative professions.

Total Enrollments

A similar trend is also observed in total enrollments in M.S.W. degree programs for the nation as a whole. Total enrollments climbed steadily throughout the decade without declines in any single year. On the other hand, total enrollment for black MSW students peaked in 1972-73 but it has continued downward since that time. Black students enrollment held parity or above parity levels through 1975-76 and was near parity in 1976-77. However, the trends set in motion in 1977 may bode serious problems in the 1980s. This situation may not only signal a loss of parity but

Table 42. First Year Enrollment in Schools of Social Work by Race and Year: All Institutions Combined, 1969-79
(MSW Programs)

Year	Total Enrolled in MSW Programs First Year Only	Total Number of Black Students in First Year	Per cent Black of Total
1969-70	6,241	888	14.2
1970-71	6,699	975	14.5
1971-72	7,137	1,067	14.9
1972-73	7,788	1,226	15.7
1973-74	8,235	1,070	12.9
1974-75	7,935	1,004	12.6
1975-76	7,840	933	11.9
1976-77	7,951	836	10.5
1977-78	8,510	864	10.1
1978-79	8,197	844	10.3
1979-80	8,056	799	9.3

Source: *Statistics on Social Work Education.* New York: Council on Social Work Education (See Reports for 1970 through 1979-80). Reprinted with permission of publisher.

Table 43. Total MSW Enrollment in Schools of Social Work by Race and Year: All Institutions Combined, 1969-79

Year	Total MSW Enrollment	Total Number Black Students	Per cent Black of Total
1969-70	12,061	1,495	12.4
1970-71	13,205	1,847	13.9
1971-72	14,456	2,116	14.5
1972-73	15,596	2,352	15.8
1973-74	16,716	2,217	13.2
1974-75	17,238	2,210	12.8
1975-76	17,388	2,149	12.3
1976-77	17,638	1,969	11.1
1977-78	18,399	1,921	10.4
1978-79	18,493	1,894	10.2
1979-80	17,397	1,722	9.9

Source: *Statistics on Social Work Education.* New York: Council on Social Work Education; 1970-79/80. Reprinted with permission of publisher.

Table 44. First year Black Student Enrollment in Twenty-Five Largest Schools of Social Work for Selected Years, 1970-1979 (MSW Programs)

Name of Institution	Black Student Enrollment By Year					
	1970	1972	1974	1976	1978	1979
Univ. of Alabama	14	15	12	5	16	16
Univ. of Calif./Berkeley	13	11	7	5	7	7
Howard Univ.	32	53	53	67	79	72
Florida State Univ.	24	18	11	2	1	3
Atlanta Univ.	61	61	61	40	53	62
Univ. of Chicago	45	44	26	17	14	6
Univ. of Kansas	15	15	25	10	11	14
Univ. of Illinois/Urbana	20	55	16	8	12	4
Boston Univ.	15	18	7	11	10	5
Univ. of Michigan	42	28	22	27	25	23
Wayne State Univ.	29	41	32	27	11	10
Univ. of Maryland	19	24	26	17	14	17
Rutgers Univ.	27	40	38	36	25	29
Adelphi Univ.	18	14	3	6	6	8
Columbia Univ.	51	50	44	33	39	27
Fordham Univ.	43	24	17	17	14	26
Hunter College	17	41	19	15	17	12
New York Univ.	45	24	26	13	17	13
Case Western Reserve	17	29	19	15	17	11
Bryn Mawr College	13	19	10	9	10	8
Univ. of Pennsylvania	44	45	31	28	17	8
Univ. of Pittsburgh	45	57	23	23	15	22
Virginia Commonwealth U.	12	24	24	14	11	9
Univ. of Southern Calif.	13	11	7	8	11	5
Univ. of Calif./Los Angeles	23	22	17	10	8	4

Source: *Statistics on Social Work Education* New York: Council on Social Work Education (See Reports for 1970 through 1979-80). Reprinted with permission of the publisher.

portend major difficulties in attracting black students to social work in significant numbers. (See Table 43).

Information on enrollment in selected institutions for selected years provide insights into the general problem of retrenchment. (See Table 44).

Using biennial data for comparative purposes, it is evident from this table that the enrollment of black students in twenty-five of the largest schools of social work was uneven at best during the decade. Over 90 per cent of these institutions do not presently enroll as many students in social work as they did in 1970. It is also evident from this table that the major thrust for recruitment and enrollment seemed to have peaked once again in 1972 through 1974 followed by a continuing downward spiral. The data provided on such institutions in this group as Florida State University, the

University of Chicago, Wayne State University, Adelphi, Columbia, UC/Berkeley, and UCLA - all illustrate this point.

These diminishing numbers and proportions of black students in first year enrollments raise significant questions about institutional behavior and external conditions which might account for the changes observed. What has happened to institutional behavior? Has the commitment of an earlier period vanished or are there uncontrollable external and internal forces that render it difficult to sustain institutional commitment in measurable or identifiable ways? How successful are black faculty in affecting decisions? Have their internal pressures paled in significance as they mobilize their energies to achieve the personal goal of gaining tenure? Where is the money needed to fund social work education? Why is this money not available?

"The Adams States"

The "Adams States" had mixed results in the enrollment of black students during the seventies. According to Table 45, the University of Oklahoma was the least successful in enrolling black students. The two largest public schools of social work in these states are located at the University of Maryland, and Virginia Commonwealth University. In general, each of these institutions sustained an enrollment of black students considerably above ten per cent of the total number of first year students enrolled in every year of the decade. The same was true at the University of North Carolina whose school of social work is considerably smaller in student capacity than the two institutions just described.

No steady trend can be discerned at Florida State University, the University of Georgia or Louisiana State University. The University of Southern Mississippi just opened its school of social work in 1974. However, it should be pointed out that in 1978, Florida State University enrolled one-twenty-fourth the number of black students as it did in 1970. The first year enrollment of black students at the University of Georgia, on the other hand, reached the ten per cent level in seven of the nine years on which data were reported. LSU achieved a similar but somewhat higher proportion in eight of the nine years. However, these proportions must be viewed in the context of the proportion of blacks in the total population of the State. When, this is done, we see that none has achieved total parity (See Table 45).

Table 45. First Year Enrollment in Primary Public Institutions
of "Adams States," by Race and Year, 1970-79/80
(Schools of Social Work)

Name of Institution	1970	1971	1972	1973	1974	1975	1976	1977	1978	1979
Florida State Univ.										
Total Enrollment	112	91	97	75	67	79	95	113	83	89
Number of Black Students	24	13	18	6	11	2	2	6	1	3
University of Georgia										
Total Enrollment	58	69	111	43	37	43	39	47	89	53
No. Black Students	5	4	15	5	8	8	7	8	20	4
Louisiana State Univ.										
Total Enrollment	92	80	112	90	106	115	106	87	93	75
No. Black Students	8	11	14	11	17	10	9	2	17	8
Univ. of Maryland										
Total Enrollment	117	211	207	198	186	173	187	169	203	153
No. Black Students	19	41	24	27	26	24	20	13	14	17
So. Mississippi										
Total Enrollment	NA	NA	NA	NA	NA	NA	51	47	57	50
No. Black Students	—	—	—	—	—	—	11	4	6	14
Univ. of North Carolina										
Total Enrollment	66	65	80	49	60	53	41	44	45	41
No. Black Students	9	6	9	6	8	6	4	7	6	2
Univ. of Oklahoma										
Total Enrollment	27	49	61	61	75	49	43	57	48	—
No. Black Students	1	5	4	2	5	2	3	3	2	—
Virginia Commonwealth										
Total Enrollment	78	73	111	119	135	101	120	123	111	91
No Black Students	10	10	24	20	24	10	14	13	11	9

Note: No data Available on the University of Arkansas;   N/A = School was not open.
Source: Statistics on Social Work Education; Council on Social Work Education-New York City 1970 through 1978/79
Reprinted with permission of Publisher

Enrollment in Doctorate (DSW) Programs

The black student enrollment in doctor of social work degree (DSW) programs increased steadily throughout the decade. The only exception to this generalization was in the 1976-78 period which showed a slight decline from the preceding year. However, a major absolute increase followed this downturn. In every year of the decade, the enrollments increased steadily. The only exception to this generalization was in the 1976-78 period which showed a slight decline from the preceding year. However, a major absolute increase following this downturn. In every year of the decade, the proportion of black students in DSW programs exceeded their proportion in the national population. In terms of percentage of total DSW students enrolled, the enrollment of black students peaked in 1974-75 when the 123 students enrolled represented 17.4 per cent of total DSW enrollment. According to the latest available data, black students comprise 15.5 per cent of these enrollments. And, this is clearly open access and above parity levels. It should also be noted that the number of institutions offering post MSW graduate work increased from 23 in 1970 to 40 in 1978. This growth in the number of doctoral degree programs helps to account for the numerical increases in student enrollment.

The description so far pertains exclusively to national totals. It is only through an analysis of institutional data that one can discern either institutional behavior or characteristics of specific states. Trends in the "Adams States" are illustrative in this regard. In 1970, when the "Adams States" comprised 13 per cent (3/23) of the doctoral degree institutions, they enrolled 11.1 per cent of black DSW students. In 1972, these States had 16 per cent (4/25) of the institutions but 13.8 per cent of black DSW students. Their proportion of these students declined in 1974 when, with 13.7 per cent of the institutions (4.27), they enrolled 12.3 per cent of black DSW students. The downward trend continued into 1976 when they comprised 14 per cent of the institutions (8/35) but only twelve of 120 or 10 per cent of the black students. Finally, an upswing occurred in 1978 in both the number of institutions (8 of 40 or 20 per cent)[15] and in the absolute number and proportion of black DSW students. In 1978, these eight institutions enrolled 23 of the 127 black students enrolled in DSW programs or 18.7 per cent of these students.

Therefore, it appears that the enrollment increases among black doctoral students in these states occurred only as the number of doctorate granting institutions increased. There is little indication that specific institutions made appreciable gains in the total number of blacks matriculated in doctorate programs.

Table 46. Enrollment of Black Students in Doctor of Social Work Degree Programs by Race and Year, 1970-1979/80: All Institutions Combined

Year	Total Enrollment	Number of Black Students	Per cent Black of Total
1970-71	462	54	11.7
1971-72	504	70	13.3
1972-73	565	87	15.4
1973-74	617	104	16.8
1974-75	648	113	17.4
1975-76	712	123	17.3
1976-77	769	120	15.6
1977-78	866	116	13.4
1978-79	821	127	15.5
1979-80	954	119	12.5

Note: For this analysis, DSW enrollments are substituted for the designation "post-MSW"

Source: *Statistics on Social Work Education.* New York: Council on Social Work Education (See years, 1970-1979/80.) Reprinted with permission of publisher.

Table 47. Total MSW (Social Work) Degrees Awarded by Year and Race: All Institutions Combined, 1970-79

Year	Total MSW Degrees	Total MSW Degrees Awarded Black Students	Per cent Black of Total
1969-70	5,866	—	—
1970-71	6,284	—	—
1971-72	6,909	—	—
1972-73	7,387	—	—
1973-74	8,005	1,246	15.5
1974-75	8,824	1,071	12.1
1975-76	9,080	1,007	11.0
1976-77	9,254	971	10.4
1977-78	9,476	881	9.3
1978-79	10,080	937	9.3

Note: Blank marks indicate the absence of usable data.

Source: *Statistics on Social Work Education.* New York: Council on Social Work Education. (See reports from 1970-1979/80). One institution not included. Reprinted with permission of the publisher.

The Production of Black Professionals
in Social Work: Degrees Granted

Most graduate students in social work are enrolled in the Masters in Social Work (MSW) degree programs. Slightly less than one-half of the 84 accredited schools of social work in the United States offer the Doctor of Social Work (D.S.W.) degree. At the beginning of the decade, only twenty-three institutions offered post-MSW studies. By 1980, that number had grown to thirty-eight. For purposes of this analysis, attention is first drawn to the production of black students with the M.S.W. degree. Unfortunately, the only usable data available are for the years 1973-79.

After an initial increase in the number of black students who earned the MSW degree, in recent years there has been a decline in both their absolute numbers as well as in their proportion of all MSW degrees earned in the United States. In 1973-74, black students earned 1,246 of the total of 8,005 or 15.5 per cent of all MSW degrees. The production rate of black students with MSW degrees continued above or near parity levels for the next three years. However, the downward trend that began in 1974-75 has continued unchecked (See Table 47).

On the other hand, although the numbers of black students who earned the DSW is relatively small, their proportion of all DSWs earned stood at or above parity for the mid-1970s. It began a process of decline in 1976-77, and dropped abruptly below black representation in the total population to the 6.7 per cent level on 1977-78. As a result, only a dozen or so black persons with the DSW degree were produced. Another upswing occurred in 1978-79 when the eighteen doctorates earned by blacks represented 10.9 per cent of the total.

These downward trends mean that a critical shortage of trained black social workers may be on the horizon in the late 1980s because the present production level cannot keep pace with retirement losses and the demand for social work personnel in teaching, administrative, state, local and federal government positions.

Black Faculty in Schools of Social Work

Of all the disciplines or fields of study offered in graduate and professional schools, social work has been the most successful in racial and ethnic integration. Although, some institutions may have only one or two black faculty members in social work, there is a visible presence of black faculty in most of the 84 institutions in the United States. In 1980, only a half-dozen

Table 48. Total Doctor of Social Work Degrees Awarded by Year and Race: All Institutuons Combined, 1970-79

Year	Total DSW Degrees Awarded	DSW Degrees Awarded Black Students	Per cent Black of Total
1969-70	89	—	—
1970-71	148	—	—
1971-72	149	—	—
1972-73	112	—	—
1973-74	157	18	11.4
1974-75	155	19	12.6
1975-76	179	23	12.8
1976-77	177	18	10.1
1977-78	178	12	6.7*
1978-79	174	18	10.9

Note: Blank Spaces indicate the absence of usable data. *This drop may be explained by the failure of 7.3 per cent of the institutions to report.
Source: *Statistics on Social Work Education*. New York: Council on Social Work Education. (See Reports from 1970-1979/80). Reprinted with permission of the publisher.

schools were without a single black faculty member but there were several schools whose total black faculty members represented a substantial proportion of total faculty. Aside from the two historically black institutions of Atlanta University and Howard University, the following institutions led all institutions in the total number of black faculty: Case Western Reserve University, the University of Pennsylvania, the University of Pittsburgh, and the University of Maryland.

During the seventies, several schools of social work hired black professionals in key administrative positions such as the Dean of the School. Examples of this group include Case Western Reserve University, Boston University, and Boston College. But, these three institutions do not exhaust the list of institutions which hired black Americans as dean or associate dean of these schools.

The number of black faculty members increased through most of the decade, (Table 49). It did not peak in absolute numbers until 1979-80 when 389 black faculty members were teaching in all schools of social work in the nation. In 1973-74, an earlier peak was achieved during the period when absolute increases in the numbers of black students were enrolled. Expanding schools of social work hired more faculty, majority and minority. A downward turn occurred for the next two or three years but this was followed by notable increases.

Table 49. Total Number of Black Faculty in Schools of Social Work by Year and Percent of Total, 1970-1979/80. (All Institutions Combined) Full-Time Faculty Only

Year	Total Faculty	Total Black Faculty	Per Cent Black Faculty of Total
1970-71	2,199	259	11.7
1971-72	2,248	276	12.2
1972-73	2,341	316	13.5
1973-74	2,309	341	14.8
1974-75	2,185	330	15.1
1975-76	2,180	323	14.8
1976-77	2,058	317	15.4
1977-78	2,149	328	15.5
1978-79	2,253	336	14.9
1979-80	2,387	389	16.3

Source: *Statistics on Social Work Education.* New York: Council on Social Work Education (See years 1970 through 1979-80). Reprinted with the permission of the publisher.

The national proportion of black faculty in the total number has always been above the proportion of blacks in the total population. It reached its highest level in 1979-80 when the 389 black faculty represented 16.3 per cent of the 2,387 faculty employed in all schools of social work in the United States. A caveat must be inserted regarding these proportions since they include black faculty employed at Atlanta University, Howard University, and Norfolk State University. When these figures, whose combined total in 1977-78 was 49, are removed from the totals, the proportion of black faculty in the total faculty population falls to about 12%. In the majority of institutions, the proportion of black faculty members does not approximate even this level.

However, there are notable exceptions to this generalization. For instance, throughout much of the decade, black Americans comprised approximately one-third of the social work faculty at the University of Pittsburgh. They were fairly close to this level at the University of Pennsylvania from 1974 onward. That proportion or higher has existed at Temple

University since 1972. On the other hand, Virginia Commonwealth University had ten black faculty out of a total of 49 in 1970. The number has fallen to eight out of a total of 51 full-time in the School of Social Work. The University of Maryland increased its black faculty from 8 of 45 in 1970 to 13 of 62 in 1978 while the University of North Carolina doubled its black faculty from 2 of 26 to 4 of 37 between 1970 and 1978.[17]

Within this time frame, the number of black faculty increased from none to three at Louisiana State University, from none to two at Florida State University, from one to two at the University of Georgia, and from two to three at the University of Oklahoma.

Although some of these proportions are unimpressive, the incontrovertable fact is that the social work profession has achieved far more success in providing role models than has any other profession. The presence of visible role models in the classroom, administrative positions, as field supervisors, and as directors of agencies in which field assignments are made, has a profoundly positive impact on the efforts to expand and retain educational opportunities in this profession. It is sound evidence of the results that may be attained when institutional commitment is strong. It suggests what results are probable when a variety of role models are present to help create a positive image of the profession as well as to attract a sufficient number of black students and black faculty.

Problems

The problems facing the recruitment of black students into social work in the 1980s are not rooted in the Bakke decision. These institutions long ago implemented modified admissions criteria with sufficient flexibility to accommodate potentially excellent social work professionals from all races. A major problem is suggested in declining enrollments. Social work has difficulties of serious proportions in competing with professionals whose earning capacity is considerably higher than that which the average social worker earns in most States. In general, social workers are underpaid for the services they perform and, in many cities, towns and states, increasing their salaries or wages is not a matter of the highest priority. This poses a major threat to efforts to attract strong individuals to the profession.

The limited or dwindling pool of black applicants suggests that recruiters can no longer afford to rely so heavily upon the pool of college students who major in the social and behavioral sciences. That pool is also dwindling. It is now incumbent upon schools of social work to establish outreach programs and implement stronger recruitment activities in the high schools so that information about the field can be disseminated at earlier and earlier levels of secondary schools studies.

There are also problems of retention in some institutions because of the lack of preparation, weak study skills, communication inadequacies, and writing deficiencies of some students. But, there are equally grave problems intrinsically associated with low retention. They encompass alienation and isolation within schools in which black students are in very small numbers and without psychological support, cultural reinforcements, social bonds, and affective ties necessary for survival in what they may perceive to be either a hostile or an unwelcoming environment. Of special note is that some faculty members are still prone to lower their expectations for black and other minority students. Regardless of the motivation for this inexcusable and intolerable behavior, the damage done to the graduate students can be all but insurmountable in future years. Each of these issues can be remediated or remedied through skillful institutional and societal intervention.

The problems of financial aid will not disappear so long as black families earn only 57 per cent of the income that white families earn. It will not vanish so long as black students are not awarded fellowships, scholarships, and adequate financial aid so that they will not be saddled with lifelong debts. High interest, repayable loans are not the answer especially when the expected earning power is comparatively low. Subsidization of graduate and professional studies should be based, of course, on both need and merit. But full, financing of students in these educational programs should be based most extensively on a commitment to serve in the underserved areas as its quid pro quo.

Footnotes

1. Kurt Reichert, "Survey of Non -Discriminatory Practices in Accredited Graduate Schools of Social Work," *Ethnic Minorities in Social Work Education* in Carl Scott (ed.), New York: Council on Social Work Education, 1970, pp. 39-51.
2. Andrew Billingsley, "Black Students in A Graduate School of Social Work," in Carl Scott, (ed.), *Ethnic Minorities in Social Work Education.* New York: Council on Social Work Education, 1970. pp. 23-38.
3. Harold I. Greenberg and Carl A. Scott, "Ethnic Minority Recruitment and Programs for the Educationally Disadvantaged in Schools of Social Work," in Carl Scott ed.), *supra*, pp. 11-22.
4. *Op. Cit.*
5. *Ibid.* pp. 28-31.
6. *Ibid.*
7. *Ibid.*

8. *Ibid.*
9. *Ibid.*
10. *Ibid.*
11. *Ibid.*
12. *Op. Cit.* p. 11.
13. *Ibid.* Pp. 12-14
14. Data provided by Western Michigan University.
15. Date Provided by University of Texas/Arlington.
16. See *Statistics on Social Work Education.* New York Council on Social Work Education, 1970-1978, Passion.
17. Institutional Breakdown on ethnicity of faculty were not reported in *Statistics on Social Work Education* for 1979-80.

Chapter 12

The Production of Black Doctorates

Few areas in the academic community have stirred as much controversy or have created a crisis of such monumental proportions as has that of the production of doctoral degrees among black Americans. The production rate and the actual number of persons in this group awarded the doctorate each year have had many serious implications. The production of "Black Ph. D.s" raised a number of questions about the "actual or potential pool" of candidates for positions. This issue is often expressed today in terms of "the available pool" which, in turn, begs the question of what should be the basis for affirmative action goals and guidelines. It created others: for instance, should the number of blacks and other minorities sought as a goal be equal to their proportion in the total population of a given area, their proportion in the State population, their representation in the national population, or should it be tied to their proportion in the "available pool" of actual doctoral degree holders?

Resolution of these questions has never been satisfactory to all sides in disputes they generated. Albeit, these issues reveal a much deeper problem which is the continuing shortage of black Americans who hold the doctorate. This fundamental and pressing problem remains in the 1980s desptie the significant increases in the absolute numbers of blacks awarded the doctorate during the seventies.

Because of the scope of the doctoral fields encompassed, it is not possible to provide a detailed analysis of enrollment data field by field. However, it is essential to note that the actual number of black Americans enrolled in doctoral degree programs is miniscule when measured by any of the traditional standards. For instance, they are underrepresented in enrollment whether or not such measure as the following are employed: number of black Americans in the doctoral age group; proportion of black Americans in either the city, State, or national population; or proportion of black Americans who are college graduates.

Further, it should be reiterated that this disproportionately low

289

number of black students enrolled in doctoral degree programs continues even as the pool of blacks who have graduated from college and who, theoretically, could be eligible for advanced degree programs expands. The existence of an expanded pool of potential candidates for doctoral degree programs among blacks with college degrees suggests that any number of barriers must be operating which effectively prevent them from enrolling in doctoral degree programs. Attention will be given to some of these impediments to enrollment in a subsequent section. However, the principal concern of this analysis is on the actual production of black doctoral degree holders. Another primary concern is with their distribution by fields of study. Inasmuch as the economic disabilities of black Americans may prevent them from pursuing the doctorate as quickly as white students generally do, it is imperative to examine differences between white and black Americans who are successful in becoming doctoral degree recipients specifically in sources of financial support while in graduate school.

Overview

The proportion of black students enrolled in graduate schools increased steadily throughout the decade of the 1970s. In 1972, black graduate students comprised 4.2 per cent of all students enrolled in U.S. graduate schools. There was a 1 per cent increase, to 5.2 per cent, in their proportion by 1973.[2] By 1975, their rate rose to 6.4 per cent of total graduate school enrollment.[3] Since 1976, it appears that the percentage of black graduate students has leveled off to about 6.0 per cent of total graduate enrollment. According to the latest available data, there are approximately 1,076,980 students enrolled in all of the nation's graduate schools. Of that number, approximately 61,923 are black.

When analyzing total enrollment of black students by State, it is necessary to keep in mind that their distributions within State or publicly controlled institutions will vary widely among institutions whenever the State has an historically black college with a graduate degree program. At present, there are 34 historically black colleges, dispersed throughout the southern and border states, which offer at least one graduate degree. Most of these institutions award degrees at the Masters level only. However, the doctorate is awarded at Howard University, Atlanta University, Meharry Medical College, Texas Southern University, and recently a doctoral degree program was established at Morgan State University and at Jackson State University. Although it is possible to obtain Masters degrees in a broad range of subjects, from Education to Sports Administration, and Social

Work, most of the Masters degree programs are in Education and its subspecializations. Significantly, State enrollments might show a relatively large number of black graduate students in States with black colleges. In reality, a great majority of these students are only enrolled in Masters level programs in the historically black Colleges. Not infrequently, the racial composition of the student body at the historically black colleges demonstrates considerably greater success in racial integration than is the case in the publicly supported traditionally white institutions.

At the doctoral level, there is a paucity of black students enrolled in every field in all States. This observation holds true even when confronted with the data on enrollment of black students in doctoral programs in Education. Even in this field, an incontrovertible fact is that black Americans are underrepresented in both enrollment and in the number of doctoral degrees they earn. The critical issue here is why our nation's graduate schools have not attracted sufficiently large enrollments of black students for them to approximate parity or to move beyond a level of tokenism in the number of doctoral degrees produced annually? Another unresolved issue is why larger numbers of black students do not actually enroll in doctoral programs?

A partial answer to these questions is that graduate schools, unlike professional schools, have not in the past mounted well-organized and systematic recruitment programs specifically designed for black and other minority students. Most of them rely upon their reputation, "old boy net-works", ill-defined programs, limited scholarships and feeble an-nouncements that their institution is "an equal opportunity" institution. Often these words are shallow, empty pronouncements, with little or no in-tent to implement their full meaning.

An indisputable fact is that many black students failed to satisfy cut-off points on the Graduate Record Examination (GRE) where it was a stan-dard admission requirement. Weaknesses on the GRE occurred sometimes when college Grade point average (GPA) suggested that the students should have performed at a substantially higher level on this test. In many of these instances, especially in the early seventies when capitation funds were distributed in part on the basis of the number of students enrolled and when grant/research money was more plentiful, some graduate departments were more willing to "take a chance" on students whom, they perhaps classified as "high gain, minimum risk" if they came close to GRE cut-off points. Undoubtedly, a few who never should have been admitted were enrolled.

The influence factors suggested by the concept of "old boy network" seems to work most effectively for black students matriculated in predominantly white institutions. Further, letters of recommendation on behalf of the black candidates help when they come from a faculty member who is either well known by scholarly reputation or is well-known to one or more members of the Department's Admissions Comittee. In this regard, it is important to keep in mind that the *number* of "admissions committees" usually approximates the *number of departments* or special programs that exist within the institution. Hence, a student may be "admissible" to the University but not necessarily "admissible" to a given department. Being "admissible" to both is a prerequisite for most fellowships and scholarships awarded through institutional funds. But the "old boy network" cuts through these problems.

To take note of the paucity of departmental minority recruitment committees is not to suggest that all departments were and are insensitive to the issue of enrolling more black students in graduate schools. Many are now and were committed to this goal during the seventies. Their admissions committees did their best to "flag" a black student when assessing applications. This process was not approached in any systematic manner. It often took the form of trying to establish racial or ethnic identity by examining the name of the institution from which the baccalaureate degree was earned: the name of the students (was this a common "black name" or not?), the identity of the persons who wrote the letters of recommendation, or membership in black oriented student groups or civic organizations.

Another explanation of this underrepresentation in enrollment lies in the reputation of apathy and disinterest, lack of sensitivity, or racism attributed to a faculty member of some departments by students who had experience in the department. A reputation of indifference or racism does not attract black students to the department. Neither does the absence of a critical mass which presumably may be explained by that indifference or racism. While it is also true that some students may misinterpret strict adherence to academic standards as racism, that alone does not disprove the presence of faculty members who do not believe that their department is the "right place" for black students. Many black students report encounters with majority professors who inform them either directly or in subtle ways that they feel that these students would be more "comfortable" in another department or institution. In any event, this reputation becomes widely disseminated in the network of black students who refuse even to apply to such departments or institutions.

Further, there is the apparent problem of lack of interest among many black college students in pursuing graduate studies. This low commitment to graduate studies may exist because it never occurs to students who are first-generation college students to think beyond obtaining a college degree.

It may occur because of improper counselling and guidance programs in their undergraduate colleges.[4] They may not have adequate knowledge of the process of applying to graduate schools, of locating adequate funding, or of how to select an area of specialization within a broad major field. Importantly, they do not think of graduate school at the moment because of staggering financial obligations accrued during four years of college. While neither of these factors is necessarily peculiar to black or other minority students, each one is undoubtedly far more serious for these groups than it is for the average middle-class white college enrollee or for poor white students whose professors may have taken a special interest in them. Hence, all of these factors may act as a strong deterrent to the pursuit of graduate studies, or at least to persuade black students to defer graduate work until a later period of time.

The Production of Doctorates Among Black Americans

According to a detailed analysis of doctorates earned by women and minorities made by Dorothy M. Gilford and Joan Snyder, between 1950 and 1954, only 139 black persons with doctoral degrees were then in the labor force. During the next five year, following the *Brown* decision and between 1954-59, that number rose to 277. During the 1965-69 period, when federal enabling legislation was enacted that fostered greater access of black students to higher education, the total number of black persons who held a doctorate and were in the labor force climbed to 586.[5]

These data are instructive in understanding the results of historical patterns of segregation, discrimination, and racism that had operated to reduce the numbers of black persons with the doctorate. At no time during this period did the number of black Ph.Ds rise above 1.0 per cent of all doctorate holders in the labor force.

Although no systematic reliable national data for the years 1970-73 are available, the data on the production of blacks with doctoral degrees between 1973 and 1979 may be especially illuminating. They show a pattern of increasing absolute numbers and the proportion of total degrees awarded that coincided with successful attacks on exclusionary barriers. For instance, the 581 doctorates produced among black Americans in 1973 alone approximated the total number in this category for the combined five year period of 1965-69. (Table 50).

An examination of Table 50 shows a near doubling of the number of doctorate degrees earned by blacks in a seven year period. A peak in the number of doctorates was reached in 1977 when this group earned 1,109

Table 50. Doctorates Awarded to Black Americans by Sex and Year 1973-79

Year	Total Doctorates Awarded Black Americans	Per cent Black of Total	Men	Per Cent	Women	Per Cent
1973a	581	2.4	—	—	—	—
1974b	846	3.8	580	68.6	264	31.4
1975	989	3.3	642	64.9	347	35.1
1976	1085	4.9	647	59.6	438	40.4
1977	1109	3.7	681	61.4	428	38.6
1978	1029	4.5	581	56.5	448	43.5
1979	1055	5.0	551	52.2	504	47.8
Total	6694a	3.9	3882b	60.2	2929b	39.8

a = No breakdown by sex and race available
b = These figures are for 1974-79 only. They may be inflated by 2 in toto after rounding off.
Source: Constructed from Data provided by the National Academy of Sciences, Commission on Human Resources.

doctoral degrees. However, a decline occurred in 1978 which resulted in a loss of 80 degrees (or 4.7 per cent). In 1979, a slight rise to 1,055 degrees awarded to black students was noted.

In 1973, the 581 degrees awarded to black Americans represented 2.4 per cent of the 27,129 doctorates conferred upon U.S. citizens in all institutions combined. The percentage of all degrees earned by black Americans increased and declined the next five years. They fluctuated from a low of 3.3 per cent to a high of 4.5 per cent of all degrees awarded U.S. citizens during this period. Blacks earned 3.9 per cent of all doctorates awarded in that seven year period.

However, between 1977 and 1979, an uninterrupted increase in these proportions occurred so that the number of doctorates earned by black Americans rose from 3.7 per cent in 1977 to 4.5 per cent in 1978 and, to 5.0 per cent in 1979.

The 1979 figure was the largest representation of blacks ever among doctorates awarded to Americans. Yet, as the decade ended, all of the doctorate granting institutions and all of their departments combined produced only slightly more than 1,000 black Americans with a doctorate degree. This is gross underrepresentation in every respect.

Distribution by Sex of Recipients

Between 1973 and 1979, black Americans were awarded a total of 6,694 doctorates. This number represented 3.6 per cent of the 186,065 degrees conferred on U.S. citizens of all racial and ethnic groups during that period. An examination of Table 50 shows that the gap between black males and black females in the number of degrees earned narrowed significantly during the seventies. For instance, in 1974, over two-thirds of all degrees conferred on black Americans at the doctoral level went to black males. However, because of the consistent entry of black women in doctoral programs and in increasing numbers, that proportion narrowed so that black women now account for almost one-half (47.8 per cent) of all doctorates earned by black citizens.

A similar pattern of increasing production of women with doctorates is observed for white and other non-white groups as well. While such increases are necessary and warranted in every conceivable manner, they should not be regarded as *displacements* for black males. Rather, since there is still such a critical shortage of blacks with doctorates in the entire labor force, it is still imperative to increase the absolute numbers of black *women and men* with these degrees. The under-representation of black women with doctorates is vividly portrayed in their limited proportion of doctorates for the decade. These data show that black men were awarded six of every ten degrees earned by black Americans as a group.

Distribution by Fields

Obviously, there are many institutions and there are many fields of study in which no doctoral degree was awarded to a black American. The result is a highly distorted distribution of black doctoral degree holders in the major fields of study.

The Commission on Human Resources of the National Academy of Sciences aggregates earned doctorates into seven or eight major fields. For most of the decade, the category "teaching fields" was added to the list. These areas are: (1) Physical Sciences; (2) Engineering; (3) Life Sciences; (4) Social Sciences; (5) Arts and Sciences or Humanities; (6) Education; (7) Professions, and (8) Teaching Fields. The numbers of degrees reported is based upon self-reported data; that is, upon the number of persons who received the doctorate in each of these fields who returned a completed and usable mailed questionnaire. It is, therefore, possible that this number may be somewhat at variance with those reported from other sources, such as

Engineering, for instance, whose data are submitted by Deans to an accrediting agency or body.

Examples of sub-fields of the *Physical Sciences* include: Mathematics, Physics and Astronomy, Chemistry, Earth and Environmental and Marine Science. *Engineering* comprises all of the specialization listed in Chapter IX. The *Life Sciences* refer to the Biological Sciences, Agricultural Sciences, and Medical Sciences (e.g., Pharmacology, Paristology, Environmental Health, and Veterinary Medicine). The *Social Sciences* encompass the traditional social science fields and Psychology with its various branches or divisions. In the social sciences are such sub-fields as Anthropology, Economics, Econometrics, Sociology, Statistics, Political Science, Urban and Regional Planning, and Public Administration.

Examples of the *Humanities* include: Art, History, Theatre, Religion, Philosophy, Linguistics, Comparative Literature, and Music. The *Professions* are represented, for instance, by Theology, Business Administration, Home Economics. Journalism, Social Work, Speech, Law and Jurisprudence, and Library Science. The *Teaching Fields* include a combination of several of the sub-fields in other larger fields mentioned above. They are represented, inter alia, by Agriculture, Art, Business, English, Foreign Languages, Industrial Arts, Mathematics, Reading, Physical Education, Vocational Education, Health, and Recreation and General Science. *Education* includes a number of specializations ranging from Elementary Education to Higher Education.

During the seventies, Education dominated the fields in which doctoral degrees were earned by blacks. As a result, the impression was often given and allegations were frequently made to the effect that there "were too many blacks in Education". While more than one-half of all doctorates earned by blacks remained consistently in Education, black Americans represented approximately 1.8 per cent or slightly more among all Education doctorates conferred in each of the seven years under review. The range in rate of all degrees conferred upon blacks was from 61.5 per cent in Education in 1976 to 52 per cent in 1973. Thus during the middle years of the 70s, approximately 6 of every 10 doctorates received by black citizens were in the field of Education.

With such a comparatively high concentration of black Americans in one field, less than four of ten degrees had to be distributed among six or seven fields. The obvious consequence of this maldistribution is a substantially greater distortion of proportions of all doctorates. There are some fields in which not one doctoral degree was received by a black American. But, between 500 and 600 degrees are awarded to blacks in Education each year.

The second most common field of study for black Americans during the 70's were the social sciences and psychology. The proportions of doctorates earned by blacks increased in every year, with one exception, during this period (See Table 51). In 1979, about two of every ten doctorates received by black citizens were in the social sciences and psychology. This translates into slightly more than 200 doctoral degrees received by Black Americans that year in this field.

The third concentration was in the Arts and Sciences or Humanities. However, the range in the actual number of degrees received by blacks in these fields was from 74 to 119. Only in 1979, did the number of these degrees received exceed a total of 100 for black Americans. The number of doctorates awarded to blacks was abysmally low in the Physical Sciences, the Life Sciences, Professions and Engineering. The number of doctorates in the professions averaged about 40 per year. In the Physical Sciences, the average number for the seven years per year was 42. The situation was somewhat better in the Life Sciences in which the average number was 68. And, as previously noted, the production of black doctorates in Engineering was unimaginably low. In all of these fields, the level of productivity is barely tokenism.

Even though percentage changes from year to year might appear impressively positive in some areas, the actual production rate must be viewed in the context of the proportion of all degrees awarded U.S. citizens and requirements for parity. It is evident that with a production rate of only about four per cent of all doctorates received by U.S. citizens during the decade, it is virtually impossible for black citizens to catch up in the forseeable future. To reach parity will require a massive influx of black students in doctoral programs and an equally all-encompassing revolution in the production rate of doctorates. Unfortunately, such profound changes in higher education do not appear to be in the offing. The future is bleak and the under-representation presently observed will continue without major and sustained intervention by relevant decision-makers unafraid to use their power, influence and resources to induce structural and policy changes.

According to the Carnegie classification of doctorate-granting institutions, universities fall into one of eight rated types. They are (1) Research I (institutions which are the most researched-oriented institutions); (2) Research II (or more moderately research-orientated institutions); (3) Doctoral I; (4) Doctoral II; (5) Comprehensive I; (6) Comprehensive II; (7) Liberal Arts I; and (8) Liberal Arts II. All other doctorate-granting institutions are categorized as (other) or as "unrated universities or colleges."[6]

According to one major research study on institutions from which black Americans received the doctoral degree, the proportion of black

Table 51. Doctorates Awarded U.S. Black Citizens by Field, Percent of Total Doctorates, Percent of Total Awarded Black Americans by Years, 1973-79

Fields	Years						
	1973	1974	1975	1976	1977	1978	1979
Physical Sciences							
*Total Degrees	4338	4892	4760	4445	4369	4193	4298
Black Citizens	45	46	36	27	41	51	48
% of Degrees/Blk	6.1	5.4	3.6	2.5	3.8	5.0	4.6
Engineering							
Total Degrees	2738	3144	2959	2791	2641	2423	2494
Black Citizens	27	16	11	12	11	9	17
% of Degrees/Blk	1.0	1.9	1.1	1.1	1.1	.9	1.6
Life Sciences							
Total Degrees	4073	4894	5022	4971	4767	4887	5076
Black Citizens	96	69	55	63	55	59	52**—
% of Degrees/Blk	13.1	8.2	5.6	5.8	5.3	6.7	5.0
Social Sciences							
Total Degrees	4796	6156	6307	6583	6504	6543	6379
Black Citizens	87	107	159	172	179	191	206**
% of Degrees/Blk	11.8	12.6	16.1	15.9	17.9	18.6	19.6
Arts & Sciences							
Total Degrees	4461	5174	5046	4883	4559	4235	4143
Black Citizens	74	75	89	91	96	75	119
% of Degrees/Blk	10.1	8.9	8.9	8.4	8.6	9.8	11.3
Education							
Total Degrees	5670	7261	7349	7727	7448	7190	7370
Black Citizens	382	501	605	667	665	583	556
% of Degrees/Blk	52.0	59.2	61.2	61.5	60.0	56.7	53.0
Professions							
Total Degrees	1151	1421	1446	1474	1340	1454	1414(1413)***
Black Citizens	24	32	34	53	41	46	52
% of Degrees/Blk	2.4	3.8	3.5	4.9	3.7	4.5	5.0
Total Degrees							
U.S. Citizens	27,129	26,827	27,009	27,195	26,007	26,529	25,369
Total Degrees-Black & %	581	846	989	1,085	1,109	1,029	1,050
Black U.S. Citizens	(2.1)	(3.1)	(3.6)	(4.0)	(4.2)	(3.8)	(4.1)
All Persons							
Combined	32,727	33,000	32,913	32,923	31,672	30,850	31,200

*Totals include degrees awarded to all students in these fields, irrespective of citizenship
**—These numbers may be slightly inflated due to percentage rounding off
***The number in parenthesis refers to degrees classified this particular year as in the teaching fields
Source: Compiled from data provided by the National Academy of Sciences; Commission on Human Resources

Americans who were awarded the doctorate from Research I institutions is smaller than is the proportion for all doctoral degrees earned. On the other hand, black Americans rank first among U.S. citizens of all racial and ethnic origins in the proportion who earned doctorates from Research II universities. They also rank high in the proportion of doctorates conferred in the "other" category.[7]

Sources of Funds for Graduate Education

Major differences exist between black and white students in how they finance their graduate education. As indicated in earlier studies, these differences lie almost entirely in the heavier utilization of loans by black students and the significantly greater proportion of white students who are awarded teaching and research assistantships.[8] However, as Table 52 is analyzed; not only are these differences apparent but other variations in financing graduate education are especially evident.

White graduate students were considerably more successful during the seventies in obtaining fellowships and trainee-grants through federally subsidized projects. It is only in the very last year of the decade that the gap between these groups narrowed by an appreciable degree. By contrast, black students received more support in the category of "other fellowships," which refer to monies granted from non-federal sources and non-educational institutional grants. The percentage point differences between the two groups are significant and favored black students, proportionately, throughout the review period.

However, in the categories of teaching and research assistantships there can be no equivocation about the disproportionate benefits and successes experienced by white students compared to the limited proportions of asistantships awarded black students. Approximately one-half of all white students financed their graduate education through teaching assistantships in every year of the decade compared to approximately one-fourth of black students who were able to use this means of financial support. Similarly, in every year under review, the proportion of white students who were awarded research assistantships more than doubled the proportion of black students who received this form of support.

Since teaching and research assistantships are invaluable for graduate students, this lack of support of black students in these two areas continues to be an issue of major concern among black educators and civil rights groups. Both forms of assistantships provide immeasurable on-the-job training and practical experience which may prove of immense value in preparation for doctoral examinations, in developing mentors among the full-time faculty, in sharpening pedagogical and research skills, as well as in becoming more sophisticated in the intricacies of laboratory research, and in demonstrating special personality traits or characteristics which can be highlighted when the doctoral recipient attempts to move into the labor force. Further, these assistantships often provide an opportunity for undergraduate students to be exposed to graduate students of their own racial and/or ethnic groups. In turn, this exposure may serve as a prime inspiration for members of that group to major in the same field or to think more positively about pursuing graduate studies. Equally important is the exposure that they provide non-minority students to minority graduate students which, in turn, may force them to overcome demeaning stereotypes they may hold about the intellectual abilities of minority groups. Hence, the salience of assistantships is a multi-faceted or multi-dimensional phenomenon. Throughout the decade, black graduate students were short-changed in these two areas.

Black graduate students fared slightly better in funds received from educational and institutional sources. The major advantage experienced by black students in this area occurred in 1975. This situation coincided both with the recession (whose effects were devastating to minorities in general) *and* with losses of federal funds to support graduate and professional education. Apparently, several educational organizations and institutions assumed more of the responsibility for assisting black students to finance graduate education at this time. However, the gap between black and white graduate students was narrowed each year thereafter.

The economic predicament of most black students is reflected in the larger proportions of black students who must resort to some form of loan in order to finance their education in contrast to a smaller proportion of white students who use this funding source. As indicated in the percentage of both groups in this category in table 52, the differences are significant. The gravity of this problem increases in severity when the loan profile is viewed in terms of how much support that a graduate student can expect to receive from family contributions. In every year, white students were considerably more successful in obtaining family contributions than were black students. The peak year for both groups was 1976 which coincided with the continuing recession and with declines in federal fellowships and trainee grants.

A high proportion of both groups receive financial assistance through their personal earnings or from their spouse's earnings. However, white students obtain more support in this area than do black students. This may reflect better job opportunities with higher remuneration among white

Table 52. Sources of Support for Doctorate Students by Race and Year, 1974-79, and Percent of Students Using This Source

Sources	Years and Per Cent					
	1974	1975	1976	1977	1978	1979
	%	%	%	%	%	%
Federal Fellow/ Trainee						
Black	33.7	28.7	30.2	24.3	19.3	19.0
White	40.5	38.1	34.1	28.3	23.8	21.8
G.I. Bill						
Black	17.3	13.9	15.9	12.9	11.7	8.7
White	12.4	13.2	13.5	12.4	11.1	9.9
Other Fellowships						
Black	26.6	26.1	25.3	28.0	24.6	24.0
White	21.7	22.8	23.0	21.0	19.9	19.8
Teaching Assistantships						
Black	28.8	30.1	27.6	25.9	23.6	23.8
White	49.1	52.0	51.5	48.7	46.8	47.2
Research Assistantships						
Black	15.1	17.2	15.8	15.0	15.2	14.9
White	30.3	34.5	35.4	33.9	33.0	34.4
Educational/ Institutional Funds						
Black	14.1	17.0	15.2	12.2	11.7	12.1
White	12.1	13.2	13.2	11.3	9.7	10.2
Own/Spouse Earnings						
Black	49.9	69.5	67.1	67.4	65.9	68.4
White	52.1	64.3	74.0	72.3	69.8	69.5
Family Contributions						
Black	3.3	9.3	10.4	9.3	7.9	8.7
White	7.2	14.8	17.5	15.6	14.0	14.4
National Direct Student Loan						
Black	—	13.2	16.9	14.8	12.9	14.3
White	—	9.1	11.6	10.2	9.4	10.0
Other Loans						
Black	18.6	15.9	14.4	14.6	12.8	13.5
White	14.0	13.6	13.9	11.5	9.9	10.0
Other						
Black	4.3	6.8	6.6	5.7	4.8	7.4
White	3.7	4.7	5.3	4.7	4.4	4.2
Unknown						
Black	5.0	2.8	2.7	2.3	3.2	2.5
White	2.3	1.2	1.1	1.2	1.2	1.0

Source: Compiled from data provided by the National Academy of Sciences: Commission on Human Resources

Americans. This enables them to save more money in pursuit of the graduate degree.

The difference between the two groups on the proportions who receive G.I. Bill benefits are relatively minor. The remaining two categories are not of major consideration.

Graduate and Professional Opportunities Program (GPOP)

Title IX, Part B of Higher Education Act, initiated the Graduate Professional Opportunities Program (GPOP) of fellowships. The purpose of these fellowships is to assist the underrepresented minorities, and women, "to prepare for academic and other fields." These fellowships, which provide a twelve months stipend of $4,500, are awarded for graduate study. Recipients are remitted institutional tuition and mandatory fees.[9] These awards were made for the first time in 1978; black students received 293 or 53 per cent of the 553 fellowships awarded.

Fewer white and other minority students received GPOP fellowships during these two years compared to the rate of success for black students. However, white students were more successful than either Hispanic, Asian American or Native American students. The 1978 and 1979 rates for white students were 26.1 per cent and 22.6 per cent of the total, respectively. This translates into a total of 204 white GPOP recipients. The 1978 and 1979 percentages of total fellowships were 16.5 per cent and 16.4 per cent, respectively, for Hispanic doctoral students. Hispanic students, therefore, received 141 of all GPOP fellowships. Asian Americans received 48 of the total number awarded during the first two years and Native Americans were awarded a total of 28 fellowships. The percentages for Asian Americans were 6.9 and 4.9 respectively, while they were 3.6 and 3.1 for Native Americans in 1978 and 1979, respectively.[10]

For 1980-81, the re-newed fellowships totaled 789 while 213 new fellowships were awarded. The number of participating institutions has more than doubled (114 in 1980-81) over the 55 who participated during the first year of the program.

The GPOP program is only one of several federal sources of financial aid for doctoral students and graduate education in general. Not only is the Office of Education involved in these efforts, so are other agencies and branches of the federal government. For instance, financial aid programs or fellowships are awarded through such agencies as the National Science Foundation, the National Institute of Education, the National Institutes of Health, the National Institute of Mental Health, and the Health Resources

Administration.

An example of an organizational program to assist black and other minority graduate students to pursue a doctoral degree is the Minority Fellowship Program of the American Sociological Association (ASA). This program was developed at the insistence of the Caucus of Black Sociologists of the ASA in 1970. By 1973, the Center for the Minority Group Mental Health Programs of NIMH gave a six year grant to establish and fund a program for the recruitment and financial support of minority students who wished to pursue a doctoral degree in Sociology. Since that time, additional support for the program came from the National Institute of Education, and the Cornerhouse Fund and through tuition remissions from some of the participating institutions.[11]

A full-time project director was hired for the program, through a national search in which the ASA and the Caucus of Black Sociologists participated. The project director is housed in the national headquarters of the Association. Not only is he responsible for initiating recruitment activities, he also works with the ASA Comittee on Minority Fellowships in the selection process, visits minority fellows at their institutional location, serves as a visible role model for the fellows, and provides important non-academic support for students who sometimes need someone to listen to their experiences as they become adjusted to the rigors of a doctoral degree program in an alien enviornment. During the first three years of the program, all 88 minority fellows were in good and regular standing within their universities. Since 1973, some 168 minority students were awarded these fellowships and about 21 were added to the labor force with Ph. D. degrees in Sociology.[12]

The American Sociological Association established, again at the insistence of the Caucus of Black Sociologists, an executive associate for minorities and women. The importance of that office is measured by both the stature of the person recruited for the position and for the overall quality of the work performed by the office. One of the most effective incumbents of this position in recent years was Dr. Doris Wilkinson, a former professor of Sociology at McAllister College. As the Executive Associate for Minority and Women Affairs, her office addressed such issues as career opportunities for minority social scientists in non-academic settings, the location of fellowship monies and training seminars and special projects of interest to minority and women sociologists. Summer Institutes in research techniques, innovations and strategies were organized. And, the practice of a column devoted to these concerns, initiated by the first person who held this position, Professor Maurice Jackson of the University of California / Riverside, was continued in *Footnotes*.[13] This office over the years initiated broad contacts wtih other foundations and governmental agencies in the Greater Washington community. The early work done by previous incumbents, Dr. Jackson, Dr. Lucy Sells and Dr. Joan Harris, was essential in this regard. Since black Americans constitute only about 3 per cent of all Sociologists in the United States, and since women sociologists, like minority group members, still encounter formidable barriers in hiring, promotions, and tenure, the continuing need for this office is both real and obvious. Similarly, there is a need for minority fellowships and minority/women affairs officers in all of the professions to stimulate the process of increasing doctoral degree production among these groups. The Minority Center of the NIMH has also assisted other Associations with similar programs, including social work and social science fields.

This list of financial contributors is by no means exhaustive. The work of the private sector, the foundations and individual philanthropists was just as important in these areas as in the the professions discussed in earlier chapters. Ford Foundation, for instance, began its minority fellowship programs in the early 60s. Its most important pre-doctoral program, began in 1967, has been instrumental in helping at least 1,800 minorities receive the doctorate through its support of about $50 million. Other foundations such as Carnegie, Mellon, Rockefeller, Exxon, IBM, and Lilly assisted blacks and other minorities in the 70s.

While all of these sources of financial assistance are of immense importace to black and other minority students as they embark on their doctorate, the size of most grants is grossly inadequate. The enormity of the problem is captured by the full realization of the economic disabilities under which most black and other minority students labor. Earlier on, it was pointed out that the median black family income in 1980 is only about 57 per cent of that of white families. It is less for Puerto Rican families and still worse for Native Americans. Rectification of these inequities demands substantially larger financial outlays to assure support for a larger cohort of students who could be recruited for all graduate fields of study. If the nation is, indeed, committed to the achievement of parity between the races in the number of doctorate holders in the labor force, it will have to be nothing less than make a gargantuan increase in financial support for graduate education. Further, it will be absolutely necessary to avoid the kinds of legal challenges to these efforts that resentful groups are likely to organize against the very existence of programs designed to assist minority students.

Time Lapse Between the Baccalaureate and the Doctorate

Unquestionably, the magnitude of the economic problems encountered by black students helps to account in a measurably significant way for the longer time lapse that black doctorate recipients take to obtain this degree. Consequently, black doctorate recipients are usually older as a group when they receive the doctorate than are all other groups, including whites, Chicanos, Asian Americans, and Native Americans.

Regarding the age at which the doctorate is granted, black recipients average about five years older in every year than both white recipients and the aggregate of all doctorate recipients (Table 53). They may delay entry into a doctoral degree program following receipt of the baccalaureate degree because of the necessity to work. Or, they may work part-time on the degree while being employed full-time and this practice in turn, inflates the median time lapse. However, the registered time it takes to earn the doctorate is not statistically different between black Americans, white Americans, and the aggregate of all American recipients. It takes from 5.8 to 6.5 years is registered time between the baccalaureate and the doctorate. One should be mindful that, in many instances, students also work toward a Masters degree in the interim, "stop-out" and work for a period of time before renewing doctoral studies. This pattern also adds to the time lapse.

The five years age difference between black and white doctoral recipients has a number of implications for post-doctoral employment behavior. The higher age level of black doctoral recipients may help to explain their lower representation in post-doctoral fellowships. However, that is not the sole explanation for this underrepresentation. The fact that many black doctorate holders of an older age search for and are attracted to higher paying positions in administration, industry, and government service may be attributed to the age factor. Their income demands may be higher given the need to provide for their children who are likely to be older than those of younger doctoral recipients. In such instances, some departments within universities encounter major problems in competing for black Americans with doctorates because of salary demands. Many are unwilling to offer substantially higher salaries to black doctorates for fear of alienating white faculty who may allege unfairness and deeply resent black faculty members who may be hired at somewhat higher salaries. Still another fact must be considered. Many black persons who receive the doctorate have already had several years of full-time teaching experience either in an historically black college, a small predominantly white liberal arts college, or in a community college. The problem faced by some departments is how to evaluate those years of "college teaching experience" in relationship to the salaries demanded by newly-minted doctorates who are five years older *and* black. Few black Americans with this type of experience are willing to enter an institution at "entry level" salaries or ranks. Thus, controver-

sies having broad ramifications may follow regardless of the decision reached unless a clearly defined institutional policy is articulated and has support
from a significant consensus.

Special Problems and Issues

Between 1977 and 1978, there was a decline of 4.7 per cent in the production of doctorates among black Americans. At the same time, there was
in increase of 16.9 per cent in the production of doctorates among
Hispanics and an increase of 15.6 per cent in the production of doctorates
among Asian Americans.[14] While these increases in the production rate of
doctorates among Hispanics and Asian Americans should be continued,
that growth should not obscure the overall plight and underproduction of
doctorate degrees among all minority groups. Nor should the success of
Asian Americans minimize the reality that less than one third of all doctoral
degrees received by Asians in the United States are awarded to Asian

Table 53. Median Age at Doctorate and Median Time Lapse from
Baccalaureate to Doctorate by Race and Year, 1973-1979

	1973	1974	1975	1976	1977	1978	1979
Age at Doctorate							
Total (All Students)	31.3	31.4	31.5	31.6	31.6	31.7	31.9
Black	—	36.7	36.3	36.2	35.9	36.0	36.4
White	—	31.3	31.4	31.5	31.4	31.5	31.7
Median Time Lapse							
Total	8.4	8.6	8.7	8.8	8.8	9.0	9.2
Black	—	8.5	8.5	8.7	8.7	8.9	9.1
White	—	12.4	12.4	12.5	11.9	12.3	12.2
Registered Time							
Total	5.8	5.9	6.0	6.0	6.1	6.2	6.3
Black	—	5.9	6.0	6.0	6.1	6.2	6.2
White	—	5.9	5.8	6.0	5.9	6.4	6.5

Source: Constructed From Table reported in *Summary Report of Doctorate Recipients from
United States Universities* (1973-1979). Commission of Human Resources, National
Research Council: National Academy of Sciences.

American citizens. The majority in this category are conferred upon
non-U.S. citizens of Asian origin.

The problem of the size of the applicant pool for graduate studies
among black students is persistent and widespread. Some dimensions of
this problem were highlighted in other segments of both this chapter and in

previous ones. Nevertheless, when discussing the applicant pool, the disparity between the rate of production of college graduates who are black from historically black and historicaly white colleges emerges as a factor of profound importance.

According to accumulated college enrollment and graduation data, the historically black colleges now matriculate only about 20 per cent of all black students in colleges across the nation. But almost half of all black students are enrolled in community or junior college (42 per cent) and many of these students are in programs not transferable for admission to a four-year college or university. Further, the majority of black students who are graduated with a baccalaureate degree each year are produced by the historically black colleges which have only about 20 per cent of the students.[15] This fact suggests that a major crisis exists in the traditionally white institutions regarding not only the production of black college degree holders but their failure to expand the pool of potential applicants among black students for graduate and professional degree programs.

The central and compelling issue here regarding actions of policy makers and other officials is that they must raise questions and examine institutional behavior and the quality of the teaching or learning environment in traditionally white institutions that render them singularly less successful in the production of black college graduates. The drop-out rate among black students is considerably higher in these institutions than is the case in the historically black colleges. Why this problem is so pervasive in white institutions is one question still begging careful scrutiny.

A related problem or issue is that of the recruitment and retention of black faculty for graduate faculty positions. Clearly, a part of the problem is the continuing underrepresentation of black Americans in all academic disiplines and fields of specialization. But, there is another equally pressing matter—that in a period of retrenchment, union shop principles take effect in the academic world, too: the last hired shall be the first fired. Since black and other minority faculty members fall in the category of the last hired, they are relatively unprotected when anything approaching the seniority principle is enforced. That situation is further exacerbated by the "revolving door" syndrome. According to this principle, black faculty members are hired and retained long enough for institutions to satisfy requirements of being "an equal opportunity employer" and, in the view of many black faculty victimized by the process, with little intention of ever retaining them regardless of the exceptional quality of their overall performance. As a result, many are retained for five or six years. They are forced out to search for another position and the process continues with another black or minority faculty member.

Tenure is theoretically denied because of weakness in one or more of three areas; teaching, research, and service. In many major institutions, service is a criterion of the lowest priority. Yet, that is where the strengths of many minority faculty lie because every conceivable committee within the university wanted them as a "minority presence". Acceeding to these requests meant that precious little time remained for scholarly research and productivity required for tenure. Some may argue that black and other minority faculty members made a "rational choice" to perform these services. That may very well be the case. However, a "minority presence" is essential on decision-making committees whose decisions affect the educational programs and the quality of learning for minority students.

There is ample evidence to support the contention that black and other minorities do not have equal access to research funds and to post-doctoral fellowships that strengthen research capabilities. If they do not have access to these funds, then, engaging in the type of research that generates thousands of dollars for the department and which may enrich a person's career or increase the probability for tenure is virtually impossible.[16]

Frequently, when tenure has been undertaken and articles published, the quality of the journal is seriously questioned. In the view of some white faculty members, an article published in a "white-oriented" or "mainstream" journal that is only two years in existence counts for more points than an article published in a journal with fifty years of history that carriers a "Black label."

Not infrequently, when books are published, the issue then focuses on the quality and prestige of the publishing house rather than the quality or merits of the book itself and the contribution that it makes to the literature in that particular field. In other words, too many of these faculty members are subjected to unfair and inappropriate tenure evaluations which escalate the revolving door syndrome. Having said all these things, it should be absolutely clear that *incompetence should never be rewarded regardless of race, color, ethnicity, sex, or creed. However, faculty members should have the same opportunities to meet the criteria for tenure in an equally wholesome environment irrespective of race, color, ethnicity, sex, or creed.* Further, and without equivocation, they should be evaluated fairly in all decisions affecting their performance as scholar-teachers.

The shortage of black faculty can be alleviated in part by more aggressive recruitment by the various departments. They can identify promising students in their own colleges, encourage them to pursue the doctorate, assist them in funding their graduate education, and, then, hire them in full-time teaching positions. This process may mean making exceptions to the general practice of not immediately hiring one's own graduates. Doing so may be one of the efficacious remedies for the problem of under production.

Coodinated recruitment necessitates enlargement in the size of the University or college-wide recruitment office. There are thousands of young people in smaller institutions who may never think that a graduate program is possible for them. This rich source of untapped talent should be sought as a way of increasing and enlarging the potential pool of appicants.

Most important of all, a renewal of commitment to firm programmatic action remains a sine qua non for increasing success in the production of doctoral degrees among black Americans in the 1980's and beyond.

Footnotes

1. The production of black graduates with Master degrees rose dramatically during the 1970s. Although this growth is recognized, a detailed attention to this level of graduate education is not within the purview of this study. Nevertheless, programs designed to increase the production of black doctorates will benefit from tapping that comparatively large pool of black Americans who hold the Masters degree and from regarding this group as a potential pool for doctoral level training.

2. Cf. *Racial and Ethnic Enrollment Data From Institutions of Higher Education, Fall, 1970*. Washington, D.C.: Office of Civil Rights, Department of Health, Education and Welfare, 72-8; 1973, *Racial and Ethnic Enrollment Data From Institutions of Higher Education, Fall, 1972*. Washington, D.C.: DHEW, ORC - 72-78, 1978; and Elaine H. El-Khawas and Joan Kinser, *Enrollment of Minority Graduate Students in Ph. D. Granting Institutions*. Washington, D.C.: American Council on Education, August 1974.
"Ph.D. Manpower Employment Demand and Supply 1972-1985," *Bulletin 1960* Washington, D.C.: U.S. Department of Labor, Bureau of Labor Statistics, 1973, p. 9.

4. Cf. U.S. Commission on Civil Rights, *Toward Equal Opportunity: Affirmative Admissions Programs at Law and Medical Schools*. Washington, D.C.: U.S. Government Printing Office, June 1978, and_____, *Social Indicators of Equality for Minorities and Women*. Washington, D.C.: U.S. Government Printing Office, Aug. 1978 and, James E. Blackwell, Maurice Jackson and Joan Moore, *The Status of Racial and Ethnic Minorities in Sociology A Footnote Supplement*. Washington, D.C.: The American Sociological Association, August 1977.

5. Dorothy M. Gilford and Joan Snyder, *Women and Minority Ph. D.s in the 1970's: A Data Book*. Washington, D.C.: Commission on Human Resources, National Research Council, National Academy of Sciences, 1977, p. 18.

6. *Ibid.,* p. 122.

7. *Ibid.*

8. Blackwell, *Op. Cit.*

9. *Program Support of Graduate Education*. Washington, D.C.: U.S. Office of Education, May 1, 1980.

11. Personal Communications with Dr. Paul Williams, Director of the Minority Fellowship Program of the American Sociological Association, July, 1980.

12. *Ibid.*

13. Cf. Doris Wilkinson, "Federal Employment for Sociologists," *Footnotes,* Washington, D.C.: American Sociological Assoication, March 1980; "Percentage of Women Doctorates in Sociology Increases", Loc. Cit., December, 1977;_____, "Careers, Minorities and Women", *Footnotes,* passim, 1977-80.

14. Cf. Commission on Human Resources, *Summary Report 1977 Doctorate Recipients From United States Universities.* Washington, D.C.: National Academy of Sciences, 1978, and _____, *Summary Report of 1978 Doctorate Recipients From United States Universities.* Washington D.C.: National Academy Sciences, 1979, and_____, *Summary Report of 1979 Doctorate Recipients From United States Universities.* Washington, D.C.: National Academy of Sciences, 1980.

15. National Advisory Committee on Black Higher Education & Black Colleges & Universities, *Access of Black Americans to Higher Education: How Open is the Door.* Washington, D.C.: U.S. Government Printing Office, January 1979, p. 20.

16. Stephen S. Wright, *The Black Educational Policy Researcher: An untapped National Resource,* Washington, D.C.: National Advisory Committee on Black Higher Education and Black Colleges and Universities, Department of Health, Education and Welfare, 1979.

Chapter 13

Policy Implications in Mainstreaming Outsiders: A Review

Few objective analysts deny the reality that some progress was made during the 1970s in increasing educational opportunities for black Americans and other racial and ethnic minority students in graduate and professional schools. Different conclusions may be drawn as to degrees of progress. There may be profound disagreements on whether or not the present status of black Americans in graduate and professional education portends a total loss of commitment to the optimistic goals of the early seventies. The analysis of data presented in the preceding chapters dramatizes how uneven and episodic were those efforts toward "mainstreaming black Americans" during that decade. These data also point to a leveling off that may signify subsequent retrogression and stagnation. Nevertheless, the objective fact remains that as the 1980s began more black professionals were indeed in the labor market than would have been the case had it not been for educational policy changes implemented during the seventies.

Asserting that some degree of progress was made toward better access to graduate and professional education is not to claim that the progress actually achieved fulfills the enunciated goals and collective aspirations. Without a doubt, the nation has not achieved educational parity. In fact, in most fields and in most insititutions, the educational establishment has fallen far short of public goals and expectations.

The seventies began with a high level of expectation and some evidence that expanding educational opportunities to include black Americans and other minorities had substantial national support. Federal interventions in educational policies had already achieved great significance. This situation was primarily a consequence of publicity given to actual enforcement of affirmative action policies, the utilization of capitation grants to facilitate minority students enrollment in professional schools, student aid programs, as well as the encouragement given to individual institutions to expand their educational facilities.

Enrollment in graduate and professional schools was on the upswing. Hence, it is important to reiterate an earlier assertion that black and other minority students benefitted by an educational policy that mandated enrollment expansion in general and that they did not, ipso facto, precipitate that expansion. White students were the primary beneficiaries of this policy since a far greater absolute number of students from this segment of the population actually enrolled in graduate and professional schools during the seventies than was the case with black and other minority students.

The corporate structure and private foundations played a profound role in expanding educational opportunities during the decade. The services rendered by them in the process of mainstreaming outsiders, already described in earlier chapters, were not only extensive but far-reaching in their implications for educational change. Incontestably, their collective involvement was stimulated in part by external pressures for positive action. It was, significantly, a manifestation also of the sensitivity among many leaders and lower-level decision makers in the corporate and foundation sectors to the moral responsibility that these groups have to create a larger pool of professionally trained persons among the minority population. Consequently, they drew upon their financial resources and professional expertise to support the establishment of an impressive array of educational programs designed to increase educational opportunity, parity, and a larger supply of professionals in the American population.

At the beginning of the decade, there was a certain amount of ambivalence among the leadership and other functionaries in graduate and professional schools about appropriate actions to accomplish equality of access. For many, whatever actions undertaken and whatever program implemented represented movement on an unchartered course. This uncertainty reflected the absence of a critical mass of minorities in such institutions as well as limited experience with minorities. In many instances, these were persons who had fought earlier against racial inclusion, but now in the 1970s, were expected to lead their institutions in the construction of programs to assure equality of access and in substantially improving the rate of producing black professionals. Consequently, many earlier programs were grounded on ambiguous educational policies.

Although laws may not change attitudes, they are a powerful instrument for transforming behavior. Responding to federal laws and to statutes to terminate discriminatory behavior, many institutions did, indeed, take a number of steps toward expanding educational opportunities in graduate and professional schools for minority group students. Steps were taken to improve the institutional climate in ways that would foster racial accommodation, reduce personal anxieties and promote a positive orientation toward learning. The motivation to engage in these expansion processes and to mount new programs was in part concrete evidence of the commitment of some faculty members, administrators, and legislators to the overarching principle of racial equality and social justice.

The success experienced during the 1970s in moving toward the production of a larger number of professionals in the black American population can be attributed to at least five interrelated factors. These include roles played by: (1) the federal government, (2) the private sector (as represented by the corporate or business world and foundations), (3) institutions themselves in seeking out black students with potential to become professionals, (4) the relentless determination of these students to succeed even in the face of various forms of adversity, and (5) sustained pressure.

Success was by no means total. Institutional involvement was by no means ubiquitous as the media would have the American public believe. On the contrary, institutional commitment was too often equivocal. Institutional policies too frequently vacillated in response to public opinion. A comparatively few institutions produced the lion's share of black Americans trained in the professions. As previously noted, when the historically black professional institutions are excluded from the analysis, it becomes especially clear that most of the historically white institutions failed to move beyond token levels regarding the admission, enrollment, and graduation of black Americans. Further, some professional schools neither enrolled as many as a dozen black students during the entire decade. Some never enrolled a single black student. While the situation may reflect the size of the black population in these States, the issue is considerably larger in scope. There is no State in the United States without *any* black citizens and no State without black school-age children. Then, why is it that these States with such a paucity of blacks in the total population cannot approximate even that representation in their graduate and professional schools?

Another illustration of the tendency for a small number of professional schools to enroll a majority of the black students may be drawn from medical education. According to data collected for this study, approximately forty-four of the nations's medical schools in existence at the beginning of the 1970s enrolled fewer than fifty black students in first year medical classes throughout the decade. That is to say, when all black students actually enrolled in first year medical schools classes were totalled for the decade, about forty-four medical schools had enrolled no more than fifty black students in first year classes. Even though the medical colleges of Howard University, Meharry Medical College, and Morehouse enroll only about 28 per cent of all black medical students in 1980 compared to over 80 per cent by Howard and Meharry in 1970, there is still a serious problem of token access to most medical institutions. As in other fields included in this study, many institutions have excellent "paper programs" and claim to make "good faith efforts" toward expanding educational opportunities, but have no real commitment to translate paper pronouncements into viable programs.

Despite these observations, as a result of the affirmative efforts of the seventies, the professional and graduate schools did produce 5,899 more black physicians, 1,436 more black dentists, 200 more optometrists, 6,967 more blacks with the first professional engineering degree, and 6,694 more blacks with a doctoral degree, for example, than the numbers present at the beginning of the decade. But, as demonstrated in earlier chapters, that apparent progress may be deceptive in some areas, especially when the proportionate enrollment of black students in certain of these fields in 1979-80 is compared to their proportions in 1970.

A sad truth is that, despite these increases in the absolute numbers of black professionals produced during the seventies, the percentages of black Americans in most professional fields are nothing less than shocking. For instance, in 1980, black Americans still comprise approximately 2 per cent of all physicians, 1.7 per cent of the veterinarians, 1 per cent of the optometrists, 2 per cent of the lawyers, 2 per cent of the dentists, 2 per cent of the engineers, 3 per cent of the pharmacists and 10 per cent of the social workers. At the rate of production of black professionals witnessed during the seventies, it will probably take another forty years before parity is reached in most professions. The nation cannot afford such a snail's pace of progress toward equality of educational opportunity.

Hence there is little reason to be completely sanguine about absolute numerical changes. Unfortunately, there are several danger signs which inform us that serious trouble lies ahead. These include such occurrences as (1) declining or stabilizing black student enrollment, (2) a decrease in commitment to aggressive recruitment strategies, (3) fall-out from the Bakke case which encourage some institutions to reduce their programs for minority students and to increase the weights assigned to objective admissions criteria when confronted with difficult choices, (4) a reluctance by the federal government to take the necessary leadership to persuade institutions to draw upon that portion of the Bakke decision which will facilitate racial heterogeneity in graduate and professional schools, (5) a shift in federal policies away from direct student grants to repayable loans and the establishment of larger numbers of financial programs which more directly benefit the wealthier students, (6) shrinkage in role models from their failure to receive academic tenure, and (7) opposition to minority student programs by the public and by influential special interest groups.

Policy Changes Needed (How to Succeed by Really Trying)

Clearly, educational parity has not been achieved. American black professionals are not produced at a rate commensurate with the total population of either the State in which they reside or in the nation as a whole. The evidence presented in earlier chapters shows that the nation is far from reaching educational parity between the races and equality of opportunity for Black and other minority group citizens in the United States. *Consequently, policy changes are necessary in three levels, at minimum: (1) the governmental levels, (2) the private and corporate level, and (3) the institutional level. In addition major changes are vital in the home, social, and school environments in order to increase the level of motivation for learning among greater numbers of black students as well as to strengthen their determination to succeed.*

In the first place, substantial improvements in the rate of access of black students to graduate and professional schools and in their production as professionals can be made if institutions re-new commitment to fulfill the real promise of equality of opportunity. That commitment encompasses a search for funding sources and maintainance of provisions for financial aid, scholarship monies, educational outreach programs, college enrichment and academic support services, programs to reduce attrition and to enable all students to complete successfully their training with confidence that they are capable of being highly competent professionals. It means *more* than so-called "good faith efforts" to recruit and enroll more black students and to hire more black faculty members. It demands actualization and success in these endeavors so that all students will benefit from a racially and ethnically heterogenous learning environment.

Institutional commitment is evidenced in receptivity to the concept of heterogeneity and to a positive orientation toward steps necessary to achieve it. That change often requires a willingness to reevaluate admissions requirements and sometimes to make them more flexible and amenable to policies of *inclusion*. An appropriate mix of both traditional or objective admissions criteria and non-traditional or subjective criteria can be developed that will satisfy the constitutional requirements or the legal boundaries imposed by the Bakke decision. This can only occur if institutions actually *desire* to enroll more minority students.

Similarly, the hiring of more black faculty members, who may be percieved as positive role models and as evidence that negative sterotypes internalized among some students are totally unfounded, can only become a reality through institutional will, determination, and creativity. Clearly, the production of enlarged numbers of black professionals is a first major step in this direction. But if the nation's graduate schools are annually producing only *ten* mathematicians, ten physicists, twenty chemists, twenty agricultural scientists, one anthropologist, two art historians and no biomedical engineers with doctorate degrees from the black American population, they have a monumental task before them if they want to realize this rekindled institutional commitment.

An expansion of educational outreach programs is imperative. These programs require joint enterprises with junior and senior high schools on a scale never before undertaken and with an intensity perhaps not fully understood. Secondary school counselors and college or professional school advisors should interact with a greater degree of regularity in order for students to become more familiar with changing career opportunities and with new requirements for admissions into a professional degree program. Outreach programs embrace motivational activities which stimulate students to learn and provide them with the personal strength to withstand peer group pressures to take the easy way out of high school or college.

The institutional climate may have to be transformed in order to increase the probability of potential role models for students and of students themselves wanting to accept a job or to enroll at the institution. Reputations of racism, institutional hostility, and prejudice in the classroom spread with extraordinary speed and are often difficult to overcome. Consequently, it is not unreasonable for the institutional leadership to enunciate policies against all forms of racial and ethnic prejudice and to specify clearly penalties for violations of these policies. Neither is it unthinkable in the 1980s for institutions to sponsor racial awareness programs in which the participants are faculty and students. These may be one instrument for understanding *how* subtle racial animosity, prejudice, and overt racial antagonisms may be attacked in order to create a more harmonious learning environment. By the same token, it must be recognized that racism may be alleged where it does not exist and when it does occur it is often difficult to substantiate.

If the institutions are receptive to change, they may welcome some form of federal intervention to assure that they will achieve their goals of equality of educational opportunity within a reasonable time frame. Similarly, they will be more likely to cooperate with corporations and private foundations regarding suggestions for changes in educational policies.

Skeptics may view the emphasis on changes in institutional behavior as something akin to a liberal's dream or believed unlikely to be realized without significant alterations of institutional power, external pressure, and internal demands from a coalition of enlightened faculty and students of all races. In view of apparent attacks on minority programs, allegations of "reverse discrimination", the insistence on the objectification of merit, and the willingness of some institutions to interpret legal mandates in ways that are deleterious to minority group interest, there is great force behind such arguments. However, some professional schools have, indeed, made good faith efforts, with and without apparent success, in opening their doors to minority students. Several achieved success despite lukewarm leadership from the top echelons of university administration and impediments created by hostile faculty and faculty who genuinely believe that the most promising students are always those who score the highest on admissions tests. For all these reasons, governmental intervention in educational policies should become even more widespread in the eighties than at any previous time in the history of American higher education. It may be absolutely necessary if States and institutions continually fail to commit themselves and their resources for the realization of educational equality.*

Hence, both the federal and State governments must commit themselves to an unyielding and uncompromising policy of equality of educational opportunity. This policy considers not only the States under litigation in the *Adams vs. Califano* case to dismantle dual systems of higher education but the nation as a whole. The commitment given by the federal government, its leadership and myriad agencies for the enforcement of sound educational policies regarding equality of opportunity cannot be any less than that given at the institutional and private levels. In fact, that commitment should be considerably more pronounced, substantially more visible to the public and more forcefully enunciated in public policy. Given the current crisis nature of the underrepresentation of black Americans in profesional schools and in doctorate degree granting programs, this is no time for the leadership in the federal govermnent to sacrifice the principles of equality of educational opportunity on the alter of political expediency.

The federal government should utilize its leverage to persuade or to compel institutional compliance to legal mandates regarding eductional opportunity. Its powers enable the federal government to have a major impact on educational policy. Using the force of law granted under Executive Order 11246, the Civil Rights Act of 1964, and the Higher Education Act,

---

*However, this situation is contingent upon the type of leadership taken by the Republican administration to *assure* equality of educational opportunities for minorities.

the federal government can influence institutional policies regarding recruitment of students, admissions policies and practices, enrollment patterns in professional schools, the distribution of student aid funds, allocation of teaching assistantships, research assistantships, other financial aid, and hiring as well as promotion practices in graduate and professional schools.

Federal intervention can facilitate equitable distribution of special scholarship monies, for instance, in ways that assure greater assistance to target groups. Its agencies can influence university researchers to train more minority researchers in connection with special grants received from the federal government. Federal powers can be invoked to cut-off federal funds to State systems of higher education and to specific private institutions that do not comply with regulations requiring discontinuation of action or practices which result in racial, ethnic, or sex discrimination.

Title III funds, as well as other resources, can be expanded and more judiciously utilized to enhance the educational programs of the historically black colleges. This effort may widen the potential pool of black students for graduate and professional school programs. Further, federal services can be more fully employed for faculty development programs which will enable faculty members to produce even more competitive graduates.

Important, the federal government must take the leadership in eliminating the debilitating social and structural barriers to equality of access among black and other minority students. The impediments to access and successful outcomes so widely encountered by blacks are imposed by an unemployment level twice the rate of the national average, or an income that is only about 57 per cent of the median family income of white families, or a teenage unemployement rate that is three or more times that of the national average. All these take a heavy toll on black Americans. They render far too many black families incapable of providing the quality of support needed by their youth to survive or succeed in post-secondary education. High drop-out and push-out rates among black high school students result and become a potent barrier which the federal and local governments can address through enrichment programs that will create and sustain interest in learning.

In each of the above ways, the federal government, especially, can have a profound impact on institutional behavior and eductional policy. That impact also extends to exacting assurances from institutions that just as they will no longer utilize set-aside programs of the type questioned in the Bakke case, their educational policy makers must be equally as committed to that portion of the Bakke decision which legitimates diversity and hetereogeneity and the use of race as one important variable in deciding who shall be *included* in graduate and professional schools.

A third level of policy concern is in the private sector. It is addressed specifically to the corporate structure, private foundations, and the professional associations. While each of these units of the private sector has contributed an immense amount of financial resources and expertise in the past to promote programmatic efforts which assure greater access and educational opportunity for minority students, this is not the time for retrenchment. Not only is it necessary to to sustain the level of commitment to this goal evidenced during the seventies, they must raise and articulate their efforts more forcefully.

There is a need, therefore, for the continued involvement of the private sector in the granting of both need-based scholarships and financial awards predicated on merit to minority students who have demonstrated exceptional academic promise. There is also a need for the private sector to assume greater leadership roles in establishing their own outreach programs, as well as cooperative enterprises with specific institutions in the form of early identification programs. There is abundant evidence that it is important for students to learn about career opportunities and their requirements as early as possible in their eductional career as well as to have concrete proof that members of their own groups can be successful in these occupations.

The need for pre-college programs, summer internships, college workshops, and informal encounters with members of the corporate structure is real. Earlier chapters have mentioned the success of some of these programs. The quality of those successes suggests that many of them should be expanded and that they should get greater financial and human resources support.

Without a doubt, minority students and their families must do something for themselves. Families have the responsibility for establishing a learning environment in their home, for stimulating an interest in learning, for imposing that type of discipline and respect for authority that creates self discipline in the person. Families, too, have a responsibility to motivate their children by whatever ways known to them and through cooperation with elementary and secondary school systems. There is ample evidence of success from generations of black persons who have achieved success despite origins of economic poverty and initial training in less than exceptional schools. They had determination, perseverance, self-dicipline, and an orientation to achievement that was instilled by parents who wanted them to be successful and by significant others interested in their welfare. The same can be done on an even broader scale today.

Family responsibilites include observations of educational activities of their children, whether or not they have homework, library assignments, or

special projects to complete. It also means that they pay attention to attendance at school and to membership in peer groups that reinforce self-
discipline, respect for the authority structure, and a desire to learn, rather
than to those who encourage opposition to learning. There is no reason to
assume that the middle-and-upper-class members of the dominant group in
American society must perforce have a corner on the market for the
development of sound skills. Just as hundreds of thousands of minority
students, all of disadvantaged socio-economic groups, have developed
these skills in the past, so can students from these origins today profit, in
larger numbers, from such practices learned and reinforced at home and at
elementary and secondary schools. Becoming more academically competitive is also a responsibility of the student themselves.

Elementary and secondary school systems have an equal responsibility
to minority pupils. It involves sound teaching, adherence to academic standards, compassion, racial awareness and understanding, and a belief in
their capacity for learning. It means helping to strengthen self-esteem and
personal confidence and a willingness to engage those subjects presumed to
be "tough." And, success should be recognized, praised, and rewarded.

These are joint efforts and joint actions that require mutual support
and understanding. But, this emphasis on home and school factors is not intended to communicate an impression that the kinds of programs, recommendations and courses of actions proposed for the institutions, government, and private sectors must await radical transformations in the overall
quality of pre-college education or major changes in orientations toward
education in the home. The seventies taught us, if not reinforced the belief,
that concerted action is a necessary precondition for advancing beyond
token levels of access to graduate and professional schools or for moving
forward in the production of doctorates among minority populations.

Beyond Tokenism: Five Year Plans for
Increasing Minority Enrollment

A standard practice in foreign aid, modernization, and development is
to persuade developing nations to construct five year development plans.
An explicit purpose of these plans is to establish carefully conceptualized
goals that may be achieved by a definite target date. These plans include a
specification of the resources required as well as their probable sources.
Methods or courses of action are articulated in a series of phases or stages
particularized for each goal. A monitoring mechanism is incorporated for
identifying obstacles and problems as well as for ascertaining success at

critical junctures in each phase. Monitoring also permits an evaluation of the methods employed and the quality of resources specified to accomplish the original purposes of each five year plan. As a result, corrective steps may be taken to eradicate problems believed to impede progress.

Obviously, academic institutions, corporations, and private foundations frequently construct such plans. During the seventies, plans were indeed developed for the purpose of increasing minority student enrollment in professional schools (e.g., Project 75 in medicine and the Engineering Effort). The implementation of these plans played a major role in stimulating the quest for equality of educational opportunity in specific professional fields. This work should be continued by the professions.

Similarly, institutions should develop new plans which involve further joint ventures and cooperation with the federal government, the business community, private foundations, and professional associations with the same purposes of producing significant numbers of professionally trained persons from minority populations. The conceptualization of these plans should likewise be a joint enterprise and should reflect greater commitment by each of the cooperating sectors to increasing educational opportunities for outsider groups.

It is hoped that now a national consensus exists that there is a compelling urgency for the country to train as many of its citizens as possible to alleviate shortages in all of the areas of human and special services identified in this study. Specifically, there should be agreement that the nation needs more people trained for the delivery of health care services, and of legal and social services, and individuals for high technology fields and research and for teaching in our colleges and universities. The issue is also one of process and methodology. That is why the cooperation of all of the levels of policy decision-making is required in order to resolve this problem of under-representation.

Inter-organizational cooperation can provide the financial and human resources for attaining specific goals during each five year phase. A mechanism for accountability should be clearly incorporated and carefully monitored. What was learned from the seventies should be a basis for determining what has actually worked or is workable if appropriate support is given to those strategies. Hence, the cooperating organizations might lend varying types of support, coordinated by the academic institution, to specific programs.

A comprehensive program necessarily embraces a positive and psychologically supportive institutional climate, a sound recruitment strategy, well-coordinated outreach programs which extend into the institutions's target area, early identification activities, assessment, selection and admission policies based upon a good mix of objective and subjective criteria, ability to attract those students admitted to the graduate or professional school, the utilization of Summer institutes for both pre-college and pre-professional programs, the continuing use of self-paced curricula as appropriate; the availability of academic and non-academic support services at both the collegiate and professional school level, and systematic, unbiased evaluation of students so that those with potential to succeed may be reassured of their progress and those without such potential can be counseled out of the program.

The financial costs for support, operation and management of programs should also be borne jointly by the federal government, the corporate sturcture, private foundations, and by the institutions themselves. For feasibility and management purposes, some 100 institutions could be selected as cooperating institutions for the first phases of the five year plans. The number allocated to a given State may be based upon the size of the minority populations in that State or jurisdiction and the quality of the plan developed by competing institutions at the State level.

Precedents for joint ventures already exist (e.g., health service centers which incorporate medicine, dentistry, optometry, pharmacy, nursing, biomedical sciences, and other health-related fields). Such models could be applied to other professional fields where individuals or independent units operate within the same institution. This will provide more effective coordination and better cost controls. Important, the result should be a substantially improved mechanism for increasing access beyond the token level, an enriched quality of training provided students, improved inter-group understandings among students, faculty, administrators, and executives of a diverse racial or ethnic heritage, as well as the production of professionally trained minorities in numbers that will reach educational parity in the not too distant future.

How soon this change occurs may very well be contingent upon profound alterations of the power structure in the educational establishment. It may depend upon both internal and external pressures exerted on State systems of higher education and specific institutions for policy changes. It may be accelerated by the determination of minority students themselves to become more competitive in largers numbers, and by their demand that graduate and professional schools create a more positive learning environment in which a greater production of black professionals can take place.

Chapter 14

Postscript 1981 and Beyond

The election of Ronald Reagan as President of the United States has given rise to a climate of fear, hysteria and uncertainty about the future among many black Americans. Despite pronouncements to the contrary made by President Reagan shortly after his inauguration, many black Americans are convinced that his administration is less than enthusiastic about civil rights, that the President himself will not exercise strong leadership to assure equality of opportunity, and that many of his supporters, including several key members of the United States Congress, are committed to the destruction of many of the programs and policies instrumental in propelling blacks toward equality of opportunity. Consequently, many black Americans fear that entering the mainstream of the American education, econcomic and political fabric is likely to be halted under the Reagan administration. From that perspective, black people, who have always looked to the federal government for support of their efforts to achieve the rights of citizenship, have little reason to be optimistic about the 1980s.

The primary explanations for the emergence of such sentiments among so many black Americans are lodged in: (1) perceptions of an anti-black and anti-civil rights mood in the nation, (2) statements of future federal policy regarding the *Adams v. Bell* case, (3) the possible negative impact of Administration proposd budget cuts for the fiscal year 1982 budget and rescissions in the 1981 budget on economic, social and educational programs of special benefit to minorities and poor Americans, and (4) the resurgence of "states rights."

On many different occasions since the 1980 election, several civil rights leaders have warned of "hysteria" in the black community. This group includes such persons as Benjamin Hooks of the National Association for the Advancement of Colored People (NAACP), Vernon Jordan of the National Urban League (UL), and Jesse Jackson of People United to Save Humanity (PUSH). They and others have reported that the November

1980 election signalled a major shift to the political right with respect to civil rights, negative attitudes toward black Americans and support for their quest for equality of opportunity. They demand protection of those gains made by black Americans during the past thirty years.

Without a doubt, many white Americans resented former President Carter for his stand on equal rights, his enforcement of federal guidelines and policies which promoted equal economic opportunity, his insistence on the implementation of affirmative action regulations, his support of the University of California/Davis in the Bakke case, and his appointment of several minorities and women to federal judgeships. For many in this group, President Carter exceeded acceptable boundaries of support for the aspirations of black Americans and President Reagan is regarded as a "protector of white privilege." Even though these assessments may be premature with respect to the Reagan administration and with regard to the presumed insensitivity of the President himself to civil rights, these sentiments are widespread.

A second reason for the suspicions of the Reagen administration among many black Americans stems from informal statements made by governmental officials about affirmative action and the *Adams v. Bell* Case (formerly called *Adams v. Richardson* and *Adams v. Califano*). A widespread belief among black Americans who support affirmative action programs is that the Reagan administration will eliminate as many affirmative action policies and guidelines as possible, undermine existing affirmative efforts to recruit more black students for graduate and professional schools, and be less than enthusiastic about the enforcement of hiring and promotion policies with respect to blacks and minorities.

In addition, neo-conservative blacks who oppose affirmative action and the dominance of the NAACP have become more outspoken. Their views are publicized and embraced by those white neo-conservatives who welcome support for their effort to undermine gains made by blacks. One possible result of this union of black and white neo-conservatives is that black Americans may become, however unwitting as it may be, collaborators in further entrenchment of second-class citizenship among their own people. Many black Americans contend that black neo-conservatives are being exploited by those groups and institutions opposed to the further mainstreaming of black Americans, and the *Adams v. Bell* case is but one manifestation of this concern.

As a result of pressure exerted by the federal government, especially the defunct Department of Health, Education and Welfare, the new Department of Education and the Office of Civil Rights, several states in which dual systems of public higher education operated have made significant improvements toward dismantling those dual systems. Consequently, increasing numbers of black students have gained access to graduate and professional education. More and more blacks have enrolled and received degrees from state-supported schools of medicine, dentistry, law, social work and other previously restricted fields. These gains were the direct consequence of federal leadership, intervention, and enforcement of policies promulgated primarily under *Adams v. Califano.*

Without the support of the federal government and insistence that post-secondary and professional educational opportunities be expanded, few of the increases in enrollment and graduation of black students reported for the Adams States in Chapters four through eleven would have occurred. For that matter, the number of blacks who received graduate and professional degrees from institutions in the non-Adams states would also have been fewer since their enrollment in many of these institutions resulted from various forms of pressure exerted by the federal government on institutions outside the Adams domain.

In the waning days of the Carter Administration, the Adams domain was extended to cover a total of eighteen states because of their history of perpetuating a dual system of higher education. As in the case of the original ten Adams states, the decision affects only public institutions operated with state funds. Private institutions are only affected tangentially by the decision. Shortly after the Adams decision was extended to include the remaining eight states in which historically black colleges are located, some states took immediate steps to demonstrate willingness to comply with court-ordered guidelines and federal policy. However, statements attributed to Secretary of Education, T.H. Bell, have heightened suspicion that Secretary Bell will attempt withdrawal of federal enforcement of desegregation policies.

The implications of a weakened governmental posture with respect to the Adams case are far-reaching. It creates the potential for reduced state support for the historically black public institutions and for lowering the quality of education provided by these institutions. Lack of enforcement may dilute ongoing efforts to equalize educational opportunity in these states since state systems may not move aggressively toward the implementation of existing desegregation plans. They may engage in delaying tactics regarding the construction of plans acceptable to the Federal District Court of Washington, D.C. and the NAACP-LDF. The LDF will once again be forced to file time-consuming financially burdensome litigation to force compliance. States may not be compelled to demonstrate equity in the allocation of scholarship funds for black undergraduate, graduate and professional school students. Universities may not be compelled to demonstrate the lack of discrimination in the distribution of teaching and research assistantships at public institutions.

Consequently, the number of black graduate and professional school students may be reduced in proportion to the loss of these forms of financial support. State-supported colleges and universities may be even less aggressive in the recruitment of black faculty who may serve as significant others or role models for minority students and whose presence may help to eliminate demeaning stereotypes learned by "other-race" students during childhood socialization.

Attacks on the goals of the Adams case have already been made with renewed vigor by neo-conservatives who oppose affirmative action in higher education and by such groups as the conservative Heritage Foundation. The Reagan administration appears to be influenced by positions taken by such groups with respect to education issues. The position of the federal government relative to the implementation of Adams mandated policies under the Carter administration was viewed by conservatives as an unwarranted intrusion by the federal government into what they believe to be essentially autonomous prerogatives of universities and the states. Supporters of the Adams decisions perceive federal pressure as absolutely critical for the advancement of equal educational opportunity. Without the power to withhold federal funds from states that do not comply with Adams guidelines, the prevailing sentiment is that public black colleges would not be "enhanced", and desegregation would be either halted or slowed to a snail's pace. For that reason, attacks against the Adams decision and anticipated actions by Secretary Bell have created immense concern among those blacks who perceive earlier governmental intervention in the Adams litigation as positive leverage for mainstreaming blacks in higher education.

The third explanation for the pessimism among blacks with respect to the Reagan administration and the future beyond 1980 centers on the perceived impact of proposed FY 1981 and FY 1982 budget cuts on black Americans. During the first three months of 1981, when the Reagan administration was formulating its monumental budgetary changes, considerable media attention was given to the potentially devastating effect that these reductions and recisions would have on the poor, the needy, the elderly, and school-aged children. Only with the release of the major components of the Reagan budget did educators become alarmed by the radical surgery being planned for agencies of direct importance to higher education. These affected areas encompassed, among others, research and training, student loans, and scholarship programs.

According to testimonies given by various NIMH officials, analyses of the Reagan budget which appeared in such publications as the *New York Times, The Chronical of Higher Education,* and the *Footnotes* of the American Sociological Association, a major casualty of the proposed Reagan cuts will be a number of programs of immeasurable benefit to minority students. For example, about 1,000 minority students have received graduate training, most of whom studied for the doctoral degree, under a special minority fellowship program operated by the National Institute of Mental Health (NIMH) since about 1972. That program could be eliminated if the Congress enacts the Reagan budget as proposed. Clinical training programs, which offer research training and assistantships to many students, including many minority students, could be virtually eliminated in NIMH. They could also be abolished in programs funded by the Basic and Applied Social Science Division of the National Science Foundation which is scheduled to lose seventy-five per cent of its budget under the Reagan Plan.

Some student-aid cuts, such as those proposed in the Guaranteed Loan Program, would affect all students but not equally. Other proposed cuts would be a major setback to efforts designed to increase the pool of black college students and the potential pool of black students for graduate and professional schools. For instance, the Administration plans to eliminate educational grants under the Social Security program because it is claimed that such grants often are awarded automatically and to many students who do not need this support. However, Allan W. Oscar, executive director of the American Association of State Colleges and Universities, was quoted in the March 30, 1981 issue of the *New York Times* as having said that an estimated 140,000 black students would be cut out of financial assistance with the elimination of these funds.* If such students are unable to obtain financial assistance from other sources, a dramatic reduction in the potential pool of black students for entry into graduate and professional studies is likely. In turn, that loss will mean further erosion of the effort to mainstream increasingly larger numbers of black professionals.

Should these budgetary reductions occur, the policy recommendations made in Chapter Thirteen with respect to an increasing role of the corporate and business world in mainstreaming outsiders take on added significance. As pointed out earlier on, a large proportion of black students in professional schools come from families whose median annual income is below $10,000. Tuition and fees are increasing by as much as fifteen per cent in some institutions. The number of undergraduate colleges whose tui-

*See Edward B. Fiske, "College Officials Push for Alternatives for Student Aid Cuts asked by Reagan," *The New York Times,* March 31, 1981, P. B-2.

tion and fees have already exceeded the $10,000 per annum range or which are now approaching that level is increasing at an alarming rate. Without financial help, the number of black and other minority students could be reduced to token levels of the pre-civil rights period.

It is incumbent upon the corporate and business world to help expand economic opportunities in general, to help reduce the economic barriers to higher education, and provide broader employment opportunities so that more students can earn sufficient finance to help defray the costs of post-secondary and professional education. Similarly, corporations can, as a shor-run device, make more of their minority employees available for teaching posts in colleges and universities who need their expertise and who can simultaneously serve as role models for minority students.

Corporations can also encourage more of their minority employees to return to graduate and professional schools to seek advanced degrees with the assistance of corporate-financed scholarship programs.

If the Reagan administration weakens federal support for educational programs of benefit to minority students and if its actions strengthen economic barriers to equal opportunity, minorities and the poor will be forced to look elsewhere for relief and assistance. Perhaps, their greatest hope lies in the willingness of the corporate world to expand economic opportunity in order to protect its industrial investment and development by expanding the pool of individuals capable of meeting corporate needs.

Finally, there is an abhorrence among black Americans for the concept of "states rights." Beginning with President Lincoln's Emancipation, the enactment of the 13th, 14th and 15th Amendments to the Constitution and the Reconstruction Civil Rights Bills, and continuing to the present time, blacks have relied heavily upon federal government for remedies to dejure and de facto segregation and discrimination. Insistence on "states rights" has been and is now regarded as injurious to the most fundamental aspirations of black Americans.

President Reagan's efforts to revive this doctrine through block grants and the abrogation of federal leadership in the areas of education, and recissions of financial support for social programs are regarded as potentially disastrous for black Americans. For one thing, blacks have little reason to believe that legislatures controlled by the majority population will be particularly responsive or sympathetic to the needs of black people without federal leadership and pressures. In most states, black Americans have limited political clout and, if Senator Strom Thurmond is successful in either repealing or weakening the Voting Rights Act of 1965, whatever political clout that blacks now have at many state and local levels may be dissipated.

This situation highlights a number of concerns about the capacity of blacks to exact demands for greater access to graduate and professional schools and for equity in the distribution of state resources. If fewer blacks are eligible to vote or if eligible blacks do not register, vote, and remove unresponsive persons from political office, their political power and influence will be minimal.

Black students have strong desires to become physicians, dentists, lawyers, engineers, architects, optometrists, social workers, and to earn doctoral degrees. They have equally strong desires to work and earn sufficient funds to finance their graduate and/or professional studies. Related goals can only be accomplished if the economic and educational barriers against access to equal schooling, jobs and wages are eliminated. That will only occur through sustained leadership, federal and state intervention, cooperative undertakings between colleges, professional schools and high schools, and the corporate world combined with a genuine realization of equality of opportunity.

Even though black American may be extremely pessimistic about the future, it is imperative for them to take control of their own fate. Blacks should become even more self-assertive and determined to gain constitutionally legitimate rights. Now is the time for strong national and grassroot leadership that is able to utilize the many strengths and resources within the black community. More so than at any time in the recent past, the need for unity, group discipline, focus on concrete objective and clearly articulated strategies for the maintenance of past civil rights achievements is paramount.

Black people must not permit the attacks against affirmative action, whether in higher education or in equal access to employment opportunity, to continue unchecked. Clearly, many of these attacks are indeed a mask for the single purpose of dismantling those economic, educational and social remedies estblished during the past three decades. Coalitions with other minority groups and with those persons within the white population committed to expanded opportunity across racial and ethnic lines may be necessary. However, the primary responsibility for the continuation of this quest for equality rests squarely on the shoulders of black people.

Finally, despite the persistence of social, economic and educational barriers to mainstreaming, black students cannot afford to acquiesce to apathy, indifference, self-centeredness and withdrawal. It is imperative for them to commit themselves in larger numbers to intellectual rigor, self-discipline, the value of learning, and to the continued development of those skills vital for incorporation into the mainstream.

Bibliography

Abarnel, Karen. "Is A Relapse Ahead for Minority Medical Education?" *Foundation News.* (March 1977).

Abt, Donald A. *Report of the Workshop Conference on Minority Representation in Veterinary Medicine.* (Sponsored by the Association of American Veterinary Medical Colleges, The American Council in Education, and The Ford Foundation), November 1-2, 1978.

American Association of Dental Schools, *Admission Requirements of U.S. and Canadian Dental Schools: 1979-80.* Washington, D.C.: American Association of Dental Schools, 1978.

_____, "Information Abstracts, *Journal of Dental Education.* 43 (1979).

_____, *Interim Report on Student Educational Finances.* Washington, D.C.: American Association of Dental Schools, 1978.

American Bar Association, *A Review of Legal Education in the United States - Fall, 1975.* Chicago: American Bar Association, Section of Legal Education and Admissions to the Bar, 1977.

_____, *A Review of Legal Education in the United States - Fall, 1976.* Chicago: American Bar Association, Section on Legal Education and Admissions to the Bar, 1977.

_____, *A Review of Legal Education in the United States - Fall, 1977.* Chicago: American Bar Association, Section on Legal Education and Admissions to the Bar, 1978.

_____, *A Review of Legal Education in the United States - Fall, 1978.* Chicago: American Bar Association, Section on Legal Education and Admissions to the Bar, 1979.

_____, *A Review of Legal Education in the United States - Fall, 1979.* Chicago: American Bar Association, Section on Legal Education and Admissions to the Bar, 1980.

American Dental Association, *Analysis of Black Applicants to Dental Schools: April, 1970 Through April, 1971.* Chicago: American Dental Association - Division of Educational Measurements, September, 1971.

_____, *Annual Report, Dental Education 1978-79.* Chicago: American Dental Association - Division of Educational Measurements, 1980.

_____, *Distribution of Dentists in the United States By States, Region, District and County.* Chicago: American Dental Association - Bureau of Economic and Behavioral Research, 1979.

_____, *Minority Report 1974-75.* Chicago: American Dental Association - Division of Educational Measurements, 1975.

_____, *Minority Report 1975-76.* Chicago: American Dental Association - Division of Education Measurements, 1976.

_____, *Minority Report 1976-77.* Chicago: American Dental Association - Division of Educational Measurements, 1977.

_____, *Minority Report 1977-78.* Chicago: American Dental Association - Division of Educational Measurements, 1978.

_____, *Minority Report 1978-79.* Chicago: American Dental Association - Division of Educational Measurements, 1979.

_____, *Minority Report 1979-80.* Chicago: American Dental Association - Division of Educational Measurements, 1980.

_____, *Trend Analysis 1977-78: Supplement II to the Annual Report 1977-78 Dental Education.* Chicago: American Dental Association - Division of Educational Measurements, 1978.

American Medical Association. "Medical Education in the United States, 1971-72." *Journal of American Medical Association,* 222 (1972), 978-985.

_____, "Medical Education in the United States, 1974-75." *Journal of American Medical Association,* 231 (1975), 17-62.

_____, "Medical Education in the United States, 1975-76." *Journal of American Medical Association,,* 236 (1976), 2960-2970.

_____, "Medical Education in the United States, 1976-77: 77th Annual Report." *Journal of American Medical Association,* 238, (1977), 2769-2799.

Annual Survey of Optometric Educational Insitutions, 1970-71. Washington, D.C.: Association of Schools and Colleges of Optometry, 1971.

Annual Survey of Optometric Educational Institutions, 1971-72. Washington, D.C.: Association of Schools and Colleges of Optometry, 1972.

Annual Survey of Optometric Educational Institutions, 1973-74. Washington, D.C.: Association of Schools and Colleges of Optometry, 1973.

Annual Survey of Optometric Educational Institutions, 1973-74. Washington, D.C.: Association of Schools and Colleges of Optometry, 1974.

Annual Survey of Optometric Educational Institutions, 1974-75. Washington, D.C.: Association of Schools and Colleges of Optometry, 1975.

Annual Survey of Optometric Educational Institutions, 1975-76. Washington, D.C.: Association of Schools and Colleges of Optometry, 1976.

Annual Survey of Optometric Educational Institutions, 1976-77. Washington, D.C.: Association of Schools and Colleges of Optometry, 1977.

Annual Survey of Optometric Educational Institutions, 1977-78. Washington, D.C.: Association of Schools and Colleges of Optometry, 1978.

Annual Survey of Optometric Educational Institutions, 1978-79. Washington, D.C.: Association of Schools and Colleges of Optometry, 1979.

Association of American Law Schools. *1975 Directory of Minority Law Faculty Members.* Washington, D.C.: Section on Minority Groups, Association of American Law Schools, 1975. (Edited by Derek Bell).

_____, *1976 Directory of Minority Law Faculty Members.* Washington, D.C.: Section on Minority Groups, Association of American Law Schools, 1976. (Edited by Derek Bell).

_____, *1977 Directory of Minority Law Faculty Members.* Washington, D.C.: Section on Minority Groups, Association of American Law Schools, 1977. (Edited by Derek Bell).

_____, *1979-80 Directory of Minority Law Faculty Members.* Washington, D.C.: Section on Minority Groups, Association of American Law Schools, 1979. (Edited by Derek Bell).

_____, *Proceedings: 1970 Annual Meeting, Part I, Section II.*

Washington, D.C.: Association of American Law Schools, 1970.

Association of American Medical Colleges. "MSAR, Minority Student Information by Individual Medical Schools." Washington, D.C.: Association of American Medical Colleges, 1978.

_____, "MSAR, Minority Student Information by Individual Medical Schools." Washington, D.C.: Association of American Medical Colleges, 1980. (Unofficial Results).

_____, "Report of the AAMC Task Force on Minority Student Opportunities in Medicine." Washington, D.C.: Association of American Medical Colleges, June 1978.

_____, "Report of the Association of American Medical Calleges Task Force on Minority Student Opportunities in Medicine." Washington, D.C.: Association of American Medical Colleges, 1978.

Atelsek, Frank J. and Irene L. Gomberg. *Special Programs for Female and Minority Graduate Students*. Washington, D.C.: American Council on Education, 1978.

Bardolph, Richard. *The Civil Rights Record*. New York: Thomas Y. Crowell, 1970.

Begun, Martin S. "Legal Considerationa Related to Minority Groups Recruitment and Admissions." *Journal of Medical Education*. 48 (1973), 556-559.

Bennett, Lerone. *Before the Mayflower: A History of Black America*. Chicago: Johnson Publishing Co., 1969.

Bierstedt, Robert. "An analysis of Social Power." *American Sociological Review*, 15 (December 1950).

Billingsley, Andrew. "Black Students in A Graduate School of Social Work." In Carl Scott, ed., *Ethnic Minorities in Social Work Education*. New York Council on Social Work Education, 1970.

Black Perspectives on Social Work Education: Issues Related to Curriculum, Faculty and Students. Atlanta: Council on Work Education Foundation, 1975.

Blackwell, James E. *Access of Black Students to Graduate and Professional Schools*. Atlanta: The Southern Education Foundation, 1975.

_____, "Social Factors Affecting Educational Opportunity for Minority Group Students." (Chapter I) in *Beyond Desegregation: Urgent Issues in the Education of Minorities*. New York: The College Board, 1978.

_____, *The Black Community: Diversity and Unity*. New York: Harper & Row, 1975.

_____, *The Participation of Black Students In Graduate and Professional Schools*. Atlanta: Southern Education Foundation, 1977.

_____, "The Power Basis of Ethnic Conflict in American Society." In Lewis A. Coser and Otto N. Larsen, eds., *The Uses of Controversy in Sociology*. New York: The Free Press, 1976.

_____, Maurice Jackson and Joan Moore, *The Status of Racial and Ethnic Minorities in Sociology A Footnote Supplement*. Washington, D.C.: The American Sociological Association, August, 1977.

_____, Philip Hart and Roberst Sharpley. *Alienation Among Metropolitan Blacks*. New Brunswick, NJ: Transaction Books, Forthcoming.

Blake, Martin L. "Graduate Enrollment Data, September 1969 and Graduate Study in Member Colleges, 1970-71." *American Journal of Pharmaceutical Education*, 35 (February 1971), 108-127.

Blalock, Hubert M. "A Power Analysis of Racial Discrimination." *Social Forces*, 39 (1960), 53-69.

Bowles, Frank and Frank A. DeCosta. *Between Two Worlds: A Profile of Negro Higher Education.* New York: McGraw - Hill Co., 1971.

Brown, Kenneth G., Arthur T. Grant and Larry L. Leslie. *Patterns of Enrollment in Higher Education: 1965 to 1977 - Research Universities.* Tuscon: University of Arizona, 1979.

"CLEO Symposium." *Howard Law Journal,* 22 (1979), 297-496.

Collins, Randall. "Some Comparative Principles of Educational Stratification." *Harvard Educational Review,* 47 (February 1977), 1-29.

Commission on Human Resources, *Summary Report of 1973 Doctorate Recipients From United States Universities.* Washington, D.C.: National Academy of Sciences, 1974.

_____, *Summary Report of 1974 Doctorate Recipients From United States Universities.* Washington, D.C.: National Academy of Sciences, 1975

_____, *Summary Report of 1975 Doctorate Recipients From United States Universities.* Washington, D.C.: National Academy of Sciences, 1976.

_____, *Summary Report of 1976 Doctorate Recipients From United States Universities.* Washington, D.C.: National Academy of Sciences, 1977.

_____, *Summary Report of 1977 Doctorate Recipients From United States Universities.* Washington, D.C.: National Academy of Sciences, 1978.

_____, *Summary Report of 1978 Doctorate Recipients From United States Universities.* Washington, D.C.: National Academy of Sciences, 1979.

_____, *Minority Groups Among United States Doctorate Level Scientists, Engineers and Scholars, 1973.* Washington, D.C.: National Academy of Sciences, 1974.

Cordoza, Michael H. *The Association Process: 1963-1973.* Washington, D.C.: Association of American Law Schools 1975, Annual Meeting Proceedings, Part I, Section 110, 1975.

Crain, Robert L. "School Integration and Occupational Achievement Among Negroes," *American Journal of Sociology* 75(1970), 593-606.

Curtis, James L. "Minority Student Success and Failure With the National Intern and Resident Matching Program." *Journal of Medical Education,* 50 (1975), 563-570.

Dillinger, Heather, *Summary Sheet, Graduate and Professional Opportunities Program.* Washington, D.C.: U.S. Office of Education, May 1, 1980.

Dube, W.F. "Applicants for the 1973-74 Medical School Entering Class," (Datagram). *Journal of Medical Education,* 49 (1974), 1070-1072.

_____, "Medical School Enrollment, 1969-70 Through 1973-74," *Journal of Medical Education,* 49 (1974), 302-307.

_____, "Medical Student Enrollment, 1969-70 Through 1974-75," *Journal of Medical Education,* 50 (1975), 303-306.

_____, " Socioeconomic Background of Minority and Other U.S. Medical Students." (Datagram). *Journal of Medical Education,* 53 (1978), 443-445.

_____, "Study of U.S. Medical School Applicants, 1972-73." *Journal of Medical Education,* 49 (1974), 849-869.

_____, and Davis G. Johnson. "Study of U.S. Medical School Applicants, 1973-74." *Journal of Medical Education,* 50 (1975), 1015-1032.

_____, "Study of U.S. Medical School Applicants, 1974-75." *Journal of Medical Education,* 51 (1976), 877-896.

DuBois, W.E.B., *The Souls of Black Folk.* Chicago: A.C. McClury Co., 1953

Dummett, Clifton O. and Louis D. Dummett. "Afro-Americans in Dentistry." *The Crisis,* (November 1979), 389-399.

Edwards, Harry, *Black Students.* New York: The Free Press, 1970.

Edwards, Harry T. "Report of the Section on Minority Groups." *Proceedings, Part I of the Annual Meeting of the Association of American Law Schools."* Washington, D.C.: Association of American Law Schools, 1976.

Eldredge, Joan F., ed., *Statistics on Social Work Education.* New York:Council on Social Work Education, 1972.

El-Khawas, Elaine H., and Joan Kinser, *Enrollment of Minority Graduate Students in Ph. D. Granting Institutions.* Washington, D.C.: American Council on Education, August 1974.

Engineering Manpower Commission, *Engineering and Technology Degrees, 1970.* N.Y.: Engineers Joint Council, 1971.

——————————————, *Engineering and Technology Degrees, 1971.* N.Y.: Engineers Joint Council, 1972.

——————————————, *Engineering and Technology Degrees, 1972.* N.Y.: Engineers Joint Council, 1973.

——————————————, *Engineering and Technology Degrees, 1973.* N.Y.: Engineers Joint Council, 1974.

——————————————, *Engineering and Technology Degrees, 1974.* N.Y.: Engineers Joint Council, 1975.

——————————————, *Engineering and Technology Degrees, 1975.* N.Y.: Engineers Joint Council, 1976.

——————————————, *Engineering and Technology Degrees, 1976.* N.Y.: Engineers Joint Council, 1977.

——————————————, *Engineering and Technology Degrees, 1978.* N.Y.: Engineers Joint Council, 1978.

——————————————, *Engineering and Technology Degrees, 1979.* N.Y.: Engineers Joint Council, 1979.

——————————————, *Engineering and Technology Enrollments, Fall 1970.* N.Y.: Engineers Joint Council, 1971.

——————————————, *Engineering and Technology Enrollments, Fall 1971.* N.Y.: Engineers Joint Council, 1972.

——————————————, *Engineering and Technology Enrollments, Fall 1972.* N.Y.: Engineers Joint Council, 1973.

——————————————, *Engineering and Technology Enrollments, Fall 1973.* N.Y.: Engineers Joint Council, 1974.

——————————————, *Engineering and Technology Enrollments, Fall 1974.* N.Y.: Engineers Joint Council, 1975.

——————————————, *Engineering and Technology Enrollments, Fall 1975.* N.Y.: Engineers Joint Council, 1976.

——————————————, *Engineering and Technology Enrollments, Fall 1976.* N.Y.: Engineers Joint Council, 1977.

——————————————, *Engineering and Technology Enrollments, Fall 1977.* N.Y.: Engineers Joint Council, 1978.

——————————————, *Engineering and Technology Enrollments, Fall 1978.* N.Y.: Engineers Joint Council, 1979.

——————————————, *Engineering and Technology Enrollments, Fall 1979.* N.Y.: Engineers Joint Council, 1980.

Epps, Anna Cherrie, "The Howard - Tulane Challenge: A Medical Education Reinforcement and Enrichment." *Journal of the National Medical Association,* 67 (1975), 55-60.

Evans, Doris A., Paul K. Jones, Richard A. Wortman, and Edgar Jackson, Jr. "Traditional Criteria as Predictors of Minority Student Success in Medical School." *Journal of Medical Education,* 50 (1975), 934-939.

Evans, Franklin R. *Applications and Admissions to ABA Acredited Law Schools: An*

Analysis of National Data for the Class Entering the Fall of 1976. Princeton, N.J.: Law Schools Admissions Council, 1977.

_____, *Law School Admissions Research*. Princeton, N.J.: Law Schools Admission Council, 1977.

Fields, Cheryl M., "SAT Scores Don't Measure Scholastic Aptitude, Can Be Influenced by 'Coaching', Report Systs," *The Chronicle of Higher* Education, (May 27, 1980).

Financial Aid Needs of Undergraduate Minority Engineering Students in the 1980s. New York: The National Fund for Minority Engineering Students, December 1978.

Franklin, John Hope. *From Slavery to Freedom*. New York: Aldred K. Knopf, 1948.

Francis, E.A. *Black Task Force Report*. New York: Council on Social Work Education, 1973.

Gilford, Dorothy M. and Joan Snyder, *Women and Minority Ph.D.s in the 1970's: A Data Book*. Washington, D.C.: Commission on Human Resources, National Research Council, National Academy of Sciences, 1977.

Golden, Deborah, Arnuff W. Pins, and Wyat Jones. *Students in Schools of Social Work*. New York: Council on Social Work Education, 1972.

Gordon, Milton. *Assimilation in American Life*. New York: Oxford University Press, 1964.

Greenberg, Harold I. and Carl A. Scott. "Ethnic Minority Recruitment and Program for the Educationally Disadvantaged in Schools of Social Work." In Carl A. Scott, ed., *Ethnic Minorities in Social Work Education*. New York: Council on Social Work Education, 1970.

Gurin, Patricia and Edgar Epps, *Black Consciousness: Identity and Achievement* New York: John Wiley and Sons, 1975.

Hanford, George H. *Minority Programs and Activities of the College Entrance Examination Board*. New York: The College Entrance Examination Board, 1976.

Henderson, Wade J. *Statement on Behalf of the Council on Legal Education Opportunity Before the Subcommittee on Post Secondary Education*. U.S. House of Representatives, June 13, 1979. (Unpublished Document.)

Hill, Robert B. *The Illusion of Black Progress*, Washington, D.C.: National Urban League Research Department, 1978.

Information for Applicants to Schools and Colleges of Optometry, Fall 1979. *Washington, D.C.: Association of Schools and Colleges of Optometry, 1978*.

Jarecky, Roy K. "Medical School Efforts to Increase Minority Representation in Medicine." *Journal of Medical Education,* 44 (1969), 912-918.

Johnson, Davis G. and William E. Sedlacek. "Retention by Sex and Race of 1968-1972 U.S. Medical School Entrants." *Journal of Medical Education,* 50 (1975), 925-933.

Johnson, Davis G., C. Vernon Smith and Stephen L. Tarnoff. "Recruitment and Progress of Minority Medical School Entrants, 1970-72." *Journal of Medical Education,* 50 (1975), 713-755.

Johnson, Davis G. and Robert L. Tuttle. "Role of GSA in Medical Education, 1957-1972." *Journal of medical Education,* 48 (1973), 231-239.

Jolly, H.P. and Thomas A. Larson. *Participation of Women and Minorities in U.S. Medical School Faculties*. U.S. Department of Health, Education and Welfare. Washington, D.C.: DHEW Publication No. (HRA) 76-91, March 1976.

Kendall, Maurice G. and A. Stuart. *The Advanced Theory of Statistics, Vol. II*. London: Charles Griffin, 1961.

Kuhli, Ralph C. "Education For Health Careers." *Journal of School Health,* 41 (January 1971), 17.

Leonard, Walter J. "Report of the Section on Minority Groups," *Proceedings: 1973 An-*

nual Meeting, Part I. Association of American Law Schools. Washington, D.C.: Association of American Law Schools, 1973.

—————————, "Report of the Section on Minority Groups," *Proceedings: 1974 Annual Meeting, Part I. Association of American Law Schools.* Washington, D.C.: Association of American Law Schools, 1974.

Lowenberg, Frank M. and Thomas H. Shey, eds., *Statistics on Social Work Education, 1970.*

—————————, *Statistics on Social Work Education.* New York: Council on Social Work Education, 1971.

Lusterman, Seymour. *Minorities in Engineering: The Corporate Role.* New York: The Conference Board, 1979.

Malcolm, Shirley M., John Cownie, and Janet Welsh Brown. *Programs in Science for Minority Students, 1960-1975.* Washington, D.C.: American Association for the Advancement of Science, 1976.

Marshall, Edwin C. "Social Indifference or Blatant Ignorance." *Journal of the American Optometric Association, 43 (November 1972),* 1261-1266.

McKay, Robert B., et.al., *"Report of the Committee on Minority Groups," Proceedings: 1971 Annual Meeting. Part I. Section I. Association of American Law Schools.* Washington, D.C.: Association of American Law School, 1971.

Medical School Admission Requirements 1981-82. Washington, D.C.: Association of American Medical Colleges, 1980.

Minorities in Engineering: A Blueprint For Action. (A Report of the Planning Commission for Expanding Opportunities in Engineering.) New York: Alfred P. Sloan Foundation, 1974.

Minorities in Medicine: Report of A Conference. New York: Josiah Macy, Jr. Foundation, 1977.

Minority Student Opportunities in United States Medical Schools 1975-76. Washington, D.C.: Association of American Medical Colleges, 1975.

Minority Student Opportunities in United States Medical Schools 1970-71. Washington, D.C.: Association of American Medical Colleges, 1970.

Minority Student Opportunities in United States Medical Schools 1971-72. Washington, D.C.: Association of American Medical Colleges, 1971.

Minority Student Opportunities in United States Medical Schools 1980-81. Washington, D.C.: Association of American Medical Colleges, 1980.

Meuller, John H., Karl F. Schuessler, and Herbert L. Costner. *Statistical Reasoning in Sociology.* Boston, MA: Houghton Mifflin Co., 1977.

National Advisory Committee on Black Higher Education & Black Colleges & Universities, *Access of Black Americans to Higher Education: How Open is the Door.* Washington, D.C.: U.S. Government Printing Office, January 1979, p. 20.

National Medical Fellowships. "Awards Approved for 1971 Through 1978." New York: National Medical Fellowships (Data Provided, 1980).

Orr, Jack E. "Report on Enrollment in Schools and colleges of Pharmacy: First Semester, Term or Quarter, 1968-69." *American Journal of Pharmaceutical Education, 33* (February 1969), 84-99.

—————————, "Report on Enrollment in Schools and Colleges of Pharmacy: First Semester, Term or Quarter, 1970-71" *American Journal of Pharmaceutical Education.* 34 (February 1970), 92-107.

—————————, "Report on Enrollment in Schools and Colleges of Pharmacy: First Semester, Term or Quarter, 1971-72." *American Journal of Pharmaceutical Education, 36* (February 1972), 120-130.

_____, "Report on Enrollment in Schools and Colleges of Pharmacy: First Semester, Term or Quarter, 1972-73." *American Journal of Pharmaceutical Education,* 37 (February 1972), 138-153.

_____, "Graduate Enrollment Data, September 1971 and Graduate Study in Member Colleges, 1972-73." *American Journal of Pharmaceutical Education,* 37 (February 1973), 138-153.

_____, "Graduate Enrollment Data, September 1972 and Graduate Study in Member Colleges, 1973-74." Bethesda: American Association of Colleges of Pharmacy, 1973.

Otto, Luther B. and David L. Featherman, "Social, Structural and Psychological Antecedent of Self-Estrangement and Powerlessness." *American Sociological Review*, 40 (December 1975), 701-719.

Peters, Henry B. "Critical Optometric Manpower Issues." *Journal of Optometric Education,* 4 (Spring 1979), 8-12.

Plaggel, James C., Robert L. Smith, Nat E. Smith, and Lawrence M. Solomon. "Increasing the Number of Minority Enrollees and Graduates: A Medical Opportunities Program." *Journal of Medical Education,* 49 (1974), 735-745.

Proceedings of the Minority Recruitment Conference: Tuskegee Institute School of Veterinary Medicine, 1974. Tuskegee, Tuskegee Institute, 1974.

Proceedings of the Minority Recruitment Conference: Tuskegee Institute School of Veterinary Medicine, 1975. Tuskegee: Tuskegee Institute, 1975.

Purvine, Margaret, ed., *Statistics on Graduate Social Work Education.* New York: Council on Social Work Education, 1973.

Racial and Ethnic Enrollment Data From Institutions of Higher Education, Fall, 1970. Washington, D.C. Office of Civil Rights, Department of Health, Education and Welfare, 72-8, 1973.

Racial and Ethnic Enrollment Data From Institutions of Higher Education, Fall, 1972. Washington, D.C.: DHEW, OCR, p. 72-78, 1978.

Reichert, Kurt. "Survey of Non-Discriminatory Practices in Accredited Graduate Schools of Social Work." In Carl Scott, ed., *Ethnic Minorities in Social Work Education.* 1970, p.39-51.

Reitzes, Dietrich C. and Hekmat Elkhanialy. "Black Students in Medical Schools." *Journal of Medical Education,* 51 (1976), 1001-1005.

Report of the AAMC Task Force on the InterAssociation Committee on Expanding Educational Opportunities in Medicine for Blacks and Other Minority Students. Washington, D.C.: Association of American Medical Colleges, April 22, 1970.

Report To The President and Congress On The Status Of Health Professions Personnel In The U.S., 1980. Washington, D.C.: Department of Health, Education and Welfare, 1980.

Report of the Special Committee on Future Direction For Minority Group Legal Education. Proceedings, Part Two, 1976 Annual Meeting of American Law Schools. Washington, D.C.: Association of American Law Schools, (1976), 155-162.

Retention of Minority Students in Engineering. Washington, D.C.: National Academy of Sciences, 1977.

Ripple, Lilian, ed., *Statistics on Graduate Social Work Education in the United States: 1973.* New York: Council on Social Work Education, 1974.

_____, *Statistics on Social Work Education in the United States: 1974.* New York: Council on Social Work Education, 1975.

Rubin, Allen and G. Robert Whitcomb. *Statistics on Social Work Education in the United States: 1977.* New York: Council on Social Work Education, 1979.

_____, *Statistics on Social Work Education in the United States: 1978.* New

York Council on Social Work Education, 1979.

Rubin, Allen. *Statistics on Social Work Education in the United States: 1979*. New York: Council on Social Work Education, 1980.

Schaegger, Ruth G. "Corporate Leadership in a National Program." *The Conference Board Record, Vol. XIII.* September, 1976.

Schlegel, John F. *Enrollment Report on Professional Degree Programs in Pharmacy, Fall 1979.* Bethesda, Md. American Association of Colleges of Pharmacy, 1980.

_____, *Graduate Enrollment Report, Fall 1976.* Bethesda: American Association of Colleges of Pharmacy, 1977.

_____, *Graduate Enrollment Report, Fall 1978.* Bethesda: American Asssociation of Colleges of Pharmacy, 1979.

Schlegel, John F. and Christopher A. Rodowskas, Jr. *Enrollment Report on Professional Degree Program in Pharmacy, Fall 1975.* Bethesda: American Association of Colleges of Pharmacy (Reprinted from the American Journal of Pharmaceutical Education, 40 (August 1976.)

Schlegel, John F., et, al., *Report of Fall 1974 Undergraduate Enrollment in Schools and Colleges of Pharmacy.* Bethesda: American Association of Colleges of Pharmacy (Reprinted from American Journal of Pharmaceutical Education, 1975).

_____, *Report on Graduate Enrollment Fall 1974 and Graduate Programs Offered in Member Colleges, 1975-76.* Bethesda: American Association of Colleges of Pharmacy (Reprinted from American Journal of Pharmaceutical Education, 1975).

Scott, Carl, ed., *Ethnic Minorities in Social Work Education.* New York: Council on Social Work Education, 1970.

Sheehan, Joseph, ed., *Statistics on Social Work Education in the United States: 1975.* New York Council on Social Work Education, 1976.

Shyne, Ann W. and G. Robert Whitcomb. *Statistics on Social Work Education in the United States: 1976.* New York: Council on Social Work Education, 1977.

Sindler, Allan P. *Bakke, DeFunis and Minority Admissions.* New York: Longman, 1978.

Slocum, Alfred A. "CLEO: Anatomy of Success." *Howard Law Jounrnal,* 22 (1979), 335-373.

Sleth, Boyd C. and Robert L. Mishell. "Black Underrepresentation in United States Medical Schools.; *The New England Journal of Medicine,* 297 (1977), 1146-1148.

Solander, Lars. "The Sex and Ethnic Compostion of Faculty at Institutions of Pharmacy Education". (An unpublished Paper Prepared for the American Association of Colleges of Pharmacy, 1979)

Sorenson, A. A., "Black Americans and the Medical Profession, 1930-1970," *Journal of Negro Education* 41 (Fall, 1972), 337-342.

Summary Information on Masters of Social Work Programs: 1978. New York: Council on Social Work Education, 1979.

The Chronicle of Higher Education. September 8, 1980, p.2.

Thompson, Vertis, M.D., "Comments", *The Montclarian* (September 24, 1980).

U.S. Commission on Civil Rights, *Toward Equal Opportunity: Affirmative Admissions Programs at Law and Medical Schools.* Washingtom D.C.: U.S. Government Printing Office, June 1978.

_____, Social Indicators of Equality for Minorities and Women. Washington, D.C.: U.S. Government Printing Office, Aug. 1978.

U.S. Department of Labor, "Ph. D. Manpower Employment Demand and Supply 1972-1985,"*Bulletin 1960,* Washington, D.C.: Bureau of Labor Statistics, p. 9, 1973.

U.S. Office of Education, *Program Support of Graduate Education,* Washington, D.C.: HEW Publication No. (OE) 17005, August, 1978.

Walwer, Frank K. "Report of the Section on Economics of Legal Education." *Proceedings: 1975 Annual Meetings of the Association of American Law Schools, Part I.* Washington, D.C.: Association of American Law School, 1975, 54-55.

Wilkinson, Doris, "Careers, Minorities and Women", *Footnotes,* Washington, D.C.: American Sociological Association, 1977-80.

_____, "Job Seeking Guides for Social Science Majors." *The Black Collegian,* 10 (February 1980), 96-97.

_____, "Federal Employment for Sociologists", *Footnotes,* Washington, D.C.: American Sociological Association, March 1980.

_____, "Percentage of Women Doctorates in Sociology Increases", *Footnotes,* Washington, D.C.: American Sociological Association, December, 1977.

Wright, Stephen J. *The Black Educational Policy Researcher: An Untapped National Resource.* Washington, D.C.: National Advisory Committee on Black Higher Education and Black Colleges and Universities, Department of Health, Education and Welfare, 1979.

Name Index

Subject Index

342